# Scientific Innovations for Coastal Resource Management

Jorge Ramos
*CinTurs, University of Algarve, Portugal*

Benjamin Drakeford
*University of Portsmouth, UK*

Francisco Leitão
*CCMAR, Portugal*

A volume in the Advances in Environmental Engineering and Green Technologies (AEEGT) Book Series

IGI Global
Publishing Tomorrow's Research Today

Published in the United States of America by
IGI Global
Engineering Science Reference (an imprint of IGI Global)
701 E. Chocolate Avenue
Hershey PA, USA 17033
Tel: 717-533-8845
Fax: 717-533-8661
E-mail: cust@igi-global.com
Web site: http://www.igi-global.com



Library of Congress Cataloging-in-Publication Data

CIP DATA PENDING

ISBN: 9798369324363
isbn: 9798369350577
eISBN: 9798369324370

British Cataloguing in Publication Data
A Cataloguing in Publication record for this book is available from the British Library.

# Table of Contents

# Detailed Table of Contents



Scientific innovation plays a pivotal role in coastal resource management, shaping strategies to foster blue economic development while mitigating environmental risks and controlling pollution. The chapter highlights the multifaceted impacts of scientific innovation in coastal areas. Innovations such as advanced monitoring technologies, artificial intelligence, and nature-based solutions are transforming coastal management practices, enabling real-time detection of pollution events, enhancing ecosystem resilience, and promoting sustainable economic activities. By leveraging scientific knowledge and interdisciplinary collaboration, stakeholders can develop holistic approaches to coastal resource management that balance economic growth with environmental sustainability. The chapter underscores the importance of integrating scientific innovation into policy-making processes and fostering partnerships among governments, industries, and communities to ensure the long-term health and prosperity of coastal ecosystems and economies.

The small-scale fisheries in São Tomé and Príncipe play a critical socioeconomic role, supporting around 30,000 people and contributing to the nation's food security. Local fishers, limited by a lack of resources and reliance on traditional techniques, face significant challenges. These have been aggravated by the increasing use of non-selective and ecologically harmful fishing methods, which have emerged as a response to fish scarcity. Here the authors detail the implementation of an artificial reef system in 2020, designed to enhance local fisheries and promote socioeconomic development. The project was led by NGOs MARAPA in partnership with Programa Tatô and involved a wide range of stakeholders. The artificial reefs, made of locally sourced inert materials and designed to foster marine life conducive to fishing, were strategically placed in the northern region on sandy bottoms to minimize environmental impact. This initiative not only aims to replenish fish resources but also serves as a tool for capacity building and co-management, promoting better decision-making processes.

Fisheries management has evolved and addressed diverse disciplines including the application of mathematical models, economic analysis, and sector behavior considerations. Models use interrelated variables that require a balanced assessment, while recent literature stresses the importance of integrating biological/ecological with social/economic aspects. This study conducts a review of 35 studies focusing on the state-of-the-art of two key currents: management strategy evaluation and bioeconomic analysis. The analysis was done for small pelagic fisheries due to their global significance. The results show the importance of socio-economic variables and emphasizes the need to address the behavioral and economic facets specific to fisheries which have been insufficiently explored. The emphasis on prioritizing analytical reality over theoretical models highlights the importance of basing policy decisions on empirical evidence and practical considerations. The study identifies areas in need of further research and includes a critical discussion of the main approaches to a common resource.



Artificial reefs (AR) are human made structures deployed in the seabed to provide multiple economic, social, and ecological benefits. AR have emerged as a highly effective strategy to provide ecosystem services, such as habitat protection and restoration. Despite their acknowledged benefits, there remains a limited understanding of how these reefs are currently being used by the local communities. Ecosystem services (EC) are tangible and intangible benefits that the ecosystems provide to humans. The present study aims to identify the EC provided by the AR located on the south coast of Portugal. To achieve this, a literature review was done to set a baseline and find out what has been published in this research area.

This chapter aims to develop an ecohydrological modelling approach to propose an environmental engineering solution based on historical nature-based information to alleviate the severe water degradation problem within Luanda Bay, Angola. This solution would improve local communities' ecosystem health and quality of life. The model suggests a simple and practical solution to solving the pollution problem by opening a cut from the bay's shallows to the sea. This will essentially reproduce an old bay opening to the sea, which was filled early in the last century. The residence time of water in the shallows of Luanda Bay would be reduced to about one week by excavating the channel over the land is straightforward. This should return the system to a more natural state and improve the water quality for recreational activities, increase the quality of marine food resources, reduce toxic algae blooms, and improve the quality of life of people in the area. Nevertheless, preventing pollutants from reaching Luanda Bay is the only long-term solution to restore a healthy environment in Luanda Bay

Artificial reefs (AR) are essentially implemented to increase marine biodiversity, which can indirectly benefit the local ecosystem and human communities. Their presence can attract a wide range of marine organisms and promote recreational activities, attracting anglers and scuba divers, improving local tourism. The aim of this work is to analyse how the Vilamoura AR (Southern Portugal) is useful to the sustainable coastal (blue) tourism in the area. Nowadays, with the help of tools such as vessels tracking intelligence and the use of direct georeferentiation, it is possible to infer about the use of artificial reefs by analysing patterns of recreational vessels around the AR area. The application of those tools reenforced the importance of combining methodologies such as not only monitoring artificial reefs but also to have interviews with the stakeholders to better understand the recreational and blue tourism potential of the Vilamoura AR.

# Preface

Coastal regions are teeming with diverse ecosystems and providing vital resources that sustain communities around the world. Yet, these precious environments face unprecedented challenges in the face of climate change, population growth, and unsustainable human activities. A large part of the world's population lives in areas along or near the coast it is important to properly manage coastal resources, which are subject to various pressures. Factors that put pressure on resources are also eventually environmental problems, such as those that can call into question the existence of these resources if they are finite (e.g. fossil fuels), or their sustainability if they are renewable resources (e.g., fish catches). The management of coastal resources by communities directly dependent on them is an approach that has received regional attention in many places around the world.

Due to complex and interconnected issues impacting our coasts, it has become clear that innovative scientific approaches are essential to effective resource management and conservation. It is precisely in the aspect of adequate management of coastal resources that the involvement of the most varied stakeholders is required, as they sometimes have different and possibly opposing positions. Science at large, as an integral part of our society and civilizational evolution, has an extremely important role as a detector of problems, an experimenter of possibilities and a leverage of solutions. Science is always being instigated through creative and collaborative processes, sometimes on a global scale. Science increasingly enters the scene and has the role of finding solutions to all the concerns that afflict us as human beings, so that our planet Earth is, and continues to be, enjoyed not only by us today, but by generations to come also. Towards this broad sense of sustainability is important to flag and learn from good practices and successful examples from different geographies.

This book aims to explore the cutting-edge research, technologies, and strategies that are transforming the way we understand, protect, and utilize our coastal resources. Through a multidisciplinary lens, the chapters that follow delve into the latest advancements in areas such as ecosystem modeling, new fishing tools, science communication, blue growth and pollution among others topics. Various areas of knowledge are covered, from natural sciences to social sciences, passing through aspects of law connected to coastal resources. Whether you are a coastal manager, a marine scientist, or simply someone with a passion for our blue planet, this book will provide you with a comprehensive understanding of the innovative approaches that are shaping the future of coastal resource management. Cases and reflections on science in Africa, Asia, Oceania and Europe are presented here. All of these chapters touch on aspects related to scientific innovation, whether of a technological nature, or just an approach and facilitation in the inclusion of stakeholders, in a lighter and more in-depth way.

## BOOK COVERAGE

The book *Scientific Innovations for Coastal Resource Management* contains in total 10 distinct chapters. According to what seemed to be the focus of the main text, the chapters are divided into three sections: "Section 1: Management and Conservation", "Section 2: Fisheries" and "Section 3: Case Studies".

Therefore, the first section is dedicated to and addresses the management and conservation of coastal resources and includes three chapters.

Chapter 1 "Impact of Scientific Innovation in Coastal Resource Management to Fostering Blue Economic Development, Environmental Mitigation, and Pollution Control" addresses in a generalized way the importance of coastal resources and their adequate management at an environmental level. It also develops reflections on the challenges of managing coastal resources and the various problems that affect them. This reflection covers various topics from marine pollution, overfishing and habitat loss, the impact of climate change, inadequate capacity and resources, and unsustainable coastal development. There is then a reflection on the role that scientific innovation – namely science and technology – plays in the management of coastal resources and the way in which scientific innovation can help mitigate previous problems. General cases of sustainable fisheries management, tourism and the blue economy, renewable energy technologies, as well as environmental monitoring and pollution control are also covered. Finally, three case studies in different geographies are presented in a very synthetic way. Future perspectives are anticipated, and a conclusion drawn.

Chapter 2 "The Effects of Marine Debris on Indonesia's Coral Reefs and Mitigation Initiatives" explores the problem of the composition of marine debris in a very expeditious way. The different types of waste that end up in the sea and their classification are categorized. Subsequently, a generalist analysis based on literature is carried out on the sources from which this waste comes. The case study of marine litter on coral reefs is explored, and the consequences on these living beings, particularly in terms of physical damage, chemical pollution, biological impact, and socioeconomic effects. Some solutions and recommendations are put forward, namely through the engagement of communities and their education, campaigns on waste in the media, a campaign to reduce the single use of plastic, beach cleaning actions, among others, including innovative monitoring technologies and marine litter tracking. The importance of cooperation between various organizations is also highlighted. It also explores a little about the importance of the legislation that exists to regulate marine litter and describes how scientific research and monitoring of marine litter is carried out.

Chapter 3 "The Green Sentinel: Impact of Ecological Constitution on Conservation of Marine Ecosystem" reflects on the Ecological Constitution in relation to the environmental challenges that occur in our times. A review of the literature on the evolution of legal regimes is made. Some specific cases in different parts of the world are analyzed. Some criticisms and challenges that exist in the implementation of the Ecological Constitution are pointed out. Marine conservation perspectives are analyzed, namely in the establishment and maintenance of marine protected areas, ecosystem management strategies and international cooperation in marine conservation based on the strengthening of environmental constitutional principles. The issues of overfishing and pollution are addressed. These are marine issues that cross borders and there is a need to design appropriate policies and engage stakeholders, as well as the availability of governments to ease these implementations. Ecological sustainability and states that have environmental protection approaches are referred. The growing importance of aquaculture is mentioned, which has already supplanted fishing that has been stagnant for some years. Other topics are also covered, such as per capita plastic emissions in the ocean and the increase in surface temperature on the Great Barrier Reef. Finally, the relevance of the Ecological Constitution is mentioned, and some conclusions are drawn.

The second section is entirely dedicated to marine and coastal resources related to fishing. This section is also made up of three chapters.

Chapter 4 "First Insight Into the Deployment of Artificial Reefs to Support Local Fisheries in São Tome and Príncipe: Artificial Reefs to Support Local Fisheries". The authors report about the first artificial reef that was deployed in sandy areas of the north of São Tomé island. Under the apparent decline of the fishing resources, that are the base of diet in this island, NGOs organizations took the lead to propose as alternative fishing tools bottom artificial reefs. This narrative chapter describes the steps for the implementation of such a project since the planning, reef selection areas, reef unit's typology etc. An important focus is given to local communities that were surveyed before the artificial reef deployment to share their perspectives about this innovative fishing tool. The work ends with some glimpse of ecological and socioeconomic findings and with some management considerations.

Chapter 5 "Same Ocean, Different Currents: A Review of Socioeconomic Variables in Small Pelagic Fisheries Assessment" stands out in a literature review in the main scientific search engines with the aim of finding out which species of pelagic fish are most studied, what are the main focuses of studies, and what are the current approaches and socioeconomic variables. A comparative analysis of socioeconomic variables and indicators is carried out, focusing on profit, net present value, gross and net additional values, behavioral, and other approaches. A transversal diagnosis is also made on the socioeconomic dimensions and fisheries management. Some conclusions are drawn from this study on the selected literature.

Chapter 6 is "Enhancing Ecosystem Services: The Role of Artificial Reefs and AIS Technology". This chapter discusses the potential that the creation of artificial reefs has to develop ecosystem services. These artificial habitats essentially develop four types of ecosystem services: provision services, which allow the production of goods that can be consumed such as food from the sea, for example; cultural services that allow users to develop scientific and cultural enrichment capacity, such as in the case of diving; support, such as when it is possible to enhance the circularity of nutrients in the marine environment; and regulatory services, such as the possibility of being able to protect the stormy coast.

Finally, the last section addresses several case studies in particular. This section contains four chapters.

Chapter 7 "Science Communication of Coastal and Marine Issues: Insights From Diverse Case Studies" focuses on communicating science to various types of audiences. The chapter begins by providing a framework for coastal and marine science communication. It then explains the methodological approach used in different case studies on the aforementioned topic. The different case studies cover salt gardens, geosciences and dance activities where the sea rolls over the sand, animation series about the coast for children, comics about aquaculture, visits to salt marshes, photographic exhibitions, plastic pollution in the sea, impacts of climate change, debate about sustainable food on the radio and podcast, and a television documentary about sardines. The case studies are analyzed comparatively. Perceptions and lessons learned are examined, namely about what the key messages were, what resources were needed and who the target audiences were. Finally, some caveats are outlined by way of conclusion.

Chapter 8 "Applying Ecohydrology to Promote a Nature-Based Solution: To Improve Quality of Life and Ecosystem Health in Luanda Bay, Angola" contextualizes the bay of Luanda in a historical perspective since its foundation some centuries ago, and the importance it has had in the trade of goods of all kinds. The fact of the rapid expansion of the city of Luanda in the twentieth century due to the rural exodus and demographic growth in three distinct periods is praised. Subsequently, a characterization of the environmental and life quality and the challenge of the bay is made, particularly with regard to the marine species that live there. Anthropogenic pressures arising from intense urbanization are mentioned. The potential of economic activities that depend on the bay are also addressed, such as tourist and recreational activities, and fishing. Next, ecohydrological modeling methods are discussed to respond to the challenges that exist in the bay. The projects developed are briefly described and the results for future solutions, in order to have a cleaner Luanda Bay. Various graphic modeling representations are made. Finally, some conclusions are drawn from the approach taken.

Chapter 9 is “Analyzing the Role of Artificial Reefs in Blue Tourism: Vilamoura Artificial Reef (South Portugal)”. In this chapter it is explored in a synthetic way the development of the tourist resort of Vilamoura, starting from the key element that was the construction of the marina. This infrastructure, due not only to its size but also to its location, allowed the development of numerous coastal and maritime activities from an early age. In addition, another type of infrastructure was also developed in the Vilamoura area, but on the seabed and these were the artificial reefs that, after colonization by well-established trophic chains in these new habitats, have allowed the development of the portfolio of some companies that work with various types of vessels, particularly for recreational fishing and companies linked to diving activities.

Chapter 10 “Sport Events' Strategic Multi-Leverage: Linking Beach Soccer to Big Waves to Stimulate Coastal Tourism” explores the meaning of maximizing events in the town of Nazaré (Portugal), which is a tourist destination internationally known for surfing in giant waves. The authors carry out a literature review that addresses sport events from the perspective of designing strategic use, whether at an economic, social, environmental, sporting, political level, and crossed with a strategy aimed at in a transversal sense with coordination between the participants. The methodological part involves a characterization and contextualization of events, a research design, data collection and analysis. The results of the events covered in various editions are presented, namely in the strategic objectives and enhanced actions, where comparisons are made between different years. The strategic objectives for the different actions planned and implemented for the different events are also presented. The discussion is a critical analysis of the strategies, stakeholders' involvement, and the potential success of implementation. Finally, a conclusion is drawn on the contribution of this research on the concept of strategic use in sports tourism events.

**Jorge Ramos**
*CinTurs, University of Algarve, Portugal*

**Benjamin Drakeford**
*University of Portsmouth, UK*

**Francisco Leitão**
*CCMAR, Portugal*

# Acknowledgment

The co-editors would like to highlight that they are very pleased to be able to bring together all the pieces of this joint endeavor thanks to the collaboration and commitment of all those who collaborated in one way or another and who come from various parts of the world. It is essential to thank all the people involved in this project for their help. Without their support, this book would not have become a reality.

Firstly, the co-editors emphasize that they would like to thank each of the authors/co-authors for their contributions. A sincere gratitude to the chapter writers who contributed their time, knowledge and effort to this book.

Secondly, the co-editors wish to recognize the valuable and essential contributions of the reviewers, who also sacrificed their precious time to improve the quality of their peers' manuscripts. Some authors also acted as referees. Your dual task is greatly appreciated. It should also be noted that some referees review more than one manuscript.

Thirdly, some of the researchers and academics contacted, despite not being authors/co-authors, provided their time and experience in the review and gave some advice on the chapters.

Fourthly, the co-editors would like to express their enormous gratitude to the IGI Global team – particularly the Assistant Development Editors of the Books Division – who were always available to provide all the technical support necessary for this book to reach a successful conclusion.

Fifthly, the editors would also express gratefulness to the Portuguese Foundation for Science and Technology (FCT), through projects UIDP/04020/2020 and UIDB/04020/2020 (DOI:10.54499/UIDB/04020/2020, https://doi.org/10.54499/UIDB/04020/2020) and 2022.01783.PTDC (DOI:10.54499/2022.01783.PTDC, https://doi.org/10.54499/2022.01783.PTDC). JR has received the stimulus of scientific employment 2021.02180.CEECIND (DOI:10.54499/2021.02180.CEEC-

IND/CP1672/CT0003, https://doi.org/10.54499/2021.02180.CEECIND/CP1672/CT0003). FL has received Portuguese national funds from FCT under the contract programme DL57/2016/CP1361/CT0008 and FCT 2022.04803.CEECIND. In addition, this study received Portuguese national support from FCT through projects UIDP/04326/2020 (DOI:10.54499/UIDP/04326/2020) and LA/P/0101/2020 (DOI:10.54499/LA/P/0101/2020).

# Introduction

The intrinsic characteristics of coastal areas have made them preferred sites for occupation from a human history perspective. These coastal areas have suffered intense development becoming among the environments most affected by human presence and activities.

The world's coasts form a narrow interface zone between marine and terrestrial areas, where a significant and growing portion of the human population and global economic activity are concentrated. These land-coast transition areas are in the low-elevation coastal zone, encompassing 2% of the earth's land area (McGranahan et al., 2007) and are a home of 10% of the total population, including high populated urban areas. The global population living in the low elevation coastal zone is substantial and growing (IPCC 2002). Estimates using 2010 and 2020 data reveal that approximately 11% of the world's population was living in low elevation areas below an elevation of 10 meters, compared to 10% based on 2000 estimates. Based on 2000 estimates, as much as 13% of the world's urban population live in the low elevation coastal zones and most of the population in these zones occur in developing areas (IPCC 2022, 2023). The number of people living in the low elevation coastal zone worldwide may reach 1 billion people by 2050. As population density and economic activity in the coastal zone increases, pressures on coastal ecosystems increase. Among the most important pressures are habitat conversion, land cover change, pollutant loads, and introduction of invasive species. The rapid expansion of socio-economic activities in coastal areas over the last few decades, including tourism, nature conservation, energy demand (fossil and renewable), coastal fisheries, and industrial and urban development, has complicated management tasks.

It is now widely acknowledged that sustainability goals require a balanced approach to managing social, economic, and environmental issues. The concept of 'integrated' or 'ecosystem-based' management is not new arising by systems thinking and the environmental movements of the 1960s and 1970s (Douvere, 2008). By the following decade, these ideas had evolved into Integrated Coastal Zone Management (ICZM) or Integrated Coastal and Marine Area Management (ICMAM). Accordingly, to IUNEP (2008) integrated coastal zone management can be defined as "...a dynamic process for the sustainable management and use of coastal zones, taking into account at the same time the fragility of coastal ecosystems and landscapes, the diversity of activities and uses, their interactions, the maritime orientation of certain activities and uses and their impact on both the marine and land parts." Recently the conceptual basis of Integrated Coastal Zone Management evolute in the way it should considered the Ecosystem Approach a concept defined in Convention for Biological Diversity as "... places human needs at the centre of biodiversity management. It aims to manage the ecosystem, based on the multiple functions that ecosystems perform and the multiple uses that are made of these functions. The ecosystem approach does not aim for short-term economic gains but aims to optimize the use of an ecosystem without damaging it ".

Both the Coastal Zone Management and Ecosystem Approach principles recognize the inherently dynamic nature of ecosystems and the uncertainties involved in managing them. Consequently, both frameworks advocate for a holistic, adaptive, and flexible approach to natural resource management. An advantage of the Ecosystem Approach is its emphasis on longer-term, sustainable perspectives, guiding decision-makers away from short-term decisions establishes that may fail to deliver lasting, cost-effective socio-economic and environmental benefits. Longer time perspectives can indeed alter cost-benefit or cost-risk assessments and thus influence decision outcomes. The same can be argued for the Coastal Zone Management and Ecosystem Approach framework (Haines-Young et al., 2011).

Coastal zone sustains sensitive ecosystems that provide critical habitats for many endangered species and essential ecosystem services, such as coastal protection, fisheries, and other living resources, rich agricultural lands, and areas of high aesthetic value. These zones are typically considered public heritage and serve as vital connections between land and sea. Additionally, in large world cities a significant portion of global tourism, increasingly crucial to national economies, are situated along the coast. Nowadays, there has been growing concern to balance steady economic growth and social development in coastal areas while preserving their natural features and ecological services. Therefore, coastal resource management should aim to consider the abundant natural capital in coastal zones, represented by ecosystems and the ecosystem services they provide, which rely on the complementary and interdependent nature of marine and terrestrial systems.

Governments are endorsing the goal of sustainable development a more sustainable coastal management strategy requires an interdisciplinary and integrated management process. There are no simple answers to what is best for a particular system from a resource management perspective. It is the task of scientists from various disciplines to provide a comprehensive picture to decision-makers. The Coastal Zone Management System aims to equip coastal authorities with the knowledge and tools for such an integrated management approach. The common goal is to achieve socially and environmentally sustainable coastal system in these areas of the world. These coastal systems present a wide spectrum of management challenges, varying significantly in their ecological state and human pressures, from more pristine areas to more heavily occupied and degraded areas, such as estuaries.

The accelerated development in many countries presents demanding challenges in the management of natural resources, particularly in coastal areas. Integrated coastal management approaches are essential, combining human, physical, and biological aspects of the coastal zone within a unified management framework. Integrated coastal management is presented here as a broad, multi-purpose endeavour aimed at improving the quality of life for communities dependent on estuarine resources and assisting local decision-makers in achieving sustainable development of coastal areas, from the headwaters of coastal watersheds and estuarine to the outer marine areas. This is why Coastal Zone Management System plays a critical role in achieving several/most sustainable development goals (SDG), particularly those related to environmental sustainability, economic development, and social well-being:

SDG 1: No Poverty: Economic Opportunities: Promotes sustainable fisheries, tourism, and other coastal livelihoods to reduce poverty in coastal communities.

SDG 2: Zero Hunger: Sustainable Fisheries: Ensures the sustainable management of fish stocks and marine resources, providing food security for coastal populations.

SDG 6: Clean Water and Sanitation: Water Quality: Protects coastal and marine water quality through pollution control and sustainable watershed management.

SDG 8: Decent Work and Economic Growth: Blue Economy: Promotes sustainable economic activities, such as ecotourism, aquaculture, and renewable energy, in coastal areas.

SDG 11: Sustainable Cities and Communities: Resilient Infrastructure: Develops resilient infrastructure and settlements that can withstand coastal hazards and climate change impacts.

SDG 13: Climate Action: Climate Adaptation: Implements strategies to adapt to climate change, including protecting coastal ecosystems that act as natural barriers against sea-level rise and extreme weather events.

SDG 14: Life Below Water: Marine Conservation: Protects marine ecosystems and biodiversity through the establishment of marine protected areas and sustainable resource management practices.

SDG 15: Life on Land: Habitat Conservation: Ensures the conservation of coastal habitats, such as mangroves, wetlands, and estuaries, which provide critical ecosystem services.

SDG 17: Partnerships for the Goals: Collaborative Efforts: Encourages partnerships between governments, local communities, the private sector, and international organizations to achieve sustainable coastal management.

Research in coastal areas has historically been divided into two largely separate streams: the natural sciences and the social sciences. Interaction between these fields has been limited, primarily occurring in the realm of engineering. However, it has become increasingly evident that a focus on interdisciplinary approaches is essential to effectively analyse the interconnected natural and social drivers of coastal change. This includes understanding the interactions and feedback loops between human systems, extending from inland watersheds to the ocean shelf (Ramesh et al., 2015). Interdisciplinary integration was the primary focus of contemporaneous coastal management. The inclusion of social sciences, human ecology, and management-oriented subjects provided the integrative principle. The core message emphasized is that management challenges are complex and multifaceted, encompassing physical forces, natural hazards, variability, vulnerability, and socio-cultural issues (Ramesh et al., 2015).

The key elements of a Coastal Zone Management System include:

1. Integrated Coastal Zone Management (ICZM): A process that brings together all stakeholders involved in the development, management, and use of coastal zones. It aims to ensure sustainable development and minimize conflicts among various coastal uses.
2. Ecosystem Approach: Focuses on managing coastal areas by considering the entire ecosystem, including human impacts, to maintain ecological integrity and provide sustainable benefits to communities.
3. Policy and Legislation: Development of legal frameworks and policies to regulate activities within coastal zones, protect sensitive areas, and promote sustainable development.
4. Planning and Zoning: Creating land-use plans and zoning regulations to manage coastal development, protect natural resources, and reduce vulnerability to hazards like flooding and erosion.
5. Monitoring and Assessment: Regular monitoring of coastal environments to assess the health of ecosystems, the effectiveness of management measures, and emerging threats.
6. Public Participation: Involving local communities, stakeholders, and the public in decision-making processes to ensure that management actions reflect diverse interests and knowledge.
7. Adaptive Management: Implementing flexible management strategies that can be adjusted based on new information and changing conditions, ensuring long-term sustainability.

8. Capacity Building and Education: Providing training and resources to local authorities, communities, and stakeholders to enhance their ability to participate in and benefit from coastal management efforts. Well informed stakeholders are a crucial element in the decision process.

A Coastal Zone Management System can be implemented at various scales, from local to national levels, and often involves collaboration between government agencies, non-governmental organizations, academic institutions, the private sector and citizens (public at large). As a result of a multidisciplinary scientific endeavour, this book will appeal to a wide audience across environmental and social disciplines. It will interest anyone working in the field of Coastal Zone Management, focus on Innovations for Coastal Resource Management, from scientists to decision-makers. Though it focuses on examples from different countries, the approach presented in this book serves as a target for scientists and practitioners worldwide facing similar challenges.

## REFERENCES


Douvere, F. (2008). The importance of marine spatial planning in advancing ecosystem-based sea use management. *Marine Policy*, 32(5), 762–771. 10.1016/j.marpol.2008.03.021

Haines-Young, R., & Potschin, M. (2011). Integrated Coastal Zone Management and the Ecosystem Approach. PEGASO Deliverable D2.1, September 2011, 11 pp. CEM Working Paper No 7. Available under: https://www.nottingham.ac.uk/CEM/WorkingPapers.html

Intergovernmental Panel on Climate Change (IPCC). (2022). The Ocean and Cryosphere in a Changing Climate: Special Report of the Intergovernmental Panel on Climate Change (1ed.). Cambridge University Press. . ISBN 978-1-009-15796-4.10.1017/9781009157964.006

Intergovernmental Panel on Climate Change (IPCC). (2023). Climate Change 2022 – Impacts, Adaptation and Vulnerability: Working Group II Contribution to the Sixth Assessment Report of the Intergovernmental Panel on Climate Change (1 ed.). Cambridge University Press. 10.1017/9781009325844.019

IUNEP/MAP/PAP. (2008). *Protocol on Integrated Coastal Zone Management in the Mediterranean*. Priority Actions Programme.

McGranahan, G., Balk, D., & Anderson, B. (2007). The rising tide: Assessing the risks of climate change and human settlements in low elevation coastal zones. *Environment and Urbanization*, 19(1), 17–37. 10.1177/0956247807076960

Ramesh, R., Chen, Z., Cummins, V., Day, J., D'Elia, C., Dennison, B., Forbes, D. L., Glaeser, B., Glaser, M., Glavovic, B., Kremer, H., Lange, M., Larsen, J. N., Le Tissier, M., Newton, A., Pelling, M., Purvaja, R., & Wolanski, E. (2015). Land–Ocean Interactions in the Coastal Zone: Past, present & future. *Anthropocene*, 12, 85–98. 10.1016/j.ancene.2016.01.005

# Section 1
# Management and Conservation

# Chapter 1
# Impact of Scientific Innovation in Coastal Resource Management:
## Fostering Blue Economic Development, Environmental Mitigation, and Pollution Control

**Sumanta Bhattacharya**
https://orcid.org/0000-0003-2563-2787
*Maulana Abul Kalam Azad University of Technology, India*

**Bhavneet Kaur Sachdev**
*Suresh Gyan Vihar University, India*

## ABSTRACT

*Scientific innovation plays a pivotal role in coastal resource management, shaping strategies to foster blue economic development while mitigating environmental risks and controlling pollution. The chapter highlights the multifaceted impacts of scientific innovation in coastal areas. Innovations such as advanced monitoring technologies, artificial intelligence, and nature-based solutions are transforming coastal management practices, enabling real-time detection of pollution events, enhancing ecosystem resilience, and promoting sustainable economic activities. By leveraging scientific knowledge and interdisciplinary collaboration, stakeholders can develop holistic approaches to coastal resource management that balance economic growth with environmental sustainability. The chapter underscores the importance of integrating scientific innovation into policy-making processes and fostering partnerships among governments, industries, and communities to ensure*

DOI: 10.4018/979-8-3693-2436-3.ch001

*the long-term health and prosperity of coastal ecosystems and economies.*

## INTRODUCTION

The blue economy, which promotes the responsible utilisation of marine resources for economic growth, benefits greatly from scientific advances in coastal resource management. Accurate monitoring and management of fisheries, aquaculture, and other coastal activities are made possible by technological advancements in remote sensing, geographic information systems (GIS), and data analytics. Marine industries may thrive for the long haul when they adopt sustainable practises backed by scientific data, which boosts the economy and protects the coastal ecosystems. Restoration of damaged coastal habitats is made easier with the use of cutting-edge scientific techniques. Artificial intelligence and satellite photography are two examples of innovative methods that aid in tracking coastal habitat shifts and gauging the success of restoration initiatives. Resource managers' ability to reduce environmental degradation, protect biodiversity, and strengthen coastal ecosystems' resistance to climate change hinges on their grasp of coastal ecosystems' underlying processes (Wenhai et. al., 2019). The monitoring and management of pollution in coastal areas rely heavily on scientific developments. Autonomous underwater vehicles (AUVs), unmanned aerial vehicles (UAVs), and other sensors can monitor water quality in real time and report back any signs of pollution or harmful algal blooms. The timely response made possible by this data protects marine life and guarantees the well-being of coastal populations that rely on clean water for their survival. Integrated Coastal Zone Management is bolstered by scientific advancements since it is a comprehensive strategy that takes into account the interdependence of diverse activities in coastal areas. The effects of urbanisation and agriculture, for example, on coastal ecosystems can be evaluated with the aid of geographic information systems and modelling tools used by planners. ICZM promotes sustainable development that strikes a balance between human demands and the preservation of coastal habitats by including ecological, social, and economic factors (de Lima et al., 2020). In addition to technology breakthroughs, scientific innovation also includes community engagement and capacity building. To include residents in decision-making, coastal resource management programmes draw on findings from the social sciences and employ participatory methods. In turn, this helps coastal communities create resilient and sustainable practises by instilling a sense of ownership and responsibility. Blue economic development efforts are more likely to succeed when they get support from empowered communities that are better able to adapt to environmental changes, such as the effects of climate change. Coastal environments can now be maintained without jeopardising blue economic growth thanks to sustainable practises made possible by recent discoveries in aquaculture. Recirculating aquaculture systems (RAS) and integrated multitrophic aquaculture (IMTA) are two examples of responsible aquaculture practises that strive to protect marine ecosystems, lessen pollution, and secure marine resources for future generations (Wilke, 2023). Allocating maritime space and resources for environmental, economic, and social benefit is what maritime Spatial Planning is all about. Suitable places for shipping, fishing, and conservation can be found with the help of data-driven improvements in MSP. This ensures that economic growth is sustainable, minimising the likelihood of conflicts and lessening the strain on fragile coastal habitats. By tapping into clean energy sources, developments in ocean energy technologies boost the blue economy. Wave and tidal energy systems, as well as offshore wind farms, fall under this category. By offering clean and sustainable energy sources, these innovations assist economic growth while minimising the environmental damage associated with traditional energy production. When it comes to reducing pollution levels in the ocean, marine biotechnology is an essential tool. Restoration of coastal ecosystems and reduction of environmental damage are aided by bioremediation methods, which employ naturally existing microbes to degrade pollutants (Issifu et. al., 2024). Tsunamis, hurricanes, and storm surges are just some of the natural hazards that threaten coastal areas. Coastal communities are equipped with the knowledge they need to evacuate and prepare for disasters thanks to scientific advancements in early warning systems like improved monitoring technologies and predictive modelling. This helps lessen the impact of natural disasters on both human populations and coastal ecosystems. In order to achieve environmental sustainability, reduce pollution, and grow the blue economy, scientists are constantly innovating coastal resource management. These developments equip decision-makers, communities, and enterprises to maintain the long-term health and resilience of coastal ecosystems by striking a balance between economic expansion and conservation efforts (Marios & Mitsch, 2015).

## Importance of Coastal Resources

The economic growth and survival of coastal towns across the world are greatly influenced by coastal resources, which are priceless assets. The fisheries, aquaculture, tourism, and renewable energy assets that make up these resources all play an important role in the lives and overall health of the coastal communities (Figure 1). For millions of people throughout the world, fisheries are not just a source of food and income, but also a vital component of coastal economies. Seafood and fish caught in coastal seas are not only economically significant, but also provide an important nutritional component. In underdeveloped nations, where coastal populations depend mostly on marine resources for their nutritional requirements, coastal fisheries play a particularly important role in ensuring food security and nutrition. In addition, people in other fields like processing, transportation, and marketing rely on the fishing sector for their lives because of the money and jobs it creates (Pascoe et. al., 2023).

*Figure 1. Importances of Coastal Management for Environment*

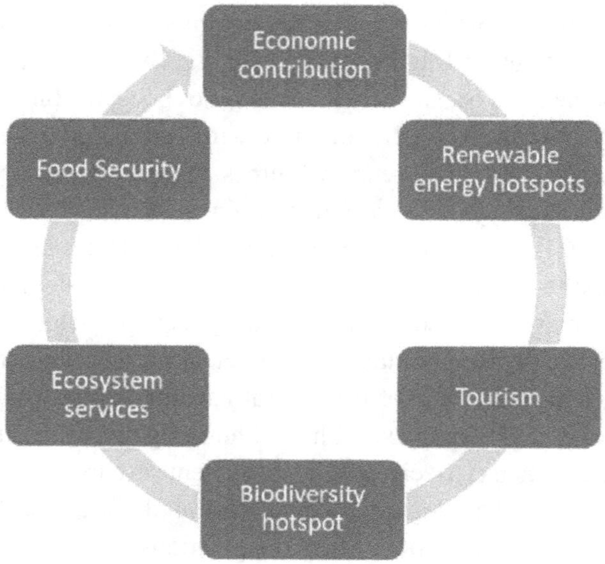


Aquaculture, which provides sustainable alternatives to wild fisheries, has become an important part of coastal resource utilisation, helping to supply the increasing demand for seafood. Because of the favourable climatic circumstances and access to nutrient-rich waterways, aquaculture enterprises, such as oyster cultivation, seaweed farming, and fish farms, flourish in coastal locations. Coastal communities benefit from aquaculture in several ways: it increases wild fish stocks, creates jobs, and generates revenue. Reducing stress on wild fish populations and encouraging responsible use of resources are two additional ways in which sustainable aquaculture techniques improve ecosystem health. (Little et. al., 2016) Another major factor boosting economies is coastal tourism, which brings in millions of tourists every year to beaches all over the globe. Tourists coming to the coast for fun things like swimming, surfing, scuba diving, and animal viewing are drawn to the stunning coastal vistas, clean beaches, and varied marine ecosystems. Businesses and towns along the coast get a financial boost from tourists spending money at restaurants, hotels, amusement parks, and other coastal attractions. In addition to boosting economic growth and development in coastal regions, tourism generates job possibilities across numerous sectors, such as the hotel, transportation, and retail industries. There is a great deal of untapped potential for renewable energy generation in coastal regions, which may tap into the forces of nature such as wind, waves, and tides to provide environmentally friendly electricity. Reducing reliance on fossil fuels and minimising greenhouse gas emissions, offshore wind farms, tidal energy installations, and wave energy converters use the abundant energy resources in coastal waterways to produce power. Investment, new infrastructure, and new jobs in the renewable energy sector are just a few ways in which renewable energy projects along the coast boost economic activity and help ensure the continued viability of the region's energy supply and the environment (Nguyen, 2021). Coastal resources are priceless assets that support the prosperity, quality of life, and economic growth of coastal towns around the globe. Coastal economies rely on industries like fishing, aquaculture, tourism, and renewable energy to generate money, create jobs, and promote sustainable development. The sustainability of coastal resources and the maximisation of their socioeconomic advantages for current and future generations depend on their effective management and protection. Communities along the coast may protect the resilience and health of coastal ecosystems while also experiencing sustainable economic growth if they value coastal resources and manage them in an integrated way.

## Challenges in Coastal Resource Management

Numerous environmental problems are plaguing coastal regions, endangering marine ecosystems and the quality of life for those who live there (Figure 2). Overfishing, habitat loss, pollution, and the effects of climate change are just a few of the threats that the coast is facing, and they all demand immediate action and coordinated strategy to combat. Overfishing, caused by factors such as rising seafood demand, improvements in fishing technology, and ineffective fisheries management, is one of the biggest problems coastal ecosystems are facing. Depleting fish populations, upsetting marine food webs, and destroying ecological harmony are all consequences of overfishing. Coastal populations who rely on fishing for their livelihoods are also put at risk. To reduce the negative effects of overfishing and bring back fish populations that have been severely reduced, it is crucial to implement sustainable fisheries management strategies. These include quotas based on scientific evidence, marine protected areas, and ecosystem-based approaches (Craig & Ruhl, 2010). The degradation of coastal habitats as a result of human activities such as urbanisation, industrialization, and coastal infrastructure projects is another pressing concern for coastal regions. A number of coastal ecosystems are particularly at risk from human interference, including mangrove forests, seagrass beds, coral reefs, and wetlands. Coastal ecosystems are under risk of losing their capacity to defend coastlines, sequester carbon, and serve as nursery grounds for marine species due to habitat loss and degradation, which in turn reduces biodiversity and ecosystem resilience. To ensure that coastal habitats and the benefits they provide are protected, it is crucial to implement integrated coastal zone management strategies that put an emphasis on restoring and conserving habitats as well as planning for sustainable land use.

*Figure 2. Challenges to Coastal Management*

Marine pollution, especially from things like oil spills and plastic trash, is a big problem and a serious danger to coastal ecosystems. The contamination of coastal waterways, beaches, and marine ecosystems by plastic pollution, which is caused by insufficient waste management procedures and excessive plastic use, endangers marine life through ingestion, entanglement, and habitat damage. In a similar vein, oil spills caused by marine mishaps, offshore drilling, or the unlawful release of pollutants may devastate coastal ecosystems through extensive pollution, the loss of habitat, and long-term ecological damage. To combat marine pollution and maintain coastal areas, it is vital to strengthen pollution control methods, enforce rigorous legislation, and promote public awareness and community involvement (Xiuzhen et. al., 2018).

The effects of sea level rise, ocean acidification, and extreme weather events are intensified in coastal locations due to climate change, which compounds pre-existing environmental concerns. Erosion, flooding, and saltwater intrusion are all consequences of sea level rise that endanger coastal populations, infrastructure, and ecosystems. The process of ocean acidification, which is caused by the absorption of carbon dioxide emissions, has far-reaching consequences for marine ecosystems, especially coral reefs and creatures that make shells. This has significant implications for biodiversity and the overall health of these ecosystems. Natural catastrophes already strike coastal areas more frequently and with more force due to increased susceptibility to climate-induced severe weather events including heatwaves, storms, and hurricanes, which can lead to coastal floods, erosion of shorelines, and destruction of habitat (Griggs & Reguero, 2021). To make coastal populations and ecosystems more resistant to the effects of climate change, adaptation strategies are essential. These include coastal zone planning, ecosystem-based methods, and climate-resilient infrastructure. Coastal people' socioeconomic security and the health of marine ecosystems are jeopardised by the myriad environmental threats that these locations confront. Coastal resource management must be approached with an integrated and multi-disciplinary mindset in order to tackle the linked problems of overfishing, habitat degradation, pollution, and the consequences of climate change. We can ensure that coastal ecosystems remain healthy and resilient and that coastal communities benefit in the long run if we tackle these concerns with conservation measures based on research, sustainable development methods, and community participation.

## Role of Scientific Innovation

However, scientific innovation is bringing forth new answers that are revolutionising the strategies that are used to manage coastal resources. Coastal ecosystems and communities are confronted with complex difficulties. The extraordinary insights into coastal habitats that have been obtained via improvements in technology, data analytics, remote sensing, and modelling approaches have made it feasible for more informed decision-making, proactive management strategies, and sustainable development efforts to be implemented. A variety of stakeholders, scientists, and policymakers can all profit from these developments (Figure 3).

One of the primary areas in which scientific innovation is making significant progress is in the monitoring and evaluation of coastal ecosystems. Coastal habitats such as mangroves, seagrass beds, coral reefs, and wetlands may be meticulously mapped and monitored via the utilisation of remote sensing technology such as satellite photography, aerial drones, and unmanned aerial vehicles (UAVs). The utilisation of these technologies enables the collection of spatial data with a high resolution. Because of these technologies, scientists are able to monitor changes in habitat size, condition, and biodiversity over time. This allows them to detect conditions such as habitat loss, degradation, and ecological collapse at an earlier stage. The utilisation of remote sensing data by coastal resource managers can facilitate the prioritisation of conservation efforts, the identification of biologically relevant places, and the development of individualised management strategies, therefore enhancing the preservation and restoration of coastal ecosystems (Rifai et. al., 2023).

*Figure 3. Role of Science and Technology in Coastal Resource Management*

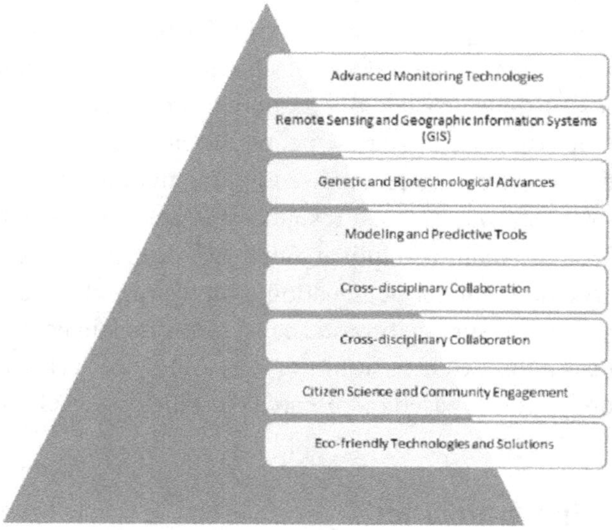


Developments in data analytics and machine learning are altering the processing, analysis, and interpretation of data pertaining to coastal resources. Big data analytics makes it possible to incorporate a wide range of datasets, which may include ecological, socioeconomic, and biological factors, in order to develop comprehensive models and forecasting tools for the management of coastal resources. In order to better allocate resources, reduce risks, and create resilience in the face of environmental change, decision-makers may use these models to anticipate ecological trends, assess ecosystem services, and study the consequences of human activities on coastal habitats. These models can allow for improved resource allocation, risk reduction, and resilience building (Ikegwu et. al., 2024).

Scientific innovation, which includes data analytics and monitoring, is allowing for the development and implementation of management plans that are based on coastal ecosystems. When stakeholders use integrated modelling frameworks such as spatial planning tools and ecosystem-based models to evaluate trade-offs, find synergies, and optimise management strategies, they are able to better manage a variety of ecosystem services. These services include fisheries, the conservation of biodiversity, and adaptation to climate change. Adaptive management approaches, collaborative decision-making, and stakeholder participation are all made possible by these technologies. The goal of these technologies is to encourage the sustainable use of coastal resources while simultaneously reducing conflicts and increasing benefits for all stakeholders (Delacamara et. al., 2020).

The development of technologies like as bioinformatics, genetic sequencing, and marine biotechnology is opening up new opportunities for the preservation of coastal ecosystems and the utilisation of their resources in a manner that is environmentally responsible. These biotechnological breakthroughs, which include genetic engineering and bio-prospecting, have the potential to enhance aquaculture operations, make aquaculture species more resistant to illnesses, and restore ecosystems that have been destroyed through ecological engineering and bio-remediation. In the realm of coastal resource management, these biotechnology advancements not only solve the problems that are now being faced, but they also open the way for future advances in the sector. The area of coastal resource management is now going through a time of revolutionary change as a result of the developments in scientific understanding that have made it feasible to develop more efficient techniques of monitoring, analysing, and managing coastal ecosystems and the resources that they contain in a manner that is sustainable. The advancement of scientific knowledge offers a plethora of chances to improve our understanding of coastal environments, to address new issues, and to guarantee the long-term viability of coastal ecosystems and populations. Remote sensing and data analytics are only two examples of the potential that are available, along with ecosystem-based modelling and biological technology developments. We can safeguard coastal resources and foster resilient and prosperous coastal communities for the benefit of future generations by accepting innovation and promoting collaboration among scientists, policymakers, and stakeholders. This will allow us to protect coastal resources and encourage coastal communities to thrive (Rotter et. al., 2021).

## Sustainable Fisheries Management

Managing fisheries in a sustainable manner is essential to the maintenance of marine ecosystems as well as the millions of people who depend on fishing for their livelihood and financial support. In recent years, technology advancements have been of significant assistance in a number of areas, including the development of fisheries management, the promotion of sustainable fishing practices, and the prevention of overexploitation of fish populations. Through the provision of information that is more precise, up-to-date, and comprehensive, modern innovations in fisheries management have significantly improved decision-making and resource utilisation. These innovations include tracking through the global positioning system (GPS), electronic monitoring (EM), and data-driven stock assessments (Bradley et. al., 2019).

One of the most significant advancements in the field of fisheries management is the widespread use of global positioning system (GPS) tracking technology. This technology makes it possible to monitor and track the whereabouts and activities of fishing vessels in real time. Using GPS tracking systems that are installed on fishing vessels, authorities are able to more effectively monitor vessel movements, quantify the amount of effort that is put into fishing, and enforce fishing regulations. In order to protect sensitive fish populations and the ecosystems in which they live, the authorities can utilise precise location data to verify that fishermen are adhering to their fishing quotas, identify instances of illegal fishing, and enforce limits on fishing in restricted areas and during specific seasons. Furthermore, stakeholders have the capacity to get data on vessel monitoring through the utilisation of GPS tracking technology, which enhances accountability and transparency in the administration of fisheries initiatives. This makes it possible for the public to have a greater say in decision-making and to be scrutinised more closely (Bradley et. al., 2019).

Electronic monitoring, sometimes known as EM, has been a game-changer in the field of fisheries management. It involves integrating remote monitoring technology with traditional onboard observer programmes. computerised monitoring systems remove the need for human observers by collecting data on fishing activities, catch composition, and fishing effort through the use of cameras, sensors, and computerised logbooks. This eliminates the need for human observers. When compared to traditional observer programmes, this technology enables authorities to dramatically reduce the cost of doing marine monitoring while simultaneously increasing the frequency and coverage of vessel observation. By using EM data to verify for compliance with fishing restrictions, measure bycatch levels, and assess conservation efforts, it is possible to make decisions regarding fisheries management that are more informed and based on evidence (van Helmond et. al., 2020).

It is vital to conduct stock evaluations that are driven by data in order to support sustainable fisheries management. The biomass, reproductive potential, and abundance of fish populations are all subject to these evaluations, which provide realistic estimates. Catch and effort statistics, which are dependent on fisheries, are utilised in the conventional methods of stock evaluation when conducting stock assessments. On the other hand, this data could be skewed and lacking in precision. Despite this, scientists are now able to obtain data that is more precise and comprehensive on fish populations and the ecosystems in which they live as a result of improvements in data collection technologies including as acoustic surveys, underwater cameras, and genetic analysis. For the purpose of determining stock characteristics, forecasting population dynamics, and assessing the potential consequences of fishing on ecosystems and species of interest, these data are employed in the construction of intricate population models and statistical algorithms (Marshall, 2009).

It is of the utmost importance to emphasise the need of combining ecosystem-based strategies into fisheries management in order to foster resilience and sustainability in marine ecosystems. Ecosystem-based fisheries management (EBFM) takes into account not only the wider dynamics of the ecosystem but also the linkages among target species, predators, prey, and habitats when making choices. Other factors that are taken into consideration include habitats. In order to make ecosystem-based forest management (EBFM) a reality, ecosystem models, marine spatial data infrastructure (MSDI), and spatial planning tools all play a part. These tools assist stakeholders in weighing the benefits and drawbacks of EBFM, identifying ecologically significant regions, and developing adaptive management plans that successfully strike a balance between conservation objectives and social and economic concerns. Ethical fisheries management (EBFM) includes ecological considerations into fisheries management in order to guarantee the sustainable utilisation of fisheries resources for future generations. The conservation of biodiversity, the resilience of ecosystems, and the provision of ecosystem services are all supported by this information (Scotti et. al., 2022).

Fisheries management activities have been substantially revolutionised as a result of breakthroughs in ecosystem-based methodologies, data-driven stock assessments, electronic monitoring, GPS tracking, and other innovations. This has led to more sustainable fishing practices and reduced overfishing. When it comes to making decisions, fishery managers are better equipped to keep an eye on fishing, enforce rules, and find a middle ground between conservation aims and social and economic concerns thanks to the use of these technologies. In order to ensure the survival of marine ecosystems and the well-being of coastal communities who are dependent on fisheries resources, we must demonstrate a readiness to embrace technological innovation and apply techniques of fisheries management that are founded on scientific principles.

The importance of aquaculture in satisfying the world's seafood demand and relieving strain on wild fish populations has grown in recent years. Aquaculture involves the growing of aquatic creatures including fish, shellfish, and plants. Habitat loss, pollution, and feeding wild fish are just a few of the environmental sustainability issues brought up by conventional aquaculture methods (Figure 4). In light of these difficulties, new approaches to aquaculture have evolved, which aim to increase productivity while decreasing negative effects on the environment and paving the way for more sustainable production practices. There have been three significant advancements in aquaculture that have helped to promote ecologically responsible practices: recirculating aquaculture systems (RAS), genetic enhancements, and sustainable feed options (Goethel et. al., 2023).

A revolutionary development in aquaculture technology, recirculating aquaculture systems (RAS) provide a closed-loop system that recycles and purifies water inside the production facility, reducing water consumption and wastewater outflow. In order to keep the water quality high, RAS facilities usually have filtration systems and water treatment technologies installed, in addition to tanks or ponds where the fish are raised. As an eco-friendly substitute for conventional open-net pen aquaculture systems, RAS continually recirculates water, which cuts down on water volume requirements and pollution discharge. By enabling farmers to precisely regulate water temperature, oxygen levels, and waste collection, RAS optimises environmental conditions for fish development, which is a major benefit. Because of this, high-quality fish may be produced in a controlled setting all year round, independent of weather or seasonality. Also, unlike monoculture, which may lead to disease outbreaks and environmental pollution, RAS allows for the production of various fish species in a single facility, which promotes diversification (Laine et. al., 2023).

*Figure 4. Sustainable Fisheries Management for Coastal Protection*

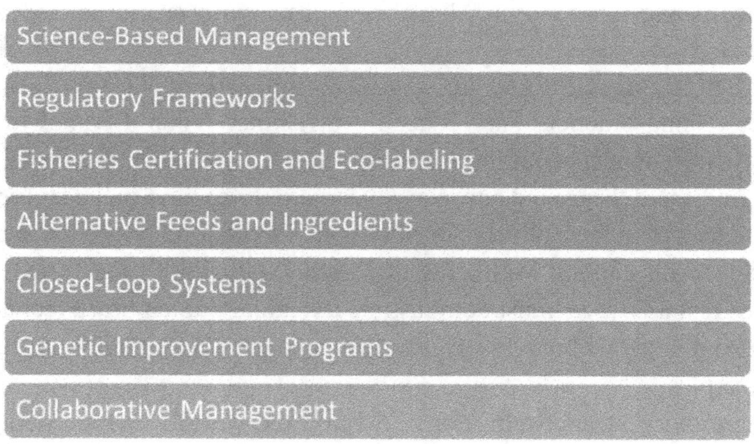


The use of selective breeding and genetic engineering to increase farmed fish species' performance, resilience, and sustainability is another cutting-edge strategy in aquaculture. Improved strains of fish that are more suited to aquaculture circumstances and need fewer external inputs can be developed by aquaculture scientists through selective breeding for desired qualities including growth rate, disease resistance, and feed efficiency. Thanks to genetic developments, aquaculture has become much more sustainable and productive, allowing farmers to raise more fish with less inputs and less of an effect on the environment. The development of fish strains with enhanced disease resistance, quicker growth rates, and higher survival rates has been made possible by selective breeding programmes. This has boosted the efficiency and profitability of aquaculture operations. Further improvements to farmed fish qualities and the creation of new species with desirable characteristics like a smaller environmental impact and better nutritional profile are both possible thanks to developments in biotechnology like gene editing and genetic engineering. Sustainable feed options are cutting down on the use of wild fish as feed and the negative impact that fish farming has on the ecosystem. This is completely changing the way aquaculture is done. Overfishing and resource depletion have long been associated with aquaculture feeds that have traditionally used fishmeal and fish oil obtained from wild-caught fish. Feeds made from fishmeal are not the only sustainable option; new feed formulations are becoming popular that include plant proteins, algae, insects, and microbial biomass as lipid and protein sources (Ibrahim et. al., 2023).

For instance, plant-based diets prioritise resource efficiency in aquaculture and alleviate stress on wild fish populations by using protein sources derived from agricultural crops like wheat, maize, and soybeans. In a similar vein, feeds made from algae provide farmed fish with protein, lipids, and other vital elements while also being sustainable and kind to the environment. This feed offers a cyclical solution to feed production while decreasing dependency on scarce resources. It is sourced from insects like mealworms and black army flies and is rich in protein and amino acids.

Microbial biomass is a potential alternative feed element that is rich in protein and has nutritional value; it is made by fermenting microorganisms including fungus, yeast, and bacteria. A sustainable and scalable answer to feed production in aquaculture may be found in microbial-based feeds, which can be made from organic waste streams such food leftovers and agricultural wastes. Sustainable feed options help aquaculture operations be more resilient and environmentally sustainable by minimising the use of wild fish as feed, increasing resource efficiency, and diversifying feed sources. They also help reduce the ecological imprint of fish farming (Sarker, 2023).

The aquaculture sector is undergoing a transformation and is becoming more ecologically conscious thanks to novel techniques such as recirculating aquaculture systems (RAS), genetic advances, and sustainable feed alternatives. Using RAS's closed-loop system, we can reduce water use and waste output while improving the resilience and performance of farmed fish species through genetic enhancements. Aquaculture has a smaller impact on the environment when it uses sustainable feed alternatives instead of wild fish. Through the adoption of these forward-thinking methods, aquaculture can maintain its crucial role in satisfying the worldwide demand for seafood while reducing its influence on marine ecosystems and fostering sustainable development for generations to come.

## Tourism and Blue Economy

Coastal regions all around the world benefit from the blue economy, which includes marine tourism as a component, in a variety of ways, including economic growth, the production of employment opportunities, and the preservation of cultural traditions (Figure 5). Coral reefs, marine habitats, and biodiversity hotspots are examples of coastal ecosystems that are especially susceptible to the adverse impacts of marine tourism, which is seeing tremendous expansion. Scientific developments in marine tourism are very necessary in order to tackle these difficulties, improve sustainable tourism practices, and protect coastal ecosystems at the same time. As a result of developments such as environmentally friendly infrastructure, coral reef restoration, and marine spatial planning, stakeholders are now in a position to strike a balance between the economic benefits of tourism and the preservation of marine biodiversity and ecosystem services (Ramos & Drakeford, 2021). Eco-friendly infrastructure is a significant breakthrough in the marine tourism industry. This type of infrastructure aims to enhance the experiences of tourists and the communities that they visit, while simultaneously lessening the negative impact that tourism structures and activities have on the environment (Bax et. al., 2022). For the purpose of adhering to these objectives, eco-lodges, resorts, and eco-tourism endeavours include concepts of cultural preservation, resource efficiency, and environmental sustainability into their design, construction, and administration. These buildings make use of water conservation strategies, renewable energy sources, and green construction technologies in order to reduce their influence on the environment. This leads to a reduction in the amount of energy used, carbon emissions produced, and waste discharged. Low-impact tourist activities, such as sustainable mobility options, community participation initiatives, and responsible waste management, are regularly implemented by eco-friendly infrastructure in order to further limit the detrimental impacts that tourism has on marine environments and local inhabitants.

*Figure 5. Blue Economy and Tourism Contribution to Coastal Management*

MSP, or marine spatial planning, is an additional innovative approach to the management of marine tourism. MSP aims to manage coastal and marine resources in an ecosystem-based manner, while simultaneously promoting sustainable development and the preservation of biodiversity. The MSP's rigorous mapping and allocation of marine space may encompass a wide variety of marine uses, including tourism, enjoyment, conservation, and resource exploitation, to name just a few of the numerous marine uses that may be included. In order to guarantee the establishment of marine protected areas (MPAs) and the regulation of human activities in a manner that minimises conflicts and maximises the benefits of marine tourism for both people and nature, the Marine Stewardship Programme (MSP) identifies areas of ecological significance, sensitive habitats, and biodiversity hotspots. The stakeholders are able to prioritise conservation activities as a result of this (Chen et. al., 2024).

Coral reef restoration is an innovative method to marine tourism conservation that aims to restore damaged coral reef ecosystems, make them more resistant to the effects of climate change, and strengthen sustainable tourism. These are all goals that would be accomplished via the process. Coral reefs are one of the most economically and ecologically vital ecosystems on the globe because they are home to a large number of marine species, they provide a source of food for fishermen, and they bring in billions of dollars annually from visitors and residents alike. Human activities that represent an increasing danger to coral reefs include overfishing, pollution, coastal development, and coral bleaching induced by climate change. These are only some of the activities that these human activities involve (Arias-Gonzalez et. al., 2022).

In the face of these threats, researchers, environmentalists, and tour operators are employing cutting-edge techniques in order to restore coral reefs and marine ecosystems that have been destroyed. For instance, in the practice of coral gardening, fragments of coral are grown in nurseries and then transplanted to parts of reefs that have been damaged in order to speed up the healing process of coral and to strengthen the resilience of the ecosystem. In a similar manner, artificial reef structures such as reef balls and modular reef systems are utilised in areas where natural reefs have been damaged or destroyed in order to boost biodiversity by providing habitat complexity. Moreover, emerging technologies such as micro fragmentation, genetic engineering, and three-dimensional printing provide the possibility of overcoming technical and logistical challenges and advancing efforts to restore coral reefs. Utilising the technology of three-dimensional printing, it is now possible to manufacture intricate coral structures and substrates (Bostrom-Einarsson et. al., 2020). These products fulfil the function of a suitable replacement for the natural ecology of reefs and may be utilised by corals in order to establish themselves and flourish. In order to make recovered coral populations more robust, genetic engineering is being utilised to generate strains of coral that are better able to endure stress and adapt to changing environmental situations such as ocean acidification and harsh temperatures. This is being done in order to make recovered coral populations more resilient. In order to speed up the process of reef regeneration, a technique known as micro fragmentation is utilised. This technique involves the division of healthy coral colonies into very small fragments. After that, these pieces are cultivated in nurseries and then transported to the reefs that have been impacted or damaged.

Some of the scientific developments in marine tourism that are helping to encourage sustainable tourist behaviours and safeguard coastal ecosystems include the repair of coral reefs, the design of marine spatial areas, and the construction of infrastructure that is favourable to the environment. In the event that stakeholders embrace these novel strategies, marine tourist destinations and coastal communities may be able to realise the economic benefits of tourism while also safeguarding marine biodiversity and ecosystem services. In order to accommodate the growing demand for marine tourism while also safeguarding coastal ecosystems for future generations and encouraging sustainable tourism practices, it is essential to make investments in research, technology, and capacity-building programmes.

## Renewable Energy Technologies

As the world moves towards a low-carbon energy future, renewable energy sources including offshore wind, tidal, and wave power are becoming more important. Renewable energy sources are a boon to economic growth, environmental protection, and energy security because these cutting-edge technologies convert the energy of the wind, tides, and ocean waves into usable electricity. Renewable energy technologies have come a long way thanks to scientific advancements that have increased their efficiency, dependability, and cost-effectiveness. Now we can tap into their vast potential to change the world's energy landscape by reducing emissions of greenhouse gases. One of the most promising emerging areas for renewable power generation, offshore wind has enormous unrealized potential to harness the constant and powerful winds that blow across the world's oceans. Unlike on land, offshore wind farms may install bigger turbines and achieve better energy outputs by positioning themselves in shallow or deep oceans, where wind speeds are stronger and more constant. Offshore wind energy development presents a number of technical and logistical hurdles, such as those pertaining to turbine design, foundation engineering, and installation procedures. However, these obstacles have been greatly helped by scientific innovation (Rahman et. al., 2022).

The advancement of bigger and more efficient wind turbine technology is one of the main advancements propelling the expansion of offshore wind power. Modern turbines can harness more wind energy and have greater capacity factors thanks to advancements in aerodynamics, materials science, and design that have made it possible to build taller hubs and wider rotor blades. In addition, offshore wind farms in deeper seas cannot be installed using fixed-bottom foundations. Floating wind turbine technology has therefore become an attractive alternative. The use of mooring devices to attach floating wind turbines to the seabed allows for their deployment in deeper seas away from the coast, creating new possibilities for offshore wind energy development in areas rich in wind resources (Asim et. al., 2022).

Another cutting-edge renewable energy source that uses the force of ocean tides to create power is tidal energy. The kinetic energy of the ebbing and flowing tidal currents is converted into electrical power in tidal power plants by means of underwater turbines or barrages. It is possible to reliably anticipate tidal patterns—which are controlled by the moon's and sun's gravitational pull—years in advance, which is a benefit of tidal energy. Optimisation of tidal energy conversion technologies, improvement of turbine efficiency, and reduction of environmental problems associated with tidal power generation have all been greatly aided by scientific innovation. The evolution of state-of-the-art turbine designs and materials adapted to the severe marine environment is a critical innovation that is propelling the expansion of tidal energy. Strong and long-lasting building materials and protective coatings are necessary for tidal turbines so that they can endure the corrosive saltwater, high flow velocities, and turbulence. Turbine blade design, rotor combinations, and hydrodynamic modelling advancements have made it possible to create dependable and efficient tidal energy systems that can work in a variety of tidal circumstances. The deployment of tidal energy projects in tough coastal areas, such as those with strong currents, shallow seas, and delicate ecosystems, has been made possible by advancements in marine engineering and installation techniques. Another exciting renewable energy source is wave power, which converts the kinetic energy of the ocean into usable electricity. When waves travel over the ocean's surface, wave energy converters (WECs) collect that motion and transform it into usable mechanical or electrical energy. The renewable, plentiful, and predictable nature of wave energy makes it an attractive option for coastal areas worldwide looking for a substantial boost to their clean energy production. Improvements in device efficiency, dependability, and capacity to withstand severe marine conditions have been made possible by scientific advancements in wave energy conversion technology. The evolution of distinct wave energy converter (WEC) designs that are optimised for certain wave conditions and deployment situations is a critical innovation that is propelling the expansion of wave energy. diverse types of wave energy converters (WECs) use diverse ideas for capturing and converting wave energy; some examples include attenuators, oscillating water columns, and point absorbers. More efficient and cost-effective wave energy converters that can operate in a variety of maritime habitats and capture a broader range of wave situations have been developed as a result of advancements in control systems, materials science, and WEC design. Additionally, advancements in grid integration technologies, mooring systems, and deployment methods have made it easier to install and integrate wave energy projects into the current energy infrastructure. This has allowed for the larger-scale commercialization and deployment of wave energy (Soukissian et. al., 2017).

Renewable energy sources like offshore wind, tidal, and wave power have come a long way thanks to scientific advancements, which have increased their efficiency, dependability, and cost-effectiveness. These technologies have the potential to revolutionise the world's energy system by cutting down on emissions of greenhouse gases. Innovative technologies are enhancing energy security, promoting economic development, and safeguarding the health and resilience of marine ecosystems for future generations by utilising the power of natural phenomena like wind, tides, and waves. These sources of energy are clean, dependable, and abundant.

## Environmental Monitoring and Pollution Control

Sustainable environmental management relies on environmental monitoring and pollution control to maintain ecosystem health, human health, and soil, water, and air resource quality. Modern environmental monitoring has been radically altered by the advent of cutting-edge monitoring technology like sensor networks, drones, and satellite images, which allow for the real-time identification of pollution incidents and the implementation of more efficient pollution control methods. These technologies enable stakeholders, regulators, and policymakers to take swift and focused action to reduce pollution and safeguard the environment by collecting, analysing, and interpreting environmental data in a way that has never been possible before. This data will shed light on the origins, pathways, and impacts of pollution. Drones, which can fly by themselves, gather high-resolution images, and track environmental variables in real time; they are also referred to as unmanned aerial vehicles (UAVs) or unmanned aerial systems (UAS). Drones have recently become an effective tool for pollution control and environmental monitoring. The employment of drones equipped with various monitoring equipment allows for the rapid and precise assessment of environmental conditions, the detection of pollution sources, and the monitoring of changes in land use, plant cover, and water quality. Environmental scientists and regulators can save time and money by using drones to access previously inaccessible areas. This improves the spatial and temporal resolution of environmental monitoring efforts and reduces the need for expensive and time-consuming field surveys (Hariram et. al., 2023).

Offering worldwide coverage and multispectral imaging capabilities, satellite photography has transformed environmental monitoring and pollution management by revealing changes in the environment over vast geographic regions. With the use of remote sensing instruments such as high-resolution cameras and radar sensors, scientists are able to track changes in land cover, deforestation, urbanisation, and pollution levels with a level of precision and timeliness never seen before. These satellites also take detailed pictures of the Earth's atmosphere and oceans. To minimise environmental impacts and protect human health, satellite imagery allows the detection of pollution sources like industrial facilities, power plants, and agricultural activities. It also tracks pollution plumes, spills, and environmental disasters, allowing for rapid response and mitigation efforts (Ma et. al., 2023).

Continuous monitoring of pollution levels and trends is made possible by sensor networks, which are networks of linked sensors and monitoring devices spread out throughout a region. These networks give real-time data on environmental factors including water and air quality as well as soil contamination. Often referred to as the "Internet of Things" (IoT), these sensor networks collect data from a wide range of sources, such as gas, particulate matter, water quality, and soil sensors, and then send that information to a central database or cloud platform for further processing. Early detection of pollution events, identification of pollution sources, and assessment of pollution impacts on human health and the environment are all made possible by sensor networks, which provide real-time information on pollution levels and trends. This information empowers stakeholders, regulators, and policymakers to implement targeted interventions and control measures (Popescu et. al., 2024). Drones, satellite images, and sensor networks are just a few examples of the cutting-edge monitoring technologies that are revolutionising environmental monitoring and pollution control. These tools can give detailed, multi-dimensional information about pollution pathways, pollution sources, and overall environmental conditions. Environmental regulators and scientists can better understand environmental processes, spot new pollution dangers, and focus pollution control efforts on the most critical issues when they combine data from different platforms. Furthermore, cutting-edge monitoring tools allow for stakeholder engagement and real-time data exchange, which in turn helps with environmental problem solving, human health protection, and adaptive management techniques (Manfreda et. al., 2018).

Where conventional monitoring equipment and resources are few, as is often the case in underprivileged areas and developing nations, new monitoring technologies hold great promise for transforming environmental monitoring and pollution management. These technologies can help local communities, citizen scientists, and NGOs monitor pollution levels, advocate for environmental protection, and hold polluters accountable. They are cost-effective, scalable, and user-friendly solutions for environmental monitoring and pollution control. Air, water, and marine pollution, as well as pollution flows across international borders, are examples of transboundary pollution issues that can be better addressed and mitigated through international cooperation and collaboration made possible by cutting-edge monitoring technologies (Mamun & Natcher, 2023).

New monitoring technologies like sensor networks, satellite imagery, and drones are changing the game when it comes to pollution control and environmental monitoring. These tools make it possible to detect pollution events in real-time, aid in the implementation of effective pollution control measures, and give stakeholders, regulators, and lawmakers the tools they need to safeguard human and environmental health. More resilient and long-lasting environmental management practices will be possible in the 21st century thanks to these technologies, which provide detailed, multi-dimensional information on environmental conditions, pollution sources, and pollution pathways. This allows for a more integrated and holistic approach to monitoring and controlling pollution.

# CASE STUDIES

## Case Study 1: Managing the Coastal Zone Integrated in Australia's Great Barrier Reef

Under the auspices of the Great Barrier Reef Marine Park Authority (GBRMPA), an all-encompassing Integrated Coastal Zone Management (ICZM) strategy has been put in place to safeguard and environmentally manage the world-renowned Great Barrier Reef (GBR). In order to tackle the many environmental and socioeconomic issues that the GBR is currently confronting, this effort brings together several government agencies, local communities, industrial players, and research institutions. In order to keep an eye on and control the maritime environment around the GBR, scientists have used cutting-edge technologies including satellite imaging, spatial planning tools, and remote sensing. Authorities may use remote sensing data to identify problem regions, monitor changes over time, and prioritise conservation efforts based on coral reef health, water quality, and marine biodiversity. In order to promote coastal activities like fishing and tourism while minimising human impacts on vulnerable ecosystems and species, spatial planning tools make it easier to designate regions as marine protected areas, create zoning plans, and establish sustainable use zones. Environmental sustainability, pollution management, and ecosystem resilience have all seen considerable advances since ICZM was implemented in the GBR. The GBRMPA has improved coral reef resilience to climate change impacts like coral bleaching events, decreased overfishing, and reduced pollutant discharge from coastal development by taking a comprehensive and scientifically-based strategy to managing coastal resources. Both the environment and the region's thriving tourist sector, which pumps billions of dollars into Australia's coffers every year, have been protected thanks to these initiatives (Harvey, 1999).

## Case Study 2: Restoring Oyster Reefs in the Chesapeake Bay, USA

In order to restore oyster reefs, boost biodiversity, and ensure sustainable fishing in the Chesapeake Bay watershed, the United States' Chesapeake Bay Programme (CBP) has launched massive restoration initiatives. Oyster reefs are vital to marine ecosystems because they filter water, stabilise sediments, and provide a home for many marine animals. Oyster stocks in the Chesapeake Bay have been devastated by decades of overfishing, habitat loss, and water pollution. To revive oyster populations and repair damaged ecosystems in the Chesapeake Bay, scientists have used cutting-edge aquaculture methods, genetic enhancements, and ecosystem modelling. Genetic advances seek to generate disease-resistant and fast-growing oyster strains to promote restoration success, while aquaculture technologies like spat-on-shell production and oyster gardening allow for the proliferation of oyster larvae and juveniles for transplantation onto restoration sites. In order to optimise resource allocation and direct restoration efforts, ecosystem modelling methods are utilised to forecast the ecological effects of oyster reef restoration on water quality, biodiversity, and fisheries production.

Improved water clarity, higher fisheries yields, and more recreational options for local communities are just a few of the environmental and socioeconomic advantages that have resulted from the restoration of oyster reefs in the Chesapeake Bay. Supporting the region's economy and cultural legacy, the CBP has helped restore oyster populations and the ecosystem services they provide, which in turn has helped the Chesapeake Bay's ecology recover and be more resilient (Carey, 2021).

## Case Study 3: Protecting Mangroves and Promoting Ecotourism in Bocas del Toro, Panama

Extensive mangrove forests in Panama's Bocas del Toro Archipelago safeguard coastal towns from erosion and hurricanes while also providing a home for many marine animals. These woods also help local people make a living. But pollution, deforestation, and irresponsible development have all posed risks to Bocas del Toro's mangrove ecosystems.

In order to foster environmental sustainability, economic development, and community resilience, Bocas del Toro has introduced innovative techniques to sustainable tourism and mangrove protection. To evaluate the state of mangroves, pinpoint places where they have degraded, and track the progress of restoration initiatives, scientists have used cutting-edge tools including ecological monitoring, drone imaging, and satellite photography. Community-based ecotourism projects, such educational programmes, kayak excursions, and guided mangrove tours, help locals make a living while drawing attention to the need to preserve mangroves and other ecosystems.

Sustainable tourism in Bocas del Toro, which includes mangrove conservation, has benefited the local community and the environment. Stakeholders have protected mangrove forests and advocated for sustainable tourist practices, which has strengthened coastal ecosystems, helped conserve species, and given coastal communities more economic options. Environmental sustainability and socio-economic development may be achieved in coastal locations across the world via science-based conservation and community participation, as shown by these efforts. These case studies show how advancements in science have improved coastal resource management in many parts of the world. Stakeholders can safeguard coastal ecosystems' health and resilience for future generations, combat environmental concerns, and advance sustainable development via the use of cutting-edge technology, multidisciplinary cooperation, and community involvement (Yingst, 2016).

## FUTURE PERSPECTIVES

The future of sustainable blue economic growth, environmental risk mitigation, and coastal pollution management depends on our ability to tap into scientific innovation. To accomplish these goals, moving forward should centre on making the most of new technology, encouraging cooperation across disciplines, and giving local people the tools they need to be an active part of coastal resource management.

AI and Big Data Analytics Integration: Moving forward, it would be wise to use AI and big data analytics to sift through mountains of environmental data, spot patterns, and foretell how coastal ecosystems will evolve. Algorithms powered by artificial intelligence may sift through data from sensors and satellite images to identify instances of pollution, track the deterioration of habitats, and evaluate the effects of human activities on coastal ecosystems. Stakeholders may improve pollution control and ecosystem preservation by effectively anticipating environmental threats, implementing proactive management methods, and optimising resource allocation through the integration of AI-driven predictive modelling with real-time monitoring systems.

Creating State-of-the-Art Monitoring Systems: In order to better monitor coastal habitats, new monitoring technologies including autonomous sensor networks, underwater drones, and unmanned underwater vehicles (UUVs) are constantly being developed. These innovative tools can monitor water quality measures, gather data from hard-to-reach places, and evaluate the state of habitats with greater precision and coverage over longer periods of time. Stakeholders can generate real-time data streams on ecosystem dynamics, pollution levels, and environmental conditions by deploying a network of interconnected monitoring devices with AI capabilities. This will allow for effective pollution control measures and rapid response to new threats.

Applying Solutions Derived from Nature: Restoring coastal wetlands, reforesting mangroves, and constructing living shorelines are all examples of nature-based solutions that may help manage coastal resources and reduce pollution in a sustainable way. In addition to protecting coastal areas, sequestering carbon, and preserving biodiversity, these natural ecosystems also make coastal towns more resistant to the effects of climate change. To maximise the benefits of nature-based solutions for both people and nature, future efforts should focus on restoring and conserving coastal wetlands, mangroves, and other natural habitats as part of integrated coastal zone management strategies. This can be achieved by leveraging scientific knowledge and community engagement.

Reducing pollution, minimising waste, and promoting sustainable resource use may be achieved by transitioning coastal areas to a circular economy model, which we will discuss in the fourth point. Coastal communities can lessen their impact on marine litter and plastic pollution by adopting innovative waste management strategies like recycling, composting, and waste-to-energy conversion. These initiatives also offer economic opportunities, such as green job creation and the development of circular economy industries. In order to ensure the long-term viability of coastal ecosystems and economies, scientific advancements can lend a hand in the development and execution of circular economy programmes by offering answers to problems like resource recovery, sustainable consumption habits, and waste management.

Enhancing Governance and Policy Frameworks: To foster long-term blue economic growth and reduce coastal pollution, strong governance and policy frameworks are needed. To solve complicated environmental problems, future work should encourage multi-stakeholder cooperation, establish regulatory frameworks, and improve enforcement mechanisms. In order to achieve a balance between environmental conservation and socio-economic development goals in coastal resource management, it is important to incorporate scientific evidence into policy-making processes and encourage dialogue amongst local communities, scientists, politicians, and industry stakeholders.

Coastal communities, decision-makers, and stakeholders must be empowered to actively participate in coastal resource management and pollution control activities. This can only be achieved via capacity development and information sharing programmes. Building local capacity in environmental monitoring, pollution control, and sustainable resource management techniques should be the priority of future initiatives that prioritise investment in education, training, and awareness-raising efforts. Coastal resource management can also benefit from knowledge exchange platforms like community-based monitoring networks, online portals, and collaborative research initiatives. These can help diverse stakeholders share best practices, lessons learned, and scientific innovations.

Ultimately, if we want to foster long-term blue economic growth, lessen environmental dangers, and improve coastal pollution control, we must tap into scientific innovation. In order to preserve coastal ecosystems and ensure the continued prosperity of coastal communities for future generations, stakeholders must embrace new technology, encourage interdisciplinary cooperation, and give local communities more agency.

## CONCLUSION

Sustainable development plans must continue to incorporate scientific innovation if coastal resource management is to have any hope of a bright future. In order to tackle the complex issues, such pollution management and adaptation to climate change, that coastal communities are experiencing, stakeholders may make use of new technology like AI, sophisticated monitoring systems, and solutions derived from nature. The translation of scientific knowledge into policies and practices that promote social equity, economic prosperity, and environmental sustainability in coastal regions worldwide will also require interdisciplinary collaboration, strengthened governance mechanisms, and the empowerment of local communities. To make this transition to a blue economy that can last the test of time, we must use scientific innovation to keep coastal ecosystems strong and resilient, coastal people' means of subsistence secure, and future generations' health and prosperity guaranteed. It is impossible to exaggerate the significance of coastal resource management in this age of unparalleled environmental change and global interconnection. To fully realise coastal ecosystems' potential as climate buffers, biodiversity engines, and economic growth engines, coastal management must be science-based and holistic, drawing on innovative technological developments and interdisciplinary research. Looking ahead, let us stay dedicated to collaborating across borders, fields, and industries to create a future where our planet's coastal areas are better able to withstand environmental challenges, where thriving communities and ecosystems coexist, and where everyone shares fairly in the benefits of blue economic development.

## REFERENCES


Arias-González, J. E., Baums, I. B., Banaszak, A. T., Prada, C., Rossi, S., Hernández-Delgado, E. A., & Rinkevich, B. (2022). Editorial: Coral Reef Restoration in a Changing World: Science-Based Solutions. *Frontiers in Marine Science*, 9, 919603. 10.3389/fmars.2022.919603

Asim, T., Islam, S. Z., Hemmati, A., & Khalid, M. S. U. (2022). A Review of Recent Advancements in Offshore Wind Turbine Technology. *Energies*, 15(2), 579. 10.3390/en15020579

Bax, N., Novaglio, C., Maxwell, K. H., Meyers, K., McCann, J., Jennings, S., Frusher, S., Fulton, E. A., Nursey-Bray, M., Fischer, M., Anderson, K., Layton, C., Emad, G. R., Alexander, K. A., Rousseau, Y., Lunn, Z., & Carter, C. G. (2022). Ocean resource use: Building the coastal blue economy. *Reviews in Fish Biology and Fisheries*, 32(1), 189–207. 10.1007/s11160-021-09636-033679009

Boström-Einarsson, L., Babcock, R. C., Bayraktarov, E., Ceccarelli, D., Cook, N., Ferse, S. C. A., Hancock, B., Harrison, P., Hein, M., Shaver, E., Smith, A., Suggett, D., Stewart-Sinclair, P. J., Vardi, T., & McLeod, I. M. (2020). Coral restoration – A systematic review of current methods, successes, failures and future directions. *PLoS One*, 15(1), e0226631. 10.1371/journal.pone.022663131999709

Bradley, D., Merrifield, M., Miller, K. M., Lomonico, S., Wilson, J. R., & Gleason, M. G. (2019). Opportunities to improve fisheries management through innovative technology and advanced data systems. *Fish and Fisheries*, 20(3), 564–583. 10.1111/faf.12361

Carey, J. (2021, June). The complex case of Chesapeake Bay restoration, PNAS. https://www.pnas.org/doi/10.1073/pnas.2108734118

Chen, L., Hu, Y., Wang, R., Li, X., Chen, Z., Hua, J., Osman, A. I., Farghali, M., Huang, L., Li, J., Dong, L., Rooney, D. W., & Yap, P.-S. (2024). Green building practices to integrate renewable energy in the construction sector: A review. *Environmental Chemistry Letters*, 22(2), 751–784. 10.1007/s10311-023-01675-2

Craig, R. K., & Ruhl, J. B. (2010). Governing for Sustainable Coasts: Complexity, Climate Change, and Coastal Ecosystem Protection. *Sustainability (Basel)*, 2(5), 1361–1388. 10.3390/su2051361

de Lima, R. L. P., Boogaard, F. C., & de Graaf-van Dinther, R. E. (2020). Innovative Water Quality and Ecology Monitoring Using Underwater Unmanned Vehicles: Field Applications, Challenges and Feedback from Water Managers. *Water (Basel)*, 12(4), 1196. 10.3390/w12041196

Delacámara, G., O'Higgins, T. G., Lago, M., & Langhans, S. (2020). Ecosystem-Based Management: Moving from Concept to Practice. In O'Higgins, T., Lago, M., & DeWitt, T. (Eds.), *Ecosystem-Based Management, Ecosystem Services and Aquatic Biodiversity*. Springer., 10.1007/978-3-030-45843-0_3

Goethel, D. R., Omori, K. L., Punt, A. E., Lynch, P. D., Berger, A. M., de Moor, C. L., Plagányi, É. E., Cope, J. M., Dowling, N. A., McGarvey, R., Preece, A. L., Thorson, J. T., Chaloupka, M., Gaichas, S., Gilman, E., Hesp, S. A., Longo, C., Yao, N., & Methot, R. D. (2023). Oceans of plenty? Challenges, advancements, and future directions for the provision of evidence-based fisheries management advice. *Reviews in Fish Biology and Fisheries*, 33(2), 375–410. 10.1007/s11160-022-09726-736124316

Griggs, G., & Reguero, B. G. (2021). Coastal Adaptation to Climate Change and Sea-Level Rise. *Water (Basel)*, 13(16), 2151. 10.3390/w13162151

Hariram, N. P., Mekha, K. B., Suganthan, V., & Sudhakar, K. (2023). Sustainalism: An Integrated Socio-Economic-Environmental Model to Address Sustainable Development and Sustainability. *Sustainability (Basel)*, 15(13), 10682. 10.3390/su151310682

Harvey, N. (1999). Australian Integrated Coastal Management: A Case Study of the Great Barrier Reef. In Salomons, W., Turner, R. K., de Lacerda, L. D., & Ramachandran, S. (Eds.), *Perspectives on Integrated Coastal Zone Management. Environmental Science*. Springer. 10.1007/978-3-642-60103-3_16

Ibrahim, L. A., Shaghaleh, H., El-Kassar, G. M., Abu-Hashim, M., Elsadek, E. A., & Alhaj Hamoud, Y. (2023). Aquaponics: A Sustainable Path to Food Sovereignty and Enhanced Water Use Efficiency. *Water (Basel)*, 15(24), 4310. 10.3390/w15244310

Ikegwu, A. C., Nweke, H. F., Mkpojiogu, E., Anikwe, C. V., Igwe, S. A., & Alo, U. R. (2024). Recently emerging trends in big data analytic methods for modeling and combating climate change effects. *Energy Informatics*, 7(1), 6. 10.1186/s42162-024-00307-5

Issifu, I., Dahmouni, I., García-Lorenzo, I., & Sumaila, U. R. (2024). Economics in Marine Spatial Planning: A Review of Issues in British Columbia and Similar Jurisdictions. *Sustainability (Basel)*, 16(3), 1210. 10.3390/su16031210

Laine, C., Ollikainen, M., Kankainen, M., Setälä, J., & Vielma, J. (2023). Social net benefits from aquaculture production: A comparison of net cage cultivation and recirculating aquaculture systems. *Aquaculture Economics & Management*, 28(1), 1–31. 10.1080/13657305.2023.2222681

Little, D. C., Newton, R. W., & Beveridge, M. C. M. (2016). Aquaculture: A rapidly growing and significant source of sustainable food? Status, transitions and potential. *The Proceedings of the Nutrition Society*, 75(3), 274–286. 10.1017/S002966511600066527476856

Ma, J., Ma, R., Pan, Q., Liang, X., Wang, J., & Ni, X. (2023). A Global Review of Progress in Remote Sensing and Monitoring of Marine Pollution. *Water (Basel)*, 15(19), 3491. 10.3390/w15193491

Mamun, A. A., & Natcher, D. C. (2023). The promise and pitfalls of community-based monitoring with a focus on Canadian examples. *Environmental Monitoring and Assessment*, 195(4), 445. 10.1007/s10661-022-10841-y36877375

Manfreda, S., McCabe, M. F., Miller, P. E., Lucas, R., Pajuelo Madrigal, V., Mallinis, G., Ben Dor, E., Helman, D., Estes, L., Ciraolo, G., Müllerová, J., Tauro, F., De Lima, M. I., De Lima, J. L. M. P., Maltese, A., Frances, F., Caylor, K., Kohv, M., Perks, M., & Toth, B. (2018). On the Use of Unmanned Aerial Systems for Environmental Monitoring. *Remote Sensing (Basel)*, 10(4), 641. 10.3390/rs10040641

Marois, D. E., & Mitsch, W. J. (2015). Coastal protection from tsunamis and cyclones provided by mangrove wetlands – a review. *The International Journal of Biodiversity Science, Ecosystem Services & Management*, 11(1), 71–83. 10.1080/21513732.2014.997292

Marshall, C. T. (2009). Implementing Information on Stock Reproductive Potential in Fisheries Management: The Motivation, Challenges and Opportunities. In Jakobsen, T., Fogarty, M. J., Megrey, B. A., & Moksness, E. (Eds.), *Fish Reproductive Biology*., 10.1002/9781444312133.ch11

Nguyen, Q. H. (2021). Impact of Investment in Tourism Infrastructure Development on Attracting International Visitors: A Nonlinear Panel ARDL Approach Using Vietnam's Data. *Economies*, 9(3), 131. 10.3390/economies9030131

Pascoe, S., Paredes, S., & Coglan, L. (2023). The Indirect Economic Contribution of Fisheries to Coastal Communities through Tourism. *Fishes*, 8(3), 138. 10.3390/fishes8030138

Popescu, S. M., Mansoor, S., Wani, O. A., Kumar, S. S., Sharma, V., Sharma, A., Arya, V. M., Kirkham, M. B., Hou, D., Bolan, N., & Chung, Y. S. (2024). Artificial intelligence and IoT driven technologies for environmental pollution monitoring and management. *Frontiers in Environmental Science*, 12, 1336088. 10.3389/fenvs.2024.1336088

Rahman, A., Farrok, O., & Haque, M. M. (2022). Environmental impact of renewable energy source based electrical power plants: Solar, wind, hydroelectric, biomass, geothermal, tidal, ocean, and osmotic. *Renewable & Sustainable Energy Reviews*, 161, 112279. Advance online publication. 10.1016/j.rser.2022.112279

Ramos, J., & Drakeford, B. (2021). Blue economy and tourism: is there a potential to create sustainable jobs? In *Handbook of research on human capital and people Management in the Tourism Industry* (pp. 243–264). IGI Global. 10.4018/978-1-7998-4318-4.ch012

Rifai, H., Quevedo, J. M. D., Lukman, K. M., Sondak, C. F. A., Risandi, J., Hernawan, U. E., Uchiyama, Y., Ambo-Rappe, R., & Kohsaka, R. (2023). Potential of seagrass habitat restorations as nature-based solutions: Practical and scientific implications in Indonesia. *Ambio*, 52(3), 546–555. 10.1007/s13280-022-01811-236484926

Rotter, A., Barbier, M., Bertoni, F., Bones, A. M., Cancela, M. L., Carlsson, J., Carvalho, M. F., Cegłowska, M., Chirivella-Martorell, J., Conk Dalay, M., Cueto, M., Dailianis, T., Deniz, I., Díaz-Marrero, A. R., Drakulovic, D., Dubnika, A., Edwards, C., Einarsson, H., Erdo an, A., & Vasquez, M. I. (2021). The Essentials of Marine Biotechnology. *Frontiers in Marine Science*, 8, 629629. 10.3389/fmars.2021.629629

Sarker, P. K. (2023). Microorganisms in Fish Feeds, Technological Innovations, and Key Strategies for Sustainable Aquaculture. *Microorganisms*, 11(2), 439. 10.3390/microorganisms1102043936838404

Scotti, M., Opitz, S., MacNeil, L., Kreutle, A., Pusch, C., & Froese, R. (2022). Ecosystem-based fisheries management increases catch and carbon sequestration through recovery of exploited stocks: The western Baltic Sea case study. *Frontiers in Marine Science*, 9, 879998. 10.3389/fmars.2022.879998

Soukissian, T. H., Denaxa, D., Karathanasi, F., Prospathopoulos, A., Sarantakos, K., Iona, A., Georgantas, K., & Mavrakos, S. (2017). Marine Renewable Energy in the Mediterranean Sea: Status and Perspectives. *Energies*, 10(10), 1512. 10.3390/en10101512

van Helmond, A. T. M., Mortensen, L. O., Plet-Hansen, K. S., Ulrich, C., Needle, C. L., Oesterwind, D., Kindt-Larsen, L., Catchpole, T., Mangi, S., Zimmermann, C., Olesen, H. J., Bailey, N., Bergsson, H., Dalskov, J., Elson, J., Hosken, M., Peterson, L., McElderry, H., Ruiz, J., & Poos, J. J. (2020). Electronic monitoring in fisheries: Lessons from global experiences and future opportunities. *Fish and Fisheries*, 21(1), 162–189. 10.1111/faf.12425

Wenhai, L., Cusack, C., Baker, M., Tao, W., Mingbao, C., Paige, K., Xiaofan, Z., Levin, L., Escobar, E., Amon, D., Yue, Y., Reitz, A., Neves, A. A. S., O'Rourke, E., Mannarini, G., Pearlman, J., Tinker, J., Horsburgh, K. J., Lehodey, P., & Yufeng, Y. (2019). Successful Blue Economy Examples with an Emphasis on International Perspectives. *Frontiers in Marine Science*, 6, 261. 10.3389/fmars.2019.00261

Wilke, M. (2023). Comparing Public Participation in Coastal and Marine Planning in the Arctic: Lessons from Iceland and Norway. *Coasts*, 3(4), 345–369. 10.3390/coasts3040021

Xiuzhen, L., Richard, B., Christopher, C. E., & Widney, S. (2018). Coastal wetland loss, consequences, and challenges for restoration. *Anthropocene Coasts.*, 1(1), 1–15. 10.1139/anc-2017-0001

Yingst, A. L. (2016), The Effect of Mangrove Development on Coral Reef Fish in Bocas Del Toro; Panama And Global Mangrove Management Method [Bachelor's Dissertation, University of Pittsburgh]. http://d-scholarship.pitt.edu/27790/1/Yingst_Bachelor_of_Philosophy.pdf

## KEY TERMS AND DEFINITION

**Blue Economy:** Sustainable use of ocean resources for economic growth, improved livelihoods, and jobs while preserving the health of ocean ecosystem.

**Closed-Loop System:** A mechanical or electronic device that automatically regulates a system to maintain a desired state or set point without human interaction.

**Coastal Infrastructure:** Structures, systems, and facilities built along coastlines.

**Diversification of Economies:** The process of shifting an economy away from a single income source toward multiple sources from a growing range of sectors and markets.

**Environmental Sustainability:** The ability to maintain the balance of the Earth's natural environment and use natural resources responsibly to meet current needs without compromising the ability of future generations to do the same.

**Geographic Information Systems (GIS):** Are a technological development that combine computer hardware and software to store, manage, analyze, edit, output, and visualize geographic data.

**Industrialization:** The process of transforming the economy of a nation or region from a focus on agriculture to a reliance on manufacturing.

**Renewable Energy Technologies:** Technologies that use energy from natural sources that are replenished faster than they are consumed.

**Science-Based Management:** Using scientific and research-based approaches to help a business make decisions, improve performance, and meet goals.

**Sustainable Tourism Practices:** A framework that aims to reduce the negative effects of tourism on the environment, society, and economy.

# Chapter 2
# The Effects of Marine Debris on Indonesia's Coral Reefs and Mitigation Initiatives

**Andik Isdianto**
https://orcid.org/0000-0002-4808-7868
*Universitas Brawijaya, Indonesia*

**Oktiyas Muzaky Luthfi**
https://orcid.org/0000-0002-9550-9381
*Universitas Brawijaya, Indonesia*

**Arief Setyanto**
https://orcid.org/0000-0003-2704-7346
*Universitas Brawijaya, Indonesia*

**Aulia Lanudia Fathah**
https://orcid.org/0009-0005-0443-5241
*Universitas Brawijaya, Indonesia*

**Berlania Mahardika Putri**
https://orcid.org/0009-0006-9898-3085
*Universitas Gadjah Mada, Indonesia*

## ABSTRACT

*Marine debris severely threatens coral reefs in Indonesia. This chapter aims to examine the effects of marine debris on coral reefs in Indonesia and explore the corrective actions implemented to solve this problem. This chapter relies on secondary data*

DOI: 10.4018/979-8-3693-2436-3.ch002

*from scientific literature, government records, and environmental organisations. Land and sea waste damage Indonesian coral reefs, causing physical, chemical, and biological harm. Indonesia has worked with the government, environmental organisations, and local communities to mitigate this. These initiatives consist of beach and water cleanups, public awareness, and education through a variety of educational campaigns, the development of waste management technology, and the establishment of environmental policies and regulations regarding waste management. For Indonesian coral reef conservation to succeed, cross-sector collaboration and community participation are essential. Despite significant mitigation efforts, challenges remain that require a greater commitment to protecting coral reefs and reducing the damaging impacts of marine debris.*

## INTRODUCTION

Six ASEAN Member States—Indonesia, Thailand, Vietnam, the Philippines, Malaysia, and Singapore—produced 243 million metric tonnes of waste in 2016 (Smith, 2022; ASEAN, 2021). Java and Sumatra account for two-thirds of Indonesia's 346.5 ktons of plastic debris from land to ocean. Indonesia's 87 coastal cities discharge 0.68–0.86 million metric tonnes of debris into the ocean annually (World Bank, 2021). Human-made or used products are intentionally dumped into water, rivers, or seas; others are transferred indirectly via drainage channels, runoff, wind, or rivers; or they are lost in the ocean, including fishing gear (Guzman, 2022).

Marine debris that has entered the sea will be very difficult to control because it will spread due to the movement of ocean currents and waves. Waste will spread or sink to the ocean floor, polluting marine life, especially coral reefs. This will have a negative impact on coastal communities that rely on the ocean for their livelihoods. Aquaculture, shipping, and maritime tourism are impacted by marine pollution, especially plastic waste. Production losses, marine life loss, vessel damage, beach cleanup costs, and tourism reductions due to resource degradation are some examples. Moreover, marine pollution ultimately contributes to climate change (Smith, 2022).

The largest climate change carbon sinks are coastal and marine habitats. From the industrial revolution to the 1990s, the ocean absorbed 30–50% of anthropogenic carbon emissions, delaying the disastrous implications of a warming world (Smith, 2022). Plastic production will triple by 2050, according to a report by the Ellen MacArthur Foundation and the World Economic Forum, plastic production will triple by 2050, making it important to limit the release of plastic threats. An increase in such factors will undoubtedly result in adverse consequences, including a decline in the marine economy, an increase in climate change, and the worsening of disasters that represent a threat to the future of the planet. The settling sediment produces CH4 and H2S (methane gas). Ridha et al. (2019) found that methane gas drives climate change. Presently, Indonesia has the most coral reefs and the second-highest unmanaged waste of plastic (Sweet et al., 2019). The aim of sustainability and marine protection is of the highest priority at all levels, including local, national, regional, and international governance. Institutions must support this effort, along with individuals, the corporate and public sectors, and academics must all contribute to this effort (Smith, 2022).

## ORGANIZATION BACKGROUND

The making of the book "The Effects of Marine Debris on Indonesia's Coral Reefs and Mitigation Initiatives" was initiated by lecturers and students from Brawijaya University and Gadjah Maday University, namely Andik Isdianto, Ade Yamindago, Wanda Suryani Pangestu, Aulia Lanudia Fathah, and Berlania Mahardika Putri. The book focuses on marine debris from plastic and other debris in marine environments. This book examines marine debris types, land and ocean sources, coral reef impacts, and Indonesian climate change mitigation.

This book aims to educate the public about science. Academics, students, and interested readers learn about marine pollution caused by plastic and other debris. Scientific studies and accurate data show the negative effects of marine pollution and successful mitigation techniques. This book might encourage readers to take marine debris-reduction actions. This could involve changing one's lifestyle, supporting environmental groups, or contributing to marine debris management research and innovation.

## CASE DESCRIPTION

Addressing marine debris and climate change must minimize the dangers they pose to marine biodiversity and society. Climate change can alter the origin, distribution, and position of marine debris. Climate change has a significant impact on the accumulation of waste in marine ecosystems through storms, landslides, and water runoff, as well as increasing floods and storms. The deposition and distribution of marine debris have been linked to climate change, according to researchers. This shows how climate change can threaten marine ecosystems. Public perception of the link between marine debris and climate change is influenced by this information.

However, interactions and impacts vary in each marine area and depend on certain aspects of the marine environment (Lincoln et al., 2022).

According to the World Bank (2021), 87 coastal cities in Indonesia dump 0.68 to 0.86 million metric tons of waste into the sea annually. Bettim et al. (2021) predict that this number will increase in the future because of the growing population in coastal areas. Rivers and water systems account for 80% of land-based marine waste contamination. Marine waste also comes from marine activities in certain areas, apart from land sources. Around 4000 ships in Indonesian waters produce bait boxes, fishing lines and nets, buoys, and household waste (Purba et al., 2019). Then, in several areas, trash moves freely following sea currents (Lincoln et al., 2022). Marine debris and climate change have an impact on biodiversity and ecosystems, as well as society and the economy (Lincoln et al., 2022), as well as marine organisms such as coral reefs (Ford et al., 2022). The accumulation of debris leads to coral disease, species composition, and the extinction of several local species. Therefore, society currently faces various challenges affecting the resilience of coral reefs (Baumann et al., 2019; Ortiz et al., 2018).

## CASE

### Composition of Marine Debris

Marine debris consists of consumer products discarded into the ocean and on beaches, including but not limited to cans, glass or plastic bottles, bags, rubber, balloons, metal, fibreglass, and cigarettes. Fishing implements, such as lines, buoys, hooks, and fishing lines, are also included, along with any other items that are carelessly discarded into the sea or lost on or near land (Pawar et al., 2016). Marine debris is categorised as follows in accordance with the United Nations Environment Programme (UNEP) (Table 1) (Prajanti et al., 2020):

*Table 1. Categories of Marine Debris*

| No. | Type | Classification of Debris |
|---|---|---|
| 1 | Plastic | Plastic bottles; drink and food package; plastic cutlery; plastic bags; toys; gloves; sheeting; fishing gear (traps, lures & pots); fishing net; monofilament line; rope; raffia ropes; and others |
| 2 | Plastic Foam | Foam sponge; foam; and others |
| 3 | Fabric | Clothing & others accessories; bags; canvas; rope & string; Carpet & furnishing |
| 4 | Glass and Ceramics | Construction and material; Bottles & jars; cutlery; light globes & tubes; Glass or ceramic fragments and others |

continued on following page

*Table 1. Continued*

| No. | Type | Classification of Debris |
|---|---|---|
| 5 | Metal | Metal cutlery; alumuniun drink cans; other cans; Gas bottles; drums & buckets; wire; foil wrappers; fishing related (sinkers, traps, lures, hooks, & pots); and other |
| 6 | Paper and Cardboard | Paper; cardboard boxes; cups&drink containers; food wrappers, cigarette packs; and others |
| 7 | Rubber | Ballon, balls & toys; tyres; gloves; rubber bands; inner-tubes and rubber sheet; and other |
| 8 | Wood | Fishing traps and pots; Corks; wood cutlery & ice cream sticks; wooden tools; and other |
| 9 | Other items | Paraffin & wax; sanitary; Appliances & Electronics; Batteries; and other |

Sources: modification of from Prajanti et al., 2020; Tim Koordinasi Nasional Penanganan Sampah Laut, 2024.

Several studies have found differences in the type, composition, and origin of marine debris between areas near and far from human activities (for example, ports, ghost fishing, aquaculture, and ship loading and unloading). Organic waste, especially branches, accumulates in uninhabited areas in the Sempu Strait, Malang. This is different from areas filled with inorganic waste near ports, residential areas, and tourist destinations. Inorganic waste accumulates in ports due to the loading and unloading of fish catches, refuelling, markets, ship waste, port waste, and household waste. Cloth, tires, fishing lines, and ghost fishing are often found in the Sempu Strait near tourist areas, residential areas, and ports. The lack of rubbish bins and waste disposal facilities near residential areas most likely causes the large amount of plastic waste found in the Sempu Strait. Waters connecting residential areas with beaches also contribute to the movement of waste (Amri et al., 2023). Other research shows that fishing produces more marine waste than plastic waste on land (Putra et al., 2021)

## Source of Marine Debris

Marine waste can come from land or sea, both of which contribute greatly to the production of marine waste.

## Land-Based Sources

*Figure 1. The Discharge of Land-Based Debris into the Marine Environment*

*Source: Personal Illustration*

As much as 90% of the marine debris derived the waters from land sources (Nama et al., 2023), which end up in the sea (Figure 1). Waterways and rivers can carry waste that is not properly managed on land, eventually leading it to end up in the sea. Rivers carry and dispose of 83% of the annual plastic that leaks from land into the sea. Meanwhile, only 17% of plastic waste is disposed of directly in coastal areas (World Bank, 2021). In recent years, attention has turned to rivers as the main route for plastic waste entering the sea. However, researchers still have a limited understanding of the transport mechanisms and behavior of plastic waste in waterways. The first global study conducted Lebreton et al. (2017) placed four rivers in Indonesia among the 20 most polluted rivers (World Bank, 2021). The following are several mechanisms that allow waste to enter the sea (Pawar et al., 2016):

### Storm Water Discharges

During rains, runoff and waste that are collected by floodwaters or rainwater enter rivers at the same time. This water channel will transport waste to the river or sea that is closest to it. The currents of the river discharge the wastewater into the ocean, where it eventually ends up.

## Combined Sewer Overflows

Combined water disposal is a way to collect wastewater and rainwater. During periods of heavy rain, wastewater treatment systems may exceed their capacity to handle wastewater and stormwater, leading to direct release of water into nearby rivers or oceans.

## Littering

Residents and tourists who are visiting the beach and leaving behind waste have the potential to become marine debris. The trash that tourists usually leave behind is cigarette butts, food and drink packaging, and plastic beach toys. Not only that, fishermen also have the potential to be involved in contributing to marine debris because they often leave their fishing gear behind. This waste is not only in coastal areas but also in urban areas, which deliberately create waste in rivers. Waste from urban workers in industries such as forests, construction, agriculture, and mining also has the potential to become marine waste if it enters river flows.

## Solid Waste Disposal and Landfills

Runoff from waste dumps located on beaches or near rivers has the potential to enter the marine environment. Additionally, coastal and marine debris is commonly caused by improperly disposed industrial or domestic waste.

## Industrial Activities

If industrial goods are improperly disposed of on land, lost during transportation, or loaded or dumped at port facilities, they have the potential to become marine debris (US EPA, 2002). Plastic resin pellets, which have a diameter of around 2 to 6 millimetres, have accidentally entered the marine environment via breaches that transpired throughout the stages of manufacturing, processing, shipment, and management. Marine wildlife has ingested these pellets, which have also spread across sediments and beaches.

## Case Study in Indonesia

On Java, the North Coast (50.4 million) and South Coast (31.4 million) population densities have an impact on marine debris discharge. Jakarta Bay collects $23 \pm 7.10$ tonnes of debris per day from Jakarta, Tangerang, and Bekasi on Java's north coast over nine river pathways. Between June 2015 and June 2016, the Bekasi

River discharged 6.67 ± 2.20 tonnes of debris every day, while the Dadap River near Tangerang dumped 7.92 ± 2.71 tonnes. Plastic waste accounts for 8.32 ± 2.44 tonnes, followed by wood and paper, and textile and fiber waste, which is the most wasteful in Bekasi (Cordova et al., 2023). Beach tourism in Pasir Putih Losari, in Brebes, North Java, increases marine debris. Plastic debris contributes to 35% of marine debris from tourists, fishermen, and industry (Amri et al., 2023). Through the Cimandiri River, human activities contribute 307 ± 192.69 kg of debris every day to Pelabuhan Ratu Bay every day. The average everyday debris includes 6.84 ± 10.98 kg of paper, metal, glass, newspapers, tissues, and other paper products. Thin plastic bags, wrappers, containers, and sachets are popular. Every year, the Cimandiri River discharges Pelabuhan Ratu Bay with 323.20 metric tonnes of plastic waste (Meijer et al., 2021).

Debris accumulation could impact Bali, Losari Beach in Makassar, and Java, attracting tourists for their natural beauty. Researchers detected straws, glasses, and trash bags on 14 Bali beaches. Marine debris averages 0.356–0.366 items/m$^2$ and weighs 4.057–3.892 g/m$^2$ (Suteja et al., 2021). Losari Beach has a lot of plastic waste, including bottles and food. Mounds of neglected debris frequently settle and float on the beach, causing unpleasant odors and reducing the beach's aesthetics and comfort of the beach. This reduces the appearance of the beach and may deter visitors (Kaharto et al., 2023).

## Ocean-Based Sources

Offshore industrial platforms, vessels, and boats of all varieties all have the potential to produce marine debris. This waste may be the result of unplanned loss, careless disposal, or illegal dumping. Additionally, historical waste disposal practices may have contributed to this situation. The following are examples of marine debris sources that originate in the ocean (Pawar et al., 2016):

### Commercial Fishing

The generation of marine debris is the responsibility of commercial fishermen who intentionally discard fishing gear or other detritus they dump on board or are unable to retrieve fishing gear. Commercial fishermen frequently generate waste comprising various items such as gill nets, netting, fishing lines, ropes, bait boxes, bags, and shrimp trawls. Fishing gear today consists of plastics like nylon, polyethylene, and polypropylene. The fishing gear may be abandonment-induced or lost as a result of cyclones.

## Recreational Boaters

Recreational vessels and their occupants consistently have the capacity to discharge refuse into the ocean, including sacks, food packaging, and fishing gear (Pawar et al., 2016). Private yachts and cruise ships donate trash, including electric cables and light bulbs, as well as new plastic and glass beverage bottles (Duhec et al., 2015).

## Merchant, Military, and Research Vessels

Research, military, and commercial vessels are capable of intentionally or unintentionally dumping debris overboard. However, it is also possible for the ships to inadvertently discharge waste. A considerable proportion of the crew members on these vessels generate solid waste on a daily basis; therefore, improper management could result in this waste entering the marine environment and transforming into marine debris (Pawar et al., 2016). Commercial cargo ships produce waste such as containers, large PVC plastic tarpaulins, and polystyrene items (Duhec et al., 2015).

## Offshore Exploration and the Operation of Oil and Gas Platforms

Operations on oil and gas platforms may discharge byproducts into the marine environment, either intentionally or unintentionally. Personal waste, hard helmets, gloves, and 55-gallon storage containers are among these items. Additionally, resource extraction and underwater exploration contribute to marine debris.

## Case Study in Indonesia

Unsustainable fishing gear (2%) endangers marine biota and resources by being lost, thrown away, or abandoned in the ocean (Abdillah, 2023). Nets, traps, and rampoons can destroy coral reefs in the ocean (antaranews.com, 2022). Many crab fishermen in Banten Bay lose fish traps ≥ 5 times a month and gillnets 3–4 times a month (Susanto et al., 2022). Additionally, passenger ships throw away Indonesian seas. Werinama-Banda Neira passengers faced accusations of polluting the Banda Sea in January 2024. In August 2017, a passenger ship from Tanjung Priok, Jakarta, to Natuna, Riau Islands, did the same. Workers threw away the debris in this situation (Liputan6.com, 2018).

During specific seasons, local activities on islands and countries in Indonesia during particular seasons cause marine debris. Currents, waves, and wind carry light trash such as leaves, twigs, and plastic (Kusumawati et al., 2019). Marine debris from surrounding nations contaminates most Natuna beaches in December, January, and February. Plastic drink bottles, shampoo, soap, oil containers, detergents, and floor cleaners dominate this debris. Debris from China, Malaysia, Thailand, Cambodia, and Vietnam is the most commonly found (antaranews.com, 2023). At Huntete Beach, Wakatobi receives cargo from Malaysia, Vietnam, the Philippines, Java, Ambon, and Kalimantan (Kompas Cyber Media, 2020).

## Case of Marine Debris on Coral Reefs

Social media often showcases the underwater beauty of the current digital era. However, on the other hand, NGOs and influencers have also raised the issue of marine debris on coral reefs in Indonesian waters through social media. Marine debris pollution and warming from climate change are both harming coral reefs, according to experts. Researchers have repeatedly raised the issue of marine debris on coral reefs as a potential research topic. Marine pollution has become a global problem that requires attention.

Bali, a popular religious and marine tourism destination, has coral reefs damaged by the buildup of marine debris. Marine debris frequently litters coral reefs and underwater habitats in Nusa Dua, Bali. Disposable diapers, sanitary napkins, plastic bags, food packaging, straws, drinking glasses, shampoo packaging, and rice sacks are inorganic waste that is difficult to decompose (Ramadhani & Maharani, 2018). Researchers also observed marine debris on Mules Island, Sawu Sea National Park (TNP), and NTT coral reefs. Researchers believe that organic waste, comprising 64.3% wood and its by-products, often originates from river or sea flows along the coast. Plastic waste and fishing line contribute 35.7% of coral reef inorganic waste (Idris et al., 2023). Lost, abandoned, or discarded fishing gear was found trapped in coral reefs in Karimunjawa National Park (TNKJ) in 2020. This phenomenon occurs due to ocean currents, climate, habitat type, and the composition of aquatic biocoenosis (Wibowo, 2021). Similar incidents were also reported on Harapan Island, Kelapa, and Head Dua Island, Seribu Islands. These three islands are densely populated and have lots of tourists, so the most commonly found plastic is fishing line and rope used to tie floating net cages (Putra et al., 2021).

## The Effects of Marine Debris Deposition on Coral Reefs

Marine debris in coral reef areas in several regions of Indonesia negatively impacts coral reefs, both directly and indirectly. Marine debris can directly cause damage.

## Physical Damage

Macro debris contributes significant marine debris damage, including fragmentation (Putra et al., 2021). One example is macroplastics, which can cause physical obstacles to coral growth and development. Macroplastics hinder coral growth because coral polyps expend energy cleaning themselves rather than growing and reproducing. Storms or intentional abandonment can result in the loss of ghost gear (traps) along with metal and glass fragments that physically damage coral reefs (Sweet et al., 2019; Putra et al., 2021) (Figures 2A and B).

Marine debris impacts six types of growth on the Kelapa, Harapan, and Kelapa Dua Islands. Foliose and branching corals often trap metallic trash (cables and cans) and cloth. For example, cloth, waste rope, and ship anchors in the Sempu Strait, Malang Regency, damage coral reefs, especially Millepora (Putra et al., 2021). Millepora and branching corals have a skeletal structure that is quite fragile and has spaces to trap debris (Zurba, 2019) (Figure 2D). Sheets of foliose coral, with cavities between them, easily trap marine debris. Large (massive) corals have a surface that tends to be flat, but that does not rule out the possibility that debris will get stuck. Coral encrusting, or tabulate coral, frequently contains glass and plastic debris. Low current velocity (Figure 2C) can trap debris on the smooth surface of massive corals. People also use them to anchor floating net cages (Putra et al., 2021).

*Figure 2. Physical Damage to Coral Reefs: (A) Ropes are entangled in the branching coral; (B) Fishing nets influence some corals; (C) A glass bottle is stuck in a massive coral; and (D) A woven bag affected coral.*

*(Putra et al., 2021; Wibowo, 2021; Mueller et al., 2022)*

## Chemical Pollution

Marine debris (macro and microplastics), especially plastics, can decompose and release toxic chemicals into the water. Marine debris has the potential to transmit disease to coral reefs (Hamka, 2020). If coral comes into contact with plastic waste, the chance of coral contracting disease increases from 4% to 89% % (Lamb et al., 2018; Edward et al., 2020). Polyvinyl chloride (PVC) in plastic carries the disease-causing bacteria Rhodobacterales. Polypropylene can also be a breeding ground for the genus Vibrio, a bacterial pathogen that causes whiteout syndrome and other serious diseases in many corals. Plastic waste disrupts the coral-zooxanthellae symbiosis, which compromises coral immunity (Nama et al., 2023). Pathogenic bacteria and viruses can enter coral tissue, causing coral disease symptoms (Tang et al., 2018; Utami et al., 2021). It is possible that bleaching occurs as oxygen components accumulate. This toxic compound can cause coral zooxanthellae to come out (Fujioka, 2015). Plastics such as monofilament fishing line often cause corals to bleach (Sweet et al., 2019).

Researchers found that the accumulation of marine debris in the Sempu Strait causes coral bleaching and the spread of disease. Researchers found inorganic waste like cloth, black plastic, red plastic, and white plastic on Badi Island, Spermonde Islands, impacting coral health (Nurdin et al., 2019). Researchers discovered that Tissue Loss With Algal Growth (TLAG) affected Kelapa Dua Island, Harapan Island, and Kelapa Island. Coral susceptibility to diseases, such as Black Band Disease (BBD), increases with the abundance of abnormal algae (Putra et al., 2021).

## Biological Impact

Marine debris, including plastic, fabric, and ghost gear, blocks sunlight for coral reef symbiotic algae. Rotting debris threatens marine life and reduces coral reef oxygen (Figure 3). Researchers found fabric and raffia covering several coral reefs near Malang, East Java's Sempu Strait. In 2018, foliose, millepora, and branching coral were covered by fabric in Pulau Tiga, Seribu Islands. In TNKJ (Karimunjawa National Park)., polyamide gill nets are ghost gear for Acropora (branching), Montipora (foliose), and Porites (massive) coral species. Ghost traps can also block sunlight, just like plastic and fabric. Ghost gear, such as plastic waste, disturbs production and damages heterotrophic organisms (Sweet et al., 2019). For photosynthesis and rapid calcification, coral reefs require light transmission through seawater for photosynthesis and rapid calcification. These two processes—photosynthesis and scleractinian calcification—have been the focus of the most serious research (Edmunds et al., 2018). These two processes are responsible for the light's impact of light on corals. Zooxanthellae provide coral reef-building coral nutrients through photosynthetic activity. The majority of organic molecules (20–25% glucose, amino acids, and so on) are supplied to the host as fuel for requesting metabolic processes such as tissue growth and calcium carbonate synthesis in corals' framework. Calcium carbonate, which gives corals structure, is affected by photosynthesis disturbances (LaJeunesse, 2020).

Changes in coral reefs affect organisms nearby. Coral reefs can attract bottom organisms that consume marine debris. This may influence coral reef species distribution and complex ecological interactions. Marine debris also supports invasive algae, which may threaten coral reef ecosystems.

*Figure 3. Biological Impact to Coral Reefs*

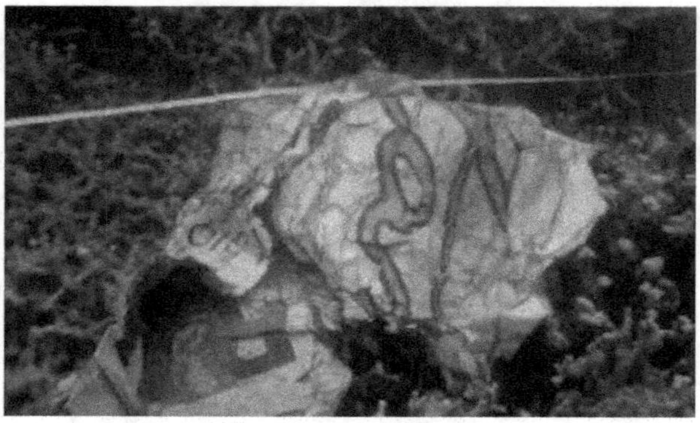

*(Putra et al., 2021)*

## Social-Economic

In addition to the impacts mentioned above, marine debris can have far-reaching economic and social consequences for surrounding communities. Marine debris on coral reefs causes significant losses to the fishing, shipping, recreation, and tourism industries, thereby reducing the environmental services provided and perceived benefits (Nama et al., 2023). Ghost gear can significantly impact the economy through the loss of resources (both non-target and commercial species), increased management costs, and potential disruptions to the fishing industry, all of which can have far-reaching effects on communities. Marine debris stuck in coral reefs diminishes the appeal of the marine environment for tourism and the economy. In addition, marine debris indirectly reduces the benefits to the ecosystem and its services. These include direct impacts on biodiversity and short-term problems like injuries, entanglements, and navigational hazards, as well as indirect and long-term impacts on quality of life (such as loss of aesthetic value, reduced recreational opportunities, and loss of unutilized value).

# SOLUTIONS AND RECOMMENDATIONS

The problem of marine waste in Indonesia, both organic and inorganic, that is difficult to decompose threatens marine ecosystems and environmental health. Research on marine waste in Indonesia shows that failure to manage waste is the main cause of pollution. Population density on the island of Java influences this (Amri et al., 2023). Marine waste, especially plastic waste, damages Indonesia's coral reefs, which support marine biodiversity and fisheries. The marine debris mitigation program for Indonesia's coral reefs includes:

## Community Engagement and Education

Education and awareness campaigns in Indonesia to prevent marine litter are critical to tackling plastic pollution. Various organizations have launched projects to raise awareness of marine debris and encourage sustainable behavior. Children and communities that interact with coastal waters, particularly fishermen who produce waste such as fishing gear, can pose a threat to marine biota. Indonesia has made the following efforts, including examples:

### Social Media Waste Campaign

The government must address the promotion of waste on social media, especially in campaigns to reduce plastic use. Rapada et al. (2021) found that social media influences plastic use, as evidenced by increasing awareness of plastic pollution among the people of Mexico, Peru, Spain, Chile, and Colombia (Tabuenca et al., 2019). Social media can help solve the problem of the ineffectiveness of laws due to a lack of oversight. Research has shown that knowledge about the marine environment and community participation improves the implementation of public policies and the relationship between the government and society's social control mechanisms through social media (Hung et al., 2022). According to Article 70 Paragraph 1 of Law Number 32 of 2009, "The community has equal and full rights and opportunities to play an active role in environmental protection and management." (Hermawan et al., 2022).

### A Single-Use Plastic Reduction Campaign

This project aims to reduce single-use plastic waste, including plastic straws and bags. The prohibition on plastic bags in several major Indonesian cities, the dissemination of information to retailers and consumers about the negative effects of single-use plastics, and the promotion of environmentally friendly alternatives are all attempts to reduce single-use plastic consumption. Several big cities are implementing this approach to prevent plastic pollution. The government prohibits plastic bags and implements a payment system for such bags. In packaged beverages, some manufacturers are using paper straws instead of plastic (Irawati, 2020).

## Beach Clean-Up

Various nations, including Indonesia, have carried out a significant number of beach cleanup efforts. The aim of this beach cleaning effort is to prevent waste from entering the ocean by preventing it from entering the water. Significant contributions to the conservation of the environment are made by a number of communities and organisations in Indonesia that are participating in this programme, including Universitas Brawijaya students. The Student Executive Board of the Faculty of Engineering carried out a clean-up of Gatra Beach in Tambak Rejo Village, Sumbermanjing Wetan District, Malang Regency, East Java, in March 2023 (humasft, 2023). The Ministry of Maritime Affairs and Fisheries carried out the "Menghadap Laut" beach clean-up on a large scale in 2018. Students from the Faculty of Fisheries and Marine Sciences participated in this programme (Figure 4).

*Figure 4. Beach clean-up actions in Malang Regency (A) Faculty of Engineering in Gatra Beach (B) Faculty of Fisheries and Marine Sciences in Pondokdadap*

*Source: humasft, 2023 and Faculty of Fisheries and Marine Sciences, 2018*

## Waste Bank Program

Waste banks are facilities developed and managed by communities, business entities, and/or local governments. Waste banks educate the public, change waste management thinking, and encourage a circular economy (Minister of the Environment Regulation No. 14 of 2021). The waste bank manages waste by "reducing, reusing, and recycling." People receive incentives to exchange plastic waste for goods or money in this project. The Indonesian Ministry of Environment and Forestry has created a mechanism to integrate the national waste bank database. https://simba .menlhk.go.id/portal hosts Indonesia's waste bank registration. This action is a good start to increasing awareness of marine debris in Indonesia and encouraging sustainable behavior. Through community engagement, education, and awareness campaigns, Indonesia may be able to reduce plastic pollution in its waters and protect the marine environment, which is a very valuable thing.

## Waste Management and Recycling

As much as 80% of marine waste comes from antropogenic on land, which are mobilised into the ocean, while 20% comes from coastal and marine activities such as shipping, transportation, and tourism (Prasetiawan, 2018). Addressing the volume of waste in Indonesia's oceans means having to manage land waste well. Sustainable waste management involves implementing practices that minimise the environmental impact of waste production and disposal while encouraging resource conservation. The following are the main components and principles of sustainable waste management:

### Reducing the Use of Single-Use Plastic

Jakarta and Bali can inspire other cities to reduce waste (Figure 5). The Bali Provincial Government issued Governor Regulation Number 97 of 2018 on December 21 to reduce single-use plastic waste and protect the environment. Production, distribution, and commercial actors are forbidden to use single-use plastic bags, styrofoam, or straws. In Bali, traditional markets, supermarkets, and restaurants employ banana leaf food wrappers and paper bags supported by the Bali Provincial Environmental Service (Department of Housing, Settlement Areas and Land, 2020). In DKI Jakarta Province, Governor Regulation (Pergub) Number 142 of 2019 prohibited single-use plastic in supermarkets, shopping malls, and public markets (Department of the Environment, 2022). Until October 2022, Jakarta's traditional marketplaces' sellers and purchasers cannot abandon single-use plastic bags Environmental KBRL laws are not being properly implemented at Cidodol Market in Kebayoran Lama, South (Belekubun, 2022). Reusable shopping bags, water bottles, and food containers may help Indonesia decrease ocean plastic pollution.

*Figure 5. Sellers and buyers use plastic bags (A) and environmentally friendly shopping bags (B) in Traditional Market, Malang, Indonesia*

*Source: Personal Documentation*

## Reuse

An extremely significant step towards protecting marine ecology and the environment is the implementation of a reuse programme in Indonesia with the goal of resolving the issue of marine debris. Within the scope of this discussion, the reuse programme encompasses initiatives aimed at lowering the quantity of plastic trash and marine debris through the reduction of single-use items, recycling, and the production of reusable products from marine waste.

## Recycle

Plastic waste may damage the environment, but managing it properly will transform it into a valuable commodity that supports the local economy. Residents of Bentiring Permai Village, Bengkulu City, are among the many Indonesian people who like to make plastic handicrafts into hats, fruit baskets, tissue boxes, and other crafts using plastic waste from food, drink, and detergent packaging (KompasTV Bengkulu, 2023). Several local UMKMs recycle plastic waste into commercial products. Kresek Project, a UMKM project, crafts plastic bags into bags, jars, baskets, plant pot covers, coasters, footwear, key chains, and paintings (Kompas Cyber Media, 2023). The UMKM Mina Tenun Lombok in NTB, Central Lombok Regency, crafts wallets, bags, pillowcases, and tablecloths from plastic waste (Sofian, 2023). In addition to plastic, craftspeople from Pandes Hamlet, Special Region of Yogyakarta, create tables, seats, art light containers, and other art decorations using hardwood sticks and branches (Sidik, 2016). As it produces recyclable plastics into economically feasible products, Surabaya, Indonesia's second-largest city, improves plastic waste management. Two-hour bus tickets cost ten plastic cups or five plastic bottles, depending on size. This programme improves Surabaya's recycling and single-use plastic waste awareness (Smith, 2022).

## Innovative Technologies

The application of technology and innovation in marine waste management in Indonesia is a key step to overcome this problem more effectively. Technology can help in monitoring, collecting, processing, and reducing marine debris. The following is an explanation of how technology and innovation are applied, along with examples in Indonesia:

### Marine Waste Collection System

Seabins collect plastic and other waste from ports and maritime waters that are waste-prone. The ocean basin's water pump collects debris and microplastics and pumps them into a container. Seabins have efficient speeds, are safe for fish and other aquatic biota, and can operate 24 hours per day (Kompasiana.com, 2023). Ports in Indonesia, such as Benoa Harbour in Bali, use this technology to purify water and reduce marine debris. Meanwhile, the Tangerang City Department of Cleanliness and Parks (DKP) has implemented a garbage pickup system (Sijesam), similar to Online Ojek, to handle residential waste (Kusumawardhani, 2016).

### Monitoring and Tracking Marine Debris

The location of marine debris is able to be monitored and reported via a mobile application. The application facilitates the ability of common people to identify locations of marine debris, capture pictures of such debris, and file reports. The data acquired via this application can be used by authorities to strategize the accumulation of marine debris and enhance their understanding of pollution patterns.

### Recycling and Processing of Marine Waste

Various technologies facilitate the recycling of plastic waste from maritime environments into functional products. This technology can transform plastic from marine environments into primary resources used in manufacturing various products, such as swimwear, apparel, and furniture. This technology achieves both the reduction of plastic pollution and the creation of economic opportunities from marine debris.

## Training and Education

Innovation is essential in the fields of education and training. The implementation of virtual simulation technology enables stakeholders, including fishermen, to complete marine debris collection, waste management, and environmental protection training. This course shows the capacity to enhance both knowledge and skills regarding the administration of marine debris.

## Data Analysis

Monitoring and managing marine debris with artificial intelligence and data analysis can provide benefits. Analysis of pollution patterns, waste classification, and sources helps improve management techniques. This technique helps identify pollution hotspots and allocate resources more efficiently. Innovative and technological solutions to marine waste pollution in Indonesia are very important, as seen above. This method can improve the management and reduction of marine debris, protect ecologically important marine habitats, and increase public awareness of the importance of a clean marine environment by using appropriate technological solutions.

## Cooperation With Various Organisations

Indonesia was involved in multifaceted international collaborations with neighbouring nations to deal with the issue of marine debris. Because the marine debris problem transcends national boundaries and demands regional coordination, cross-border cooperation is crucial. The following are examples of Indonesia's collaborative efforts:

### ASEAN Framework Collaboration

Indonesia, as a member of the Association of Southeast Asian Nations (ASEAN), has collaborated with neighbouring countries in the ASEAN region to overcome the problem of marine debris. One way is through the ASEAN Framework of Action on Marine Debris, which is an action by ASEAN countries to deal with marine debris in the Southeast Asia region. Indonesia, together with other ASEAN countries, is committed to implementing concrete actions to combat marine pollution and its impacts and emphasises the need for regional cooperation and collective action to manage marine waste effectively (Hermawan & Astuti, 2021). Several Indonesians are working to reduce marine plastic pollution (Putri & Hudaya, 2022).

1. 2012–2018: Seventeen studies on the issue of marine plastics were conducted by Indonesian and international organisations.
2. 2017: The Indonesian government established the Indonesia Plan of Action on Marine Plastic Debris 2017-2025, which outlined a target reduction of 70% in marine plastic by 2025.
3. 2018: The International Conference on Prevention and Management of Marine Plastic Pollution took place.

## Bilateral Collaboration

As an additional measure to address the issue of marine debris, Indonesia participates in bilateral collaboration with nations that are geographically close to it. Marine pollution, including marine debris, represents a significant concern along the Indonesian-Malaysian maritime border. A collaborative effort is underway between the administrations of Indonesia and Malaysia to identify an agreement on this issue. The Indonesian and Malaysian governments will provide collaborative monitoring and management of marine debris in border regions in accordance with the terms of this agreement.

## International Organizations Collaboration

Indonesia is actively participating in global environmental and marine organisations, including the United Nations (UN) and international environmental organisations, in an effort to collaboratively discover and carry out solutions to these issues. The Clean Seas Campaign initiative, led by the United Nations, is a dedicated effort that Indonesia actively supports. The Clean Seas Campaign, which aims to promote sustainable utilisation of marine resources, resolve the issue of marine plastic pollution, and encourage more clear and healthier oceans, was officially launched worldwide in 2017 (Bellou et al., 2021). This campaign emphasises the significance of collaboration and partnership between governments, non-governmental organisations, and the private sector in order to implement innovative solutions and best practices in marine debris management (Madricardo et al., 2020).

## Regional Programs Collaboration

Indonesia is a participant in the Ganges-Brahmaputra-Meghna (GBM) initiative, together with nations situated in Southeast Asia and South Asia. The Ganges-Brahmaputra-Meghna (GBM) river basin is a transboundary connected river basin spanning 1.7 million km$^2$ that crosses the nations of Bhutan, India, Nepal, China, and Bangladesh. The river basin in issue discharges an estimated 1,350 $km^2$ of freshwater annually into the Indian Ocean, positioning it as the third most significant freshwater outflow globally (Khandu et al., 2016). This initiative targets the flow of major rivers that transport plastic debris into the ocean in an effort to reduce plastic-related marine pollution. International cooperation is crucial for collective initiatives to effectively deal with the issue of marine debris, given that the consequences of plastic pollution in the ocean are worldwide and require country-to-country coordination for reaching a sustainable solution.

## Policy and Regulation

The conservation of maritime ecosystems and the sustainability of the marine environment are the primary objectives of initiatives to mitigate the negative impacts of marine debris. In considering this, the Indonesian government has formulated regulations and strategic policies to deal with and promote marine debris management. However, despite substantial efforts, there is still room for improvement in the implementation and enforcement of these regulations. The following are details of how regulations in Indonesia support marine debris management:

### Law No. 18 of 2008 Concerning Waste Management

This law establishes a strong legal basis for waste management in Indonesia as a whole, including marine debris. It includes regulations for the sorting, collection, transportation, recycling, and processing of waste. In order to conduct integrated waste management, it is necessary to take into consideration the principles of environmental conservation and sustainability.

### Government Regulation No. 81 of 2012 Concerning Management of Household Waste and Waste Similar to Household Waste

The management of waste from households is the primary emphasis of this government regulation; nevertheless, in accordance with the principles of integrated waste management, it also controls the management of waste from marine environments, which can potentially originate from activities that take place in households. The first step in minimising marine debris by preventing it from entering the water is to implement this regulation, which covers many areas of waste collection and transportation.

## Presidential Regulation No. 83 of 2018 Concerning Handling Marine Waste

In order to support activities related to handling marine waste, the Indonesian government has issued Presidential Regulation No. 83 of 2018 concerning Handling Marine Waste (Riksfardini, 2023). The Presidential Decree formed the National Action Plan for Handling Marine Waste (RAN PSL) for 2018–2025, aiming to reduce waste in the sea by 70%, particularly plastic waste (Rahman et al., 2022). RAN PSL will provide strategic directions for ministries and agencies to deal with marine waste problems. There are five strategies in the National Action Plan for Handling Marine Debris, including:

1. An effort on a national scale to raise the awareness of stakeholders
2. Land-based management of waste
3. Coast and ocean waste management
4. Mechanisms for funding, strengthening of institutions, oversight, and law enforcement
5. Development and research.

## Decree of the Coordinating Minister for Maritime Affairs No. 69 of 2019 Concerning the Implementation Team for the National Action Plan for Handling Marine Waste

The Ministry of Maritime Affairs and Investment Coordination determines the decision-making on these strategies for handling marine waste. The Coordinating Minister for Maritime Affairs of the Republic of Indonesia formed an implementation team through Decree No. 69 of 2019, which established the Implementation Team for the National Action Plan for Handling Marine Waste. As the holder of command of the Ministry of Maritime Coordination and Investment, he then coordinates tasks with his subordinates, so that both the central government and regional governments have roles according to their respective authorities. Apart from that, the National Coordination Team for Handling Marine Waste (TKN PSL) also plays a role in approaching and forming a Minister of Environment and Forestry Regulation concerning the Roadmap for Waste Reduction by Producers. You can view all activities and activity reports conducted by TKN PSL on their official website, https://sampahlaut.id/.

## Minister of Environment and Forestry Regulation No. 75 of 2019 Concerning Roadmap for Waste Reduction by Producers

The Minister of Environment and Forestry issued Regulation 75 of 2019 on the Producer Waste Reduction Roadmap. Starting in 2020, this Ministerial Regulation aims to reduce plastic, paper, glass, and aluminium waste by 30% of the total amount of products and/or product packaging produced and marketed in 2029. This ministerial decree prohibits several single-use plastics until December 31, 2029. As of May 2022, 145 Indonesian businesses promoted a waste-free lifestyle through bulk stores, product refill businesses, food container rental businesses (reuse), waste collector businesses, and upcycle businesses. Businesses that make plastic-free alternatives are also growing. This can also assist the public in discovering environmentally friendly products and packaging, such as those without packaging and those with compostable, reusable, refillable, and recyclable packaging. The Directorate General of Waste, Waste and B3 Management (2022) requires using returnable packaging that can be returned to the maker. Indonesian maritime waste management policies are important, but implementing and enforcing them is challenging. Public education, waste management infrastructure, and law enforcement need improvement. More effective and sustainable marine debris management requires cooperation between government, society, the commercial sector, and international organisations.

## Scientific Research and Monitoring

Researchers and monitors assess marine debris damage to coral reefs. Data collection facilitates marine debris data collection to inform future activities. Indonesia's Environment and Forestry Ministry established marine debris monitoring guidelines. These guidelines analyze its weight, density, types, spatial distribution, and risk to coastal and marine ecosystems. Researchers categorize marine debris from the ocean as coast, floating, and seabed debris (Prajanti et al., 2020). Long-term coral reef biodiversity and sustainability monitoring is underway. Long-term, researchers are monitoring coral reef biodiversity and diversity. The Ministry of Environment and Forestry issued the Coral Reef Health Monitoring Guidelines in 2014, which involve monitoring coral reefs, coral fish, and magabenthos (Giyanto, 2014).

## REFERENCES


Abdillah. (2023, August 15). *Reducing Marine Debris, Deputy Jodi Expects Innovation to Fight ALDFG*. Kemenko Bidang Kemaritiman dan Investasi. https://maritim.go.id:443/

Amri, R., Kholifiyanti, C., Wijayanti, E. S., Bayan, S., Hidayat, R. R., & Hidayati, N. V. (2023). Composition and Distribution of Marine Debris at Pasir Putih Losari Beach, Brebes, Central Java. *Jurnal Kelautan Tropis, 26*(1), 135–147. 10.14710/jkt.v26i1.15770

Baumann, J. H., Ries, J. B., Rippe, J. P., Courtney, T. A., Aichelman, H. E., Westfield, I., & Castillo, K. D. (2019). Nearshore coral growth declining on the Mesoamerican Barrier Reef System. *Global Change Biology*, 25(11), 3932–3945. 10.1111/gcb.1478431456305

Belekubun, R. A. (2022, April 10). *Traditional Markets in Jakarta Cannot Get Rid of Plastic Bags*. kompas.id. https://www.kompas.id/baca/humaniora/2022/10/04/pasar-tradisional-di-jakarta-belum-bisa-lepas-dari-kantong-plastik

Bellou, N., Gambardella, C., Karantzalos, K., Monteiro, J. G., Canning-Clode, J., Kemna, S., Arrieta-Giron, C. A., & Lemmen, C. (2021). Global assessment of innovative solutions to tackle marine litter. *Nature Sustainability*, 4(6), 516–524. 10.1038/s41893-021-00726-2

Bettim, M., Krelling, A. P., Di Domenico, M., Cornwell, T. O., & Turra, A. (2021). Daily environmental variation influences temporal patterns of marine debris deposition along an estuarine outlet in southern Brazil. *Marine Pollution Bulletin*, 172, 112859. 10.1016/j.marpolbul.2021.11285934418710

Cordova, M. R., Bernier, N., Yogaswara, D., Subandi, R., Wibowo, S. P. A., Kaisupy, M. T., & Haulussy, J. (2023). Land-derived litter load to the Indian Ocean: A case study in the Cimandiri River, southern West Java, Indonesia. *Environmental Monitoring and Assessment*, 195(10), 1251. 10.1007/s10661-023-11831-437768383

Department of Housing. Settlement Areas and Land. (2020, August 1). *Bali Becomes the First Province in Indonesia to Implement Plastic Waste Free Rules*. Dinas Perumahan, Kawasan Permukiman dan Pertanahan | Pemerintah Kabupaten Buleleng. https://disperkimta.bulelengkab.go.id/informasi/detail/artikel/bali-menjadi-provinsi-pertama-di-indonesia-yang-menerapkan-aturan-bebas-sampah-plastik-13

Department of the Environment. (2022, March 21). *Plastic Bag Free Jakarta Policy*. Portal Resmi Provinsi DKI Jakarta. https://jakarta.go.id/kebijakan-jakarta-bebas-kantong-plastik

Directorate General of Waste. Waste and B3 Management. (2022, May 24). *National Dialogue: Waste Reduction by Producers, Opportunities and Challenges towards Sustainable Business*. Direktorat Jenderal Pengelolaan Sampah, Limbah Dan B3 Kementrian Lingkungan Hidup Dan Kehutanan. https://pslb3.menlhk.go.id/portal/read/dialog-nasional-pengurangan-sampah-oleh-produsen-peluang-dan-tantangan-menuju-bisnis-yang-berkelanjutan

Duhec, A. V., Jeanne, R. F., Maximenko, N., & Hafner, J. (2015). Composition and potential origin of marine debris stranded in the Western Indian Ocean on remote Alphonse Island, Seychelles. *Marine Pollution Bulletin*, 96(1), 76–86. 10.1016/j.marpolbul.2015.05.04226024564

Edmunds, P. J., Tsounis, G., Boulon, R., & Bramanti, L. (2018). Long-term variation in light intensity on a coral reef. *Coral Reefs*, 37(3), 955–965. 10.1007/s00338-018-1721-y

Edward, J. K. P., Mathews, G., Raj, K. D., Laju, R. L., Bharath, M. S., Kumar, P. D., Arasamuthu, A., & Grimsditch, G. (2020). Marine debris—An emerging threat to the reef areas of Gulf of Mannar, India. *Marine Pollution Bulletin*, 151, 110793. 10.1016/j.marpolbul.2019.11079332056588

Faculty of Fisheries and Marine Sciences. (2018). Faculty of Fisheries and Marine Sciences. https://fpik.ub.ac.id/?page_id=3452

Ford, H. V., Jones, N. H., Davies, A. J., Godley, B. J., Jambeck, J. R., Napper, I. E., Suckling, C. C., Williams, G. J., Woodall, L. C., & Koldewey, H. J. (2022). The fundamental links between climate change and marine plastic pollution. *The Science of the Total Environment*, 806(Pt 1), 150392. 10.1016/j.scitotenv.2021.15039234583073

Fujioka, S. N. (2015). *The Effect of Inorganic Waste on the Condition of Hard Corals*. Universitas Hasanuddin

Giyanto. (2014). *Guide to Monitoring Coral Reef Health: Coral Reefs, Reef Fish, Megabenthos, and Report Writing*. Pusat Penelitian Oseanografi, Lembaga Ilmu Pengetahuan Indonesia.

Guzman, R. (2022, March 7). *Marine Plastic Pollution and Climate Change: Regional and National Action in Southeast Asia*. Heinrich Böll Foundation | Southeast Asia Regional Office. https://th.boell.org/en/2022/03/07/marine-plastic-pollution-seasia

Hamka, A. (2020). *Impact of Plastic Waste on Bleaching Intensity and Recovery Efforts on Acropora sp coral*. Universitas Hasanuddin.

Hermawan, S., Abiyudo, B., & Isnawanjati, G. (2022). Potential Use of Social Media Platforms to Reduce Plastic Waste in Indonesian Oceans. *Kosmik Hukum*, 22(2), 132. 10.30595/kosmikhukum.v22i2.14140

Hermawan, S., & Astuti, W. (2021). Analysing several ASEAN countries' policy for combating marine plastic litter. *Environmental Law Review*, 23(1), 9–22. 10.1177/1461452921991731

humasft. (2023, March 27). *Caring for the Environment, BEM FT Beach Clean-up*. Prasetya UB. https://prasetya.ub.ac.id/peduli-lingkungan-bem-ft-bersih-pantai/

Hung, L.-Y., Wang, S.-M., & Yeh, T.-K. (2022). Collaboration between the government and environmental non-governmental organisations for marine debris policy development: The Taiwan experience. *Marine Policy*, 135, 104849. 10.1016/j.marpol.2021.104849

Idris, I., Fakhrurrozi, F., Suhendar, D. B., Rochyadi, A., Wahyudi, R., & Larasati, C. E. (2023). Distribution of Marine debris in Coral Reef Ecosystems on Mules Island, Sawu Marine National Park. *Jurnal Biologi Tropis*, 23(2), 2. Advance online publication. 10.29303/jbt.v23i2.4618

Irawati, D. (2020, September 8). *Nestle Immediately Replace Plastic Straws with Paper*. kompas.id. https://www.kompas.id/baca/nusantara/2020/09/08/nestle-segera-ganti-sedotan-plastik-dengan-kertas

Kaharto, S. W., & Hermansyah, Y. (2023). Analysis of Public Awareness of the Impact of Pollution on Losari Beach, Makassar City. *INNOVATIVE: Journal Of Social Science Research, 3*(4), 3510–3517. https://j-innovative.org/index.php/Innovative

Khandu, F., Forootan, E., Schumacher, M., Awange, J. L., & Müller Schmied, H. (2016). Exploring the influence of precipitation extremes and human water use on total water storage (TWS). *Water Resources Research*, 52(3), 1541–2375. 10.1002/2015WR018113

Kompas, T. V. Bengkulu. (2023, March 17). *Utilizing Waste into Valuable Items*. KOMPAS.tv. https://www.kompas.tv/regional/388999/pemanfaatan-sampah-menjadi-barang-bernilai

Kompas Cyber Media. (2020, March 9). *Study of Huntete Wakatobi Coastal Waste, Much of it from Neighboring Countries*. KOMPAS.com. https://www.kompas.com/sains/read/2020/03/09/181000123/kajian-sampah-pesisir-huntete-wakatobi-banyak-dari-negara-tetangga

Kompas Cyber Media. (2023, March 5). *Kresek Project, Handicraft Business from Recycling Kresek Waste.* KOMPAS.com. https://umkm.kompas.com/read/2023/03/05/192824183/kresek-project-bisnis-kerajinan-tangan-dari-daur-ulang-sampah-kresek

Kompasiana.com. (2023, January 10). *SeaBin Ocean Cleaning Technology, Available in Indonesia?* KOMPASIANA. https://www.kompasiana.com/yudiramid0862/63bb9e8d4addee77b65bfe12/seabin-teknologi-permbersih-laut-adakah-di-indonesia

Kusumawardhani, A. (2016, November 25). *Tangerang City Government-style Garbage Pickup Application.* Bisnis.com. https://jakarta.bisnis.com/read/20161125/383/606245/aplikasi-jemput-sampah-ala-pemkot-tangerang

Kusumawati, I., Nasution, M. A., & Alamsyah, A. (2019). Distribution and Composition Coastal Marine Garbage in The Sub-District Nagan Raya District. *Jurnal Laot Ilmu Kelautan*, 1(1). Advance online publication. 10.35308/jlaot.v1i1.1073

LaJeunesse, T. C. (2020). Zooxanthellae. *Current Biology*, 30(19), R1110–R1113. 10.1016/j.cub.2020.03.05833022248

Lamb, J. B., Willis, B. L., Fiorenza, E. A., Couch, C. S., Howard, R., Rader, D. N., True, J. D., Kelly, L. A., Ahmad, A., Jompa, J., & Harvell, C. D. (2018). Plastic waste associated with disease on coral reefs. *Science*, 359(6374), 460–462. 10.1126/science.aar332029371469

Lebreton, L. C. M., Van Der Zwet, J., Damsteeg, J.-W., Slat, B., Andrady, A., & Reisser, J. (2017). River plastic emissions to the world's oceans. *Nature Communications*, 8(1), 15611. 10.1038/ncomms1561128589961

Lincoln, S., Andrews, B., Birchenough, S. N. R., Chowdhury, P., Engelhard, G. H., Harrod, O., Pinnegar, J. K., & Townhill, B. L. (2022). Marine litter and climate change: Inextricably connected threats to the world's oceans. *The Science of the Total Environment*, 837, 155709. 10.1016/j.scitotenv.2022.15570935525371

Liputan6.com. (2018, November 21). *3 Cases of Throwing Garbage into the Sea that Go Viral on Social Media.* liputan6.com. https://www.liputan6.com/lifestyle/read/3697231/3-kasus-buang-sampah-ke-laut-yang-viral-di-media-sosial

Madricardo, F., Ghezzo, M., Nesto, N., Mc Kiver, W. J., Faussone, G. C., Fiorin, R., Riccato, F., Mackelworth, P. C., Basta, J., De Pascalis, F., Kruss, A., Petrizzo, A., & Moschino, V. (2020). How to Deal With Seafloor Marine Litter: An Overview of the State-of-the-Art and Future Perspectives. *Frontiers in Marine Science*, 7, 505134. 10.3389/fmars.2020.505134

Meijer, L. J. J., Van Emmerik, T., Van Der Ent, R., Schmidt, C., & Lebreton, L. (2021). More than 1000 rivers account for 80% of global riverine plastic emissions into the ocean. *Science Advances*, 7(18), eaaz5803. Advance online publication. 10.1126/sciadv.aaz580333931460

Mueller, J. S., Bill, N., Reinach, M. S., Lasut, M. T., Freund, H., & Schupp, P. J. (2022). A comprehensive approach to assess marine macro litter pollution and its impacts on corals in the Bangka Strait, North Sulawesi, Indonesia. *Marine Pollution Bulletin*, 175, 113369. 10.1016/j.marpolbul.2022.11336935144214

Nama, S., Shanmughan, A., Nayak, B. B., Bhushan, S., & Ramteke, K. (2023). Impacts of marine debris on coral reef ecosystem: A review for conservation and ecological monitoring of the coral reef ecosystem. *Marine Pollution Bulletin*, 189, 114755. 10.1016/j.marpolbul.2023.11475536905864

National Coordination Team for Handling Marine Debris. (2024, February 2). *Tim Koordinasi Nasional Penanganan Sampah Laut | TKN PSL*. Tim Koordinasi Nasional Penanganan Sampah Laut. https://sampahlaut.id/

Nurdin, M., Litaay, M., Priosambodo, D., & Moka, W. (2019). Coral Conditions on Baranglompo and Bone Batang Islands Based on Coral Health Tables. *Jurnal Ilmu Alam Dan Lingkungan, 10*(1), 15–28. https://journal.unhas.ac.id/index.php/jai2/article/view/6384

Ortiz, J.-C., Wolff, N. H., Anthony, K. R. N., Devlin, M., Lewis, S., & Mumby, P. J. (2018). Impaired recovery of the Great Barrier Reef under cumulative stress. *Science Advances*, 4(7), eaar6127. Advance online publication. 10.1126/sciadv.aar612730035217

Pawar, P. R., Shirgaonkar, S. S., & Patil, R. B. (2016). Plastic marine debris: Sources, distribution and impacts on coastal and ocean biodiversity. *PENCIL Publication of Biological Sciences, 3*(1), 40–54. www.pencilacademicpress.org/ppbs

Prajanti, A., Berlianto, M., Simamora, R. L., Imansari, M. B., & Sari, N. (2020). *Guidelines for Monitoring Marine Debris: Beach Debris, Floating Trash, and Seabed Debris*. Direktorat Pengendalian Pencemaran dan Kerusakan Pesisir dan Laut. https://www.ppkl.menlhk.go.id

Prasetiawan, T. (2018). Efforts to Overcome Plastic Waste in the Ocean. *PUSLIT BKD, 10*(10).

Purba, N. P., Handyman, D. I. W., Pribadi, T. D., Syakti, A. D., Pranowo, W. S., Harvey, A., & Ihsan, Y. N. (2019). Marine debris in Indonesia: A review of research and status. *Marine Pollution Bulletin*, 146, 134–144. 10.1016/j.marpolbul.2019.05.05731426141

Putra, M. G. A., Zamani, N. P., Natih, N. M., & Harahap, S. A. (2021). Relationship Between Characteristics of Marine Debris and Impact to Coral Reef. *Jurnal Ilmiah Perikanan Dan Kelautan*, 13(1), 11–19. 10.20473/jipk.v13i1.18896

Putri, R. A., & Hudaya, M. (2022). The Establishment of ASEAN Framework of Action on Marine Debris: The Role of Shared Knowledge. *Global: Jurnal Politik Internasional*, 24(1), 63–84. 10.7454/global.v24i1.668

Rahman, A. N. P., Fitria, D., & Sentika, H. T. (2022). *Effectiveness of Plastic Waste Management Policies in Indonesian Seas. Presidensi G-20 Indonesia Tahun 2022 'Recover Together, Recover Stronger"*. Transformasi Administrasi Dan Kebijakan Public Di Era Post Truth Dalam Mewujudkan Indonesia Tumbuh, Indonesia Tangguh.

Ramadhani, M., & Maharani, E. (2018, November 22). *Plastic Waste Damages Bali's Coral Reef Ecosystem*. Republika Online. https://republika.co.id/share/pildnu335

Rapada, M. Z., Yu, D. E., & Yu, K. D. (2021). Do Social Media Posts Influence Consumption Behavior towards Plastic Pollution? *Sustainability (Basel)*, 13(22), 12334. 10.3390/su132212334

Ridha, D. M., Hendriarti, N., Koropitan, A. F., & Lestari, S. P. (2019, March 30). *Marine Debris and Its Impact on Climate Change*. Pojok Iklim. http://pojokiklim.menlhk.go.id/read/marine-debris-and-its-impact-on-climate-change

Riksfardini, M. (2023). Analysis of the Implementation of Marine Waste Management Policies in the Muara Angke Coastal Area, North Jakarta. *PENTAHELIX*, 1(2), 217. 10.24853/penta.1.2.217-236

Sidik, H. (2016, March 5). *Bantul residents process marine waste into crafts*. Antara News Yogyakarta. https://jogja.antaranews.com/berita/338113/warga-bantul-mengolah-sampah-laut-jadi-kerajinan

Smith, A. (2022). *Managing Marine Plastic Waste in Asia and the Pacific*. Economic and Social Commission for Asia and the Pacific (ESCAP).

Sofian, H. (2023, June 12). *Economic value, plastic waste is processed into weaving crafts*. Rri.Co.Id - Portal Berita Terpercaya. https://www.rri.go.id/mataram/umkm/259600/bernilai-ekonomi-sampah-plastik-diolah-jadi-kerajinan-tenun

Susanto, A., Syafrie, H., Nurdin, H. S., Irnawati, R., Supadminingsih, F. N., Hamzah, A., & Kurniawati, V. (2022). Lost Gear on Blue Swimming Crab Fisheries in Banten Bay: Case of Abandoned, Lost or Discarded Fishing Gear. *Marine Fisheries*, 13(2), 233–241. 10.29244/jmf.v13i2.41961

Suteja, Y., Atmadipoera, A. S., Riani, E., Nurjaya, I. W., Nugroho, D., & Purwiyanto, A. I. S. (2021). Stranded marine debris on the touristic beaches in the south of Bali Island, Indonesia: The spatiotemporal abundance and characteristic. *Marine Pollution Bulletin*, 173, 113026. 10.1016/j.marpolbul.2021.11302634628344

Sweet, M., Stelfox, M., & Lamb, J. (2019). *Plastics and shallow water coral reefs. Synthesis of the science for policy-makers*. UN Environment.

Tabuenca, B., Kalz, M., & Löhr, A. (2019). Massive Open Online Education for Environmental Activism: The Worldwide Problem of Marine Litter. *Sustainability (Basel)*, 11(10), 2860. 10.3390/su11102860

Tang, J., Ni, X., Zhou, Z., Wang, L., & Lin, S. (2018). Acute microplastic exposure raises stress response and suppresses detoxification and immune capacities in the scleractinian coral Pocillopora damicornis. *Environmental Pollution*, 243, 66–74. 10.1016/j.envpol.2018.08.04530172125

Utami, D. A., Reuning, L., Konechnaya, O., & Schwarzbauer, J. (2021). Microplastics as a sedimentary component in reef systems: A case study from the Java Sea. *Sedimentology*, 68(6), 2270–2292. 10.1111/sed.12879

Wibowo, T. W. (2021). Cleaning "Ghost Nets" in Karimunjawa National Park Waters: Pilot Study. *JECE - Journal of Empowerment Community and Education, 1*(4), 345–350. https://jurnalpengabdian.com/index.php/jece/article/view/651

World Bank. (2021). *Plastic Waste Discharges from Rivers and Coastlines in Indonesia*. World Bank. 10.1596/35607

Zurba, N. (2019). *Introduction to Coral Reefs as the Main Foundation of Our Ocean*. UNIMAL PRESS.

## KEY TERMS AND DEFINITIONS

**Climate Change Mitigation:** Efforts to reduce the impacts of global climate change, such as increasing world temperatures and rising sea levels.

**Coral Reef Degradation:** The process of destroying coral reefs due to factors such as pollution, increasing sea water temperatures, or human activities.

**Cross-Sector Collaboration:** Collaboration between the public and private sectors to conserve coral reefs.

**Ghost Gear:** Fishing equipment lost, abandoned, or thrown in the ocean, which then becomes marine debris.

**Marine Debris Management:** An approach and method designed to reduce the impact of marine waste on marine ecosystems and humans in an efficient and sustainable way.

**Marine Debris:** Solid waste or plastic waste from land or sea that ends up in the sea and can damage marine ecosystems.

**Non-Governmental Organisations (NGOs):** Organisations founded by an individual or group to volunteer to assist the public without benefiting.

**Sustainability:** Principles that support the protection of natural resources for future generations.

# Chapter 3
# Impact of Ecological Constitution on Conservation of Marine Ecosystem

**Siddharth Kanojia**
https://orcid.org/0000-0002-1479-5292
*O.P. Jindal Global University, India*

## ABSTRACT

*The declining abundance of fish stocks and the deteriorating state of marine ecosystems are triggered by a number of variables such as pollution, overfishing, habitat destruction, and climate change. These factors raise serious concerns about our marine ecosystem. In light of these challenges, there is a pressing need to reframe environmental law on managing the marine resources to better address ecological concerns and promote sustainability. Hence, this chapter established the contextual relevance between the need for Ecological Constitution on the protecting the marine ecosystem from overfishing, loss of biodiversity, ocean acidification, and climate change and provide an overview of how the adaptation of Ecological Constitution can restore the deterioration of marine ecosystem.*

## INTRODUCTION

The idea of an "Ecological Constitution" has become essential in a time of unparalleled environmental challenges. The escalating gravity of ecological concerns, such as biodiversity loss, pollution, climate change, and resource depletion, emphasizes the necessity of stringent legal frameworks to guarantee the sustainability

DOI: 10.4018/979-8-3693-2436-3.ch003

of our environment (Mega 2018). The relationships between ecological problems have been found to be overwhelming to traditional legal systems that are usually characterized by fragmented and defensive strategies towards the environment (De Lucia 2013). Perhaps, this can be solved by Ecological Constitution, which provides a wide-ranging and legally enforceable Charter that sets environmental issues as the priority of the economic and social activities.

As a matter of fact, there are several global trends and theoretical approaches that have in the past shaped the idea of integrating environmentalism into the frameworks of constitutional law. The issue of the constitutional enshrinement of environmental rights has been boosted by the environmental justice movement which has an emphasis on right and justice in the distribution of environmental gains and losses(Shrader-Frechette 2002). Furthermore, the movement for constitutional environmental provisions has been impacted by the idea of sustainable development as it is expressed in renowned international accords like the Rio Declaration and the Brundtland Report (Bellah & Chiha 2017). This idea is further supported by theoretical underpinnings like ecological democracy and Earth jurisprudence (Burdon 2013). Furthermore, the ecological democracy looks for the ways to enhance people's engagement in an environmental decision. Similarly, the Earth jurisprudence reveals that the contemporary legal systems are in dire need of paying attention to natural entities and the inherent value of nature (Alves et al., 2023). Comprehensively, these perspectives offer a strong basis for recognizing, comprehending and developing the idea of an ecological constitution.

In this regard, a number of nations have achieved notable progress in embedding environmental principles in their constitutions. The Constitution of Ecuador is a trailblazing example, as it clearly acknowledges the rights of nature and mandates the state to maintain ecological balance (O'Donnell et al., 2020). Similarly, Bolivian Constitution recognizes the interdependence of humans and nature and enshrines the idea of "*Mother Earth*" (Pachamama) (Helbert 2024). These illustrations demonstrates the viability and revolutionary potential of environmental provisions found in constitutions. While, to differing degrees, other nations like South Africa, India, and Norway have also included environmental rights and obligations in their constitutions. These instances demonstrate a broader shift in societal values and priorities by reflecting an increasing recognition and importance of constitutionalizing environmental protection.

Accordingly, implementing an ecological constitution will have a significant impact on social norms, legal frameworks and government policies (Paavola 2007). By the virtues of this, the states can guarantee that environmental factors are incorporated into all facets of policy-making, planning, and development by putting environmental protection on par with constitutional rights. Nevertheless, implementing an ecological constitution into effect is not without its own challenges. To effectively transform constitutional ideals into laws, policies, and practices, significant institutional capacity, public support, and political will are needed (Nonet et al., 2017). Whereas, striking a balance between environmental preservation and other constitutional duties and rights, like social welfare and economic growth, can be arduous and divisive. Hence, the upcoming parts of this chapter will explore the perspective of scholars and researchers through the review of existing literature. Thereon, the chapter will deal with application of ecological sustainability and present status of deterioration of marine ecosystem. Followed by discussing the prospects of redressal through legislative frameworks and establishing nexus between ecological constitution and preservation of marine ecosystem.

## REVIEW OF LITERATURE

In contemporary academic discourse, there has been a growing recognition of the connection between constitutional environmental principles and the preservation of marine ecosystems. The theoretical foundations and practical implications of homogenizing ecological constitutionalism with marine conservation are examined in this part of the chapter.

Predominantly, the book titled '*The Ecological Constitution: Reframing Environmental Law*' intends to outline the essential elements of any constitution that may be referred to as "ecological" in nature (Collins 2021). It contains few chapters ranging from providing introduction to constitutionalism with reference to Ecological Law, followed by analyzing the importance of ecological sustainability, human right, intergenerational equity, privileges of nature, prudence and effect of climate change in Ecological Constitutionalism. The author contends that within the backdrop of the Anthropocene era, the principles such as sustainability, precautionary principles, and public trust doctrines can prove to be crucial instruments for altering consciousness and promoting policy reforms related to the protection of the environment. The book starts with a comprehensive overview of the current status of environmental law and its shortcomings in dealing with the current environmental problems such as climate change, loss of biological diversity and degradation of ecosystems. The author also claims that anthropocentrism, which places human interests above the value of nature, has led to the destruction of the environment and the current environmental problems. The immense need for a paradigm shift in the formulation and execution of environmental law is emphasized, and the concept of the Ecological Constitution is introduced as a potential solution. In reference to this, the author establishes an argument that the ecological basis upon which all civilizations are built should be protected by respective national constitutions everywhere. Followed by the assertion that an Ecological Constitution might orient nation-states towards more sustainable development routes, serve as a guide for environmental law, and infuse all aspects of government with an ecological mindset. Following the argument and assertion, the author discuss the concept of Anthropocene by conveying the fact of the present and future well-being of homo sapiens, plants, animals, and ecosystems is threatened by humans' eroding the natural ecosystems on a worldwide scale. However, the author underlines the optimistic viewpoint by dwelling on the scope for transition to sustainable means of livelihood. Accordingly, emphasizing on the immense need for new regulatory regimes to cure the failures of existing laws. The author then attempts to exhibit the probable reasons (such as Corporate Hegemony, Public Apathy, underfunding of environment science) for the failure existing environmental protection regimes while taking the cognizance of the direct combat between the deficient environmental law with other dominant areas of law such as Human Rights, Right to Property & Corporate Rights. Thereby, the author setup a context for re-visioning the environmental law through the invocation of constitutional rights & remedies for ensuring sustainability in the conduct of humans.

After exhibiting the evolution of Environmental Law Regimes from Stockholm Declaration of 1972 & subsequent ratification by the nation states, the author then introduces the concept of Ecological Law which is distinctive from the traditional principles of Environmental Law as it fundamentally re-establishes the human relationship with earth by including the phenomenon of Earth Jurisprudence, Sustainability Law, Green Law Theory, Wild Law & others (Blackmore 2018). The concept of Ecological Law have traces in various international charters & conventions such as Earth Charter (2000), Universal declaration on the Rights of Mother Earth (2010), Oslo Manifesto (2016) and Siena Declaration (2017) (Musolino 2018). On this basis, few scholars have contended that the concept must manifest three traits i.e., Ecocentrism, Ecological Primacy and Justice. The Author further establishes that the Ecological Law integrates legal concepts with scientific rules of ecology because it is firmly rooted in the study of how the Earth functions and of complex systemic thresholds in the global ecosystem. In addition to this, the Ecological Law results from a core normative decision to regulate human endeavours in a way that provides contexts for the application of ecology, thermodynamics, and other disciplines in a way that maintains ecological integrity and fosters the health of the connection between human and earth. The core of the book highlights into the principles of the Ecological Constitution, providing a detailed exploration of its key components, such as acknowledging the legal rights of nature, acknowledging the interconnectedness of ecosystems, applying the precautionary principle, promoting sustainability, advocating for participatory governance, ensuring environmental justice, and establishing effective enforcement mechanisms. Each principle is discussed in depth, supported by legal analysis, case studies, and examples from various jurisdictions around the world. The author also addresses potential criticisms and challenges to the implementation of the Ecological Constitution, providing a balanced and nuanced perspective.

The concept of ecological constitution is also underpinned by a number of theoretical paradigms including earth Jurisprudence, sustainable development and environmental justice (Ruhs & Jones 2016). In the same vein, in an effort to protect endangered groups that rely on marine resources, environmental justice calls for the provision of constitutional status to the rights of the environment and the equitable share of the costs as well as the gains that are incurred to the environment (Bullard, 2005). The impacts of long-term ecological considerations in constitutional law can be organized according to the Brundtland Report (1987) on sustainable development to achieve sustainable use of marine resources and socio-economic development. The notion that legal systems should acknowledge the inherent value of nature is also furthered by earth jurisprudence (Cullinan, 2011). This viewpoint supports constitutional clauses that acknowledge and protect the rights of oceans and marine life to achieve legal protection of the rights of marine environments. Each of these theoretical frameworks supports the notion that ecological concepts should be integrated into constitutional law to achieve the systematic and effective marine conservation.

As for the particular countries, the Ecuadorian Constitution that enshrines the rights of nature is an example of ecological constitutionalism (Valladares & Boelens 2017). According to the Constitution, the state has the obligation to protect and restore natural cycles, such as marine ones. This constitutional recognition has enabled local communities and environmental organisations to legally sue for the protection of marine environments and there have been many conservation gains. Likewise, Environmental rights are also protected under South Africa's Constitution, which places a strong emphasis on natural resource preservation and sustainable development (Kotzé, 2012). The constitutional provisions have made it easier to adopt the policies in integrated coastal management and the development of MPAs that are all-inclusive. This paper has revealed that, through the legal provisions, it has become possible to achieve a balance between the socioeconomic demands of the coastal people and the conservation objectives. However, with reference to India, the judges of the courts have elaborated on Constitution to mean that it contains the right to a healthy environment despite the fact that it lacks provisions on environmental rights (Divan & Rosencranz, 2001). This judicial activism in the protection of marine ecosystems has led to the enforcement of environmental laws and regulations. Thus, the National Green Tribunal has been very crucial in the fight against marine pollution and ensuring that conservation measures are complied with.

When ecological constitutional principles are used in marine conservation, there are several real-world implications. First of all, it provides a solid legal foundation for the establishment and maintenance of MPAs. Environmental rights should be included in the constitution because citizens and non-governmental organisations will be in a better position to demand for the protection of marine habitats and also ensure that governments are accountable for their promises on the protection of the environment (Kibel, 2007). Secondly, the adoption of ecosystem-based management strategies, which are necessary for the sustainable management of marine resources can be aided by an ecological constitution. Constitutional provisions that acknowledge the interdependence of marine ecosystems can encourage integrated policies aimed at mitigating the cumulative effects of human activities on marine environments (Agardy, 2010). Lastly, international cooperation on marine conservation can be strengthened by constitutional environmental principles. Overfishing and pollution are just two examples of the many transboundary marine issues. A nation's position in international negotiations can be strengthened by constitutional commitments to environmental protection, which can also encourage cooperation on marine conservation projects (Harrison, 2017).

Notwithstanding the possible advantages, there are a number of obstacles to an ecological constitution's implementation. One such challenge is the potential rivalry of the economic development and the conservation of the environment. Efforts to protect the seas are usually regarded as posing challenges to business activities such as fishing, travel, and resource exploitation. Careful policy design and stakeholder engagement are necessary to balance these conflicting interests (Gunningham, 2009). Furthermore, governments' ability and willingness to enforce constitutional environmental provisions determines their effectiveness. The implementation of constitutional commitments to marine conservation can be undermined in many countries by weak governance structures, corruption, and resource scarcity (Jones et al., 2013).

## ECOLOGICAL SUSTAINABILITY

The case for ecological sustainability as a requirement of the constitution and outline the expanding number of operative clauses that already recognise this essential underlying idea. It has been asserted that the drafting committees of the Ecological Constitution should include ecological clauses in the areas of their separate constitutions that are subjected to action, and courts should regard such clauses to be justifiable. Furthermore, the Ecological Constitutions should fabricate governmental entities that are engaged in advising the executive and legislative branches while upholding constitutional environmental rights in court to enforce uncomplicated implementation. Additionally, there is a requirement to include permanent measures that permit enforcement proceedings to be brought by any member of the public or non-governmental organization. Where, the Environmental constitutional disputes should be resolved quickly while focusing on expanding access to justice in Ecological Constitutions. In this regard, the author has based her assertions and recommendations by exemplifying the Constitutional provisions of few nations such as Brazil wherein the *Ministerio Publico* has been granted with some powers to put sanctions against the violators of laws made to protect the environment (Little 2013). While, in Ecuador, the natural and legal persons can enforce their Right of Nature by obligating the public authorities. Then, few Latin American countries and few Asian countries such as India and Philippines has attempted to reduce the cost & simplify the legal procedures in order to increase the access to justice. The focus of the book then switches to fundamental human rights in the next section, including the right to a healthy environment. Post 1970s, when the Stockholm Declaration first hinted in 1972, the right to a healthy environment has steadily come into being. The underlying principle states: "Man has the fundamental right to freedom, equality and adequate conditions of life, in an environment of a quality that permits a life of dignity and well-being". The right to a healthy environment is contained in regional human rights treaties and environmental accords binding more than 120 states. According to David Boyd, more than 100 States have constitutional protection for it. Whereas, more than 100 States have included it in their respective environmental legislation. Lately, 155 states have eventually recognized the right to a safe and sustainable environment (Aguila & Viñuales 2019).

## DETERIORATION OF MARINE ECOSYSTEM

Marine ecosystem is referred to a complex community of living organism interacting with each other in the ocean environment. It includes a variety of organisms such as phytoplankton, zooplankton, fish, sea mammals, and invertebrates, as well as non-living components like water, rocks, and minerals (Sigee 2005). This ecosystem considered to be important because it provides habitat for a broad range of marine diversity, and also plays a crucial character in balancing the Earth's climate and the planet's overall ecosystem health (Gray 2012). Similarly, the oceans play a significant role in the global carbon cycle, absorbing about a one-third of the carbon dioxide produced by human activities and helping to reduce the impacts of climate change (Sabine et al. 2004). Over the years, marine ecosystems are under threat from human activities such as pollution, habitat destruction, overfishing and climate change (Kanojia 2024). These factors has lately disrupted the delicate balance of marine ecosystems, leading to the loss of biodiversity, changes in species distribution, and even the collapse of entire marine food webs (Prakash & Verma 2022).

The United Nations Food and Agriculture Organization (FAO) has said that a mere 34 percent of the fish stocks globally are overexploited, meaning that they are fished in a manner that is unsustainable (Pauly & Zeller 2017). For instance, the population of bluefin tuna in the Mediterranean has declined by 85% over the past few decades due to overfishing. Thereon, The FAO estimated that the global fishing fleet is two to three times larger than what is needed to sustainably harvest the world's fish stocks (Heffernan 2014). This means that there is a significant overcapacity in the fishing industry, which contributes to overfishing (*seeFigure 1*). Consequently, Overfishing has evidently altered the entire marine food webs in addition to causing a detrimental impact on the populations of fishes, by removing too many predators, such as sharks, which lead to an overabundance of their prey, leading to consumption of their own food sources and disrupting the balance of the ecosystem.

*Figure 1. Data on Aquaculture and Fisheries Capture from Year 1960 to 2015*

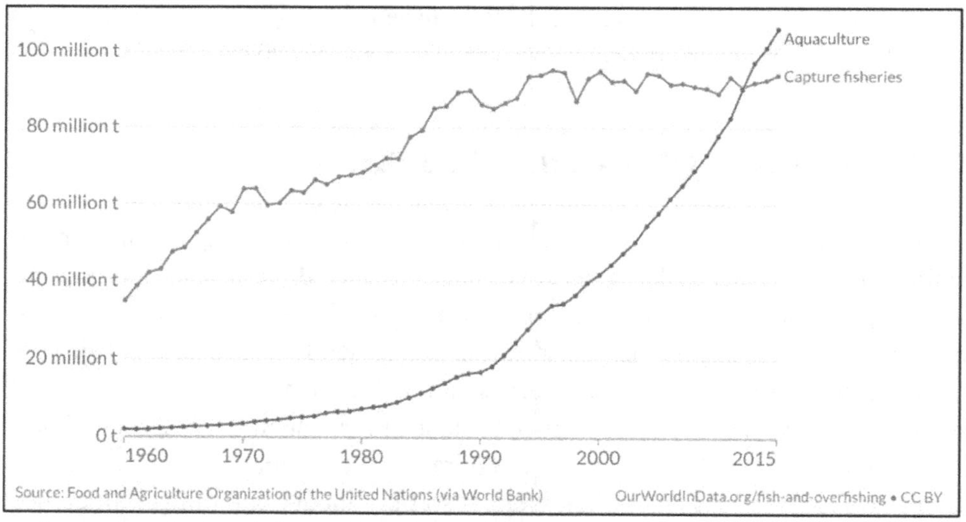


*Source: https://ourworldindata.org/fish-and-overfishing*

Similarly, According to estimates, land-based pollutants such as agricultural runoff, sewage, and plastic trash account for almost 80% of marine pollution. Every year, an estimated eight million metric tons of plastic waste enters the world's oceans, which is equivalent to dumping a garbage truck full of plastic into the ocean every minute (Hassan 2017) (*See Figure 2*). Other sources of marine pollution include oil spills, industrial discharges, and shipping activities. In the year 1989, the Exxon Valdez oil spill released 10.8 million gallons of oil into Prince William Sound, Alaska, causing extensive damage to marine ecosystems and local economies. Further, it is estimated that the cost of lost tourism revenue, property damage, and clean-up efforts from the Deepwater Horizon oil spill in 2010 was $17.2 billion. Besides, the Great Pacific Garbage Patch, an assemblage of plastic debris in the North Pacific Ocean, is gauged to be twice the size of territory of Texas and contains over 80,000 metric tons of plastic waste. Hence, it is often construed that the effects of marine pollution has been far-reaching and long-lasting (Picou & Martin 2007).

*Figure 2. Per Capita Emission of Plastic Waste in Ocean*

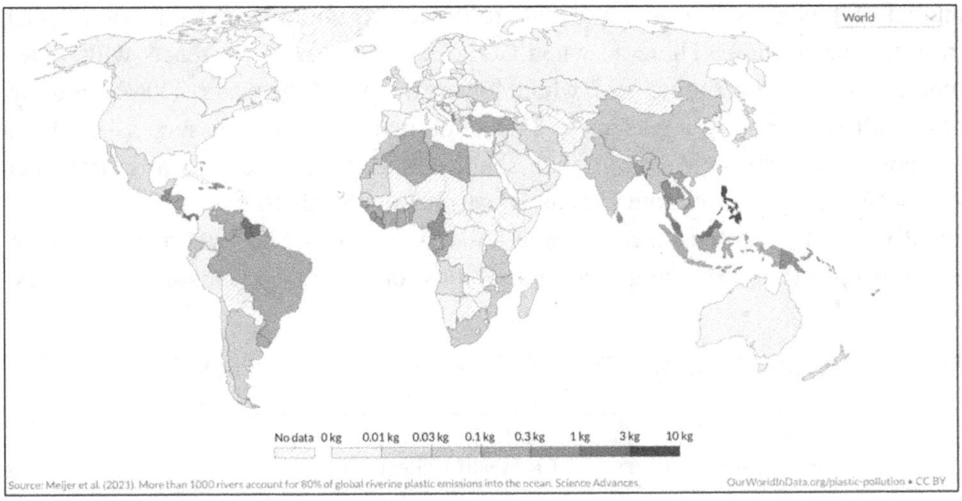


*Source: https://ourworldindata.org/plastic-pollution*

Whereas, with an estimated 25% of all marine species they sustain, coral reefs are among the biodiverse ecosystems on the globe, despite covering only 0.1% of the ocean floor. They provide habitat for countless species of fish, crustaceans, mollusks, and other marine organisms (Knowlton et al. 2010). They also tend to protect coastlines from storms and erosion by absorbing wave energy and reducing the impacts of waves and currents and they are the potential source of new medicines and treatments for a variety of diseases. Various researchers have identified compounds in coral reef organisms that have anti-inflammatory, anti-cancer, and antibiotic properties (Malve, 2016). However, they are under an acute threat from human interventions such as overfishing, excessive pollution and climate change (Tymoshenko et al. 2022). According to the report, It is estimated that around 50% of the world's coral reefs have been lost in the last 3 decades and up to 90% of the world's coral reefs may be lost by 2050 if current trends continue (Mooney et al., 2009). Coral reefs are particularly very sensitive to the increases in ocean temperatures, which lead to the bleaching and death of coral. Evidently, there is a rise of almost 2.5 degrees centigrade in the sea surface of great barrier reef over the period of last century (*SeeFigure 3).* Accordingly, in the year 2016 and 2017, two back-to-back coral bleaching events affected large portions of the Great Barrier Reef, causing consequential damage. The degradation of reefs have caused weighty economic impacts, particularly in countries that rely on tourism and fishing hence leading o reduction of income and employment opportunities, as well as the loss of important cultural and ecological resources. Other the other hand, the ocean absorbs approximately 30% of the carbon dioxide which are emitted by human activities, subsequently leading to acidification of ocean. Technically, it is a process in which the pH of oceanic water decreases due to the absorption of carbon dioxide ($CO_2$) from the atmosphere. Thereon, when $CO_2$ dissolves in water, it reacts with water molecules to form carbonic acid, which lowers the pH of the water. This eventually has resulted in causing a pernicious impact on marine organisms such as shellfish and corals, who struggle to form their shells or skeletons in acidic water (Raven et al. 2005). Lastly, the climate change is causing sea levels to rise, which can have negative impacts on coastal ecosystems and communities. Sea level rise can lead to erosion of coastlines, flooding of coastal habitats, and loss of wetlands and mangroves.

*Figure 3. Increase in Surface Temperature over Great Barrier Reef*

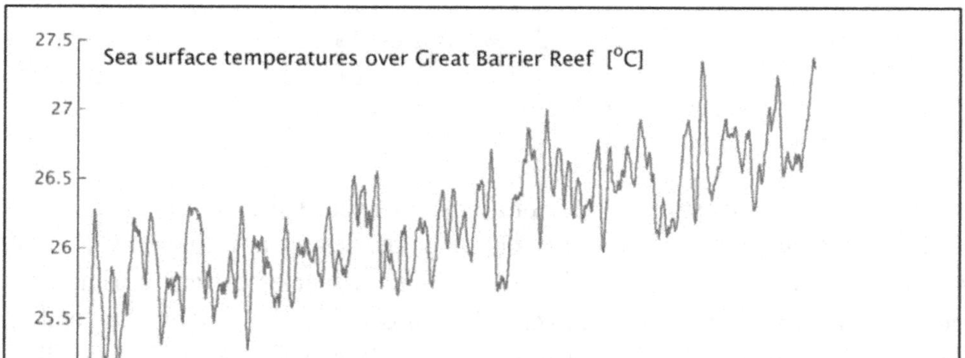


*Source:* https://www.climate-lab-book.ac.uk/2017/coral-reef-bleaching

## LEGISLATIVE REDRESSAL

The Great Barrier Reef Marine Park Act was passed by the Australian government in 1975 to protect the Great Barrier Reef, one of the world's largest and most complex marine ecosystems. This act has endeavoured to establish a clear guidelines for the management and protection of the reef while imposing the ardent regulations on overfishing, and promoting the sustainable tourism practices. However, there have been few concerns about its effectiveness in addressing some of the major ecological menaces. Hence, the stakeholders realized the need for ongoing efforts to improve the management and protection of the Great Barrier Reef and other marine ecosystems around the world (Day & Dobbs 2013). On the similar lines, the Marine and Coastal Access Act was legislated by the UK government in the year 2009 with an intention to enhance the management and protection of marine ecosystems in the territory of England. By the virtues of this act, a network of Marine Protected Areas (MPAs) around the country has been established, which are formulated to protect critical habitats and encourage the sustainable fishing practices. Yet, some stakeholders have upheaved the concerns about the appropriate implementation of this Act and its propensity to compellingly protect and conserve the marine environment (Terry et al. 2019). For example, there have been few criticisms about the lag in establishing the network of Marine Protected Areas, and concerns that these protected areas may not be large or comprehensive enough to adequately protect marine biodiversity. Additionally, some stakeholders have also implicated that act does not contain substantial provisions in addressing the pressing issues such as overfishing and marine pollution. The United States, on the other hand, has also established a network of Marine Protected Areas (MPAs) around its territorial boundaries in order to protect critical habitats and stimulate sustainable fishing practices. These MPAs are managed by a collective efforts of federal, state, and local governments, and they are designed to protect a range of marine ecosystems, from coral reefs to kelp forests. These MPAs are instituted in variety of forms such as marine reserves, marine sanctuaries, and marine parks. Some of the notable MPAs are Papahānaumokuākea Marine National Monument, Channel Islands National Marine Sanctuary, Florida Keys National Marine Sanctuary, Olympic Coast National Marine Sanctuary. While, Marine Protected Areas (MPAs) in the United States are intended to protect and conserve marine biodiversity. Subsequently, there are pitfalls and challenges associated with their implementation and management, such as the effectiveness of MPAs being limited due to inappropriate accessibility & connectivity between protected areas. This is considered as a particular challenge for migratory species, which may travel from long distances and cross international borders. Whereas, many MPAs in the United States are underfunded and understaffed which in a way affect their operational efficiency and the major challenges with MPAs is ensuring that the regulations and management plans are effectively enforced. This can be difficult, particularly in areas where there is limited law enforcement presence or where there is conflict between user groups (Megdal et al. 2015).

## RELEVANCE OF ECOLOGICAL CONSTITUTION

As discussed earlier, an Ecological Constitution is a set of guidelines and regulations that are designed to protect and restore ecosystems. Accordingly, the implementation of ecological constitution can play integral role in restoring marine ecosystem. As a matter of fact, the Ecological Constitution can assist in establishing the lucid guidelines for the protection and restoration of critical marine habitats, such as coral reefs, seagrass beds, and mangroves while establishing the regulations and policies to promote sustainable fishing practices, including the imposition of limits on fishing quotas, the use of selective fishing gear, and the protection of spawning grounds. The constitution can also be adopted to establish guidelines for the reduction of pollution in marine environments. These guidelines can include regulations for curtailing the use of plastic, other pollutants, disposal of illicit waste and hazardous chemicals. Besides, these guidelines can mandate the adoption of the necessary steps to address the effects of climate change on marine ecosystems and encourage the use of renewable energy sources. Subsequently, the implementation of Ecological Constitution can also promote collaboration between governments, NGOs, and local communities to implement conservation efforts, as well as educational initiatives to increase awareness about the importance of marine ecosystems and encouraging the public participation in restoration efforts (Da Costa et al. 2020). Most importantly, there is a need to establish protocols for monitoring and evaluating the health of marine ecosystems and the effectiveness of conservation efforts which can be achieved through the implementation of Ecological Constitution. This can further lead to certify that conservation efforts are effective and that interventions can also be adjusted as needed.

## CONCLUSION

The integration of an Ecological Constitution has an enormous and diverse effect on the conversation of marine ecosystem. The highest legal frameworks can guarantee that marine conservation is a constitutional mandate rather than just a choice in policy by enshrining environmental principles and rights. As discussed in the chapter, the constitutionalizing of environmental protection establishes a strong legal basis for sustainable marine management, enabling civil society and governmental entities to jointly promote and carry out the required conservation efforts. In reference to this, establishing and enforcing Marine Protected Areas (MPAs) can be considered as one of the most imperative effects of an ecological constitution. These regions are essential for maintaining healthy ecosystems, and preserving marine biodiversity. By giving MPAs a clear legal authority and mandate and guaranteeing that these areas are sufficiently protected from destructive activities, constitutional provisions can encourage the establishment of MPAs. Furthermore, a culture of environmental stewardship and responsibility is also promoted by the constitutional recognition of environmental rights, which empowers individuals and non-governmental organisations to hold governments responsible for their conservation pledges.

Besides, an ecological constitution may also encourage the use of ecosystem-based management strategies. These methods acknowledge the interdependence of marine ecosystems and the necessity of addressing the cumulative effects of different human activities. In principle, the Constitutional law enshrining these principles enables nations to formulate and carry out more comprehensive and integrated policies that take the sustainability and long-term health of marine environments into account. In order to address complex and interconnected threats like overfishing, pollution, bleaching of coral reefs and climate change, a shift towards comprehensive management is important. While, an ecological constitution can have a bearing on international cooperation in marine conservation. Overfishing, pollution, and climate change are just a few of the transboundary problems pertaining to marine conservation that call for cooperative solutions. A nation's credibility and leadership in international negotiations can be strengthened by constitutional commitments to environmental protection, which can lead to stronger alliances and more fruitful multilateral initiatives. Thereby, countries can support their objectives of international marine conservation and gain from pooled resources and knowledge by coordinating their national legal frameworks with international environmental agreements.

Nonetheless, there are few challenges associated with implementing an ecological constitution into effect. It's still difficult and frequently divisive to strike a balance between environmental preservation and economic growth. It is possible for economic pursuits like tourism, and resource extraction to be seen as at odds with efforts to conserve the marine environment. Carefully crafting policies, involving stakeholders, and creating viable economic alternatives that support conservation objectives are all necessary to resolve these conflicts. Furthermore, the ability and willingness of governments to enforce constitutional provisions determines their effectiveness. The execution of constitutional commitments to marine conservation may be jeopardised by weak governance frameworks, a shortage of funding, and corruption. However, the advantages of an ecological constitution in improving marine conservation outweigh these challenges. Constitutional frameworks can be strengthened by incorporating ecological principles to provide a solid legal basis for the preservation and sustainable management of marine environments. This strategy encourages a more just and equitable distribution of environmental benefits and burdens while also ensuring the preservation of marine biodiversity and ecosystems. The incorporation of constitutional environmental protection aims to offers a promising path towards a more resilient and sustainable future for marine ecosystems, particularly as nations continue to face urgent environmental challenges.

## REFERENCES


Agardy, T. S. (2010). *Ocean zoning: making marine management more effective*. Routledge. 10.4324/9781849776462

Aguila, Y., & Viñuales, J. E. (2019). A global pact for the environment: Conceptual foundations. *Review of European, Comparative & International Environmental Law*, 28(1), 3–12. 10.1111/reel.12277

Alves, F., Costa, P. M., Novelli, L., & Vidal, D. G. (2023). The rights of nature and the human right to nature: An overview of the European legal system and challenges for the ecological transition. *Frontiers in Environmental Science*, 11, 1175143. 10.3389/fenvs.2023.1175143

Blackmore, A. C. (2018). Rediscovering the origins and inclusion of the public trust doctrine in South African environmental law: A speculative analysis. *Review of European, Comparative & International Environmental Law*, 27(2), 187–198. 10.1111/reel.12239

Brundtland, G. H. (1987). Brundtland report. Our common future. *Comissão Mundial*, 4(1), 17–25.

Bullard, R. (2005). Environmental justice in the 21st century. *Debating the earth*, 3222-3356.

Burdon, P. D. (2013). The earth community and ecological jurisprudence. *Oñati Socio-Legal Series*, 3(5).

Collins, L. (2021). *The ecological constitution: reframing environmental law*. Routledge. 10.4324/9780429277320

Cullinan, C. (2011). *Wild law: A manifesto for earth justice*. Bloomsbury Publishing.

Da Costa, J. P., Mouneyrac, C., Costa, M., Duarte, A. C., & Rocha-Santos, T. (2020). The role of legislation, regulatory initiatives and guidelines on the control of plastic pollution. *Frontiers in Environmental Science*, 8, 104. 10.3389/fenvs.2020.00104

Day, J. C., & Dobbs, K. (2013). Effective governance of a large and complex cross-jurisdictional marine protected area: Australia's Great Barrier Reef. *Marine Policy*, 41, 14–24. 10.1016/j.marpol.2012.12.020

De Lucia, V. (2013). Towards an ecological philosophy of law: A comparative discussion. *Journal of Human Rights and the Environment*, 4(2), 167–190. 10.4337/jhre.2013.02.03

Divan, S., & Rosencranz, A. (2001). *Environmental law and policy in India: cases. Materials and Statistics* (2nd ed.). OUP.

El-Mohtady Bellah, A., & Chiha, I. (2017). Sustainable Development in International and Comparative Constitutional Law.

Gray, M. (2012). Valuing geodiversity in an 'ecosystem services' context. *Scottish Geographical Journal*, 128(3-4), 177–194. 10.1080/14702541.2012.725858

Gunningham, N. (2009). Environment law, regulation and governance: Shifting architectures. *Journal of Environmental Law*, 21(2), 179–212. 10.1093/jel/eqp011

Harrison, J. (2017). *Saving the oceans through law: the international legal framework for the protection of the marine environment.* Oxford University Press.

Hassan, D. (2017). *Protecting the marine environment from land-based sources of pollution: towards effective international cooperation.* Routledge. 10.4324/9781315245829

Heffernan, J. P. (2014). Dealing with Mediterranean bluefin tuna: A study in international environmental management. *Marine Policy*, 50, 81–88. 10.1016/j.marpol.2014.05.014

Helbert, M. (2024). The Space Between Motherhood and Mother Earth: An Ecofeminist Analysis of the Post-Development Model in Bolivia. In *Ecofeminism on the Edge: Theory and Practice* (pp. 201-214). Emerald Publishing Limited.

Jones, P. J., Qiu, W., & De Santo, E. M. (2013). Governing marine protected areas: Social–ecological resilience through institutional diversity. *Marine Policy*, 41, 5–13. 10.1016/j.marpol.2012.12.026

Kanojia, S. (2024). Guardians of Atolls: Examining the Paris Agreement's Role in Climate Change Mitigation and Resilience in the Maldives. In Bridging Health, Environment, and Legalities: A Holistic Approach (pp. 271-291). IGI Global.

Kibel, P. S. (2007). Grasp on water: A natural resource that eludes NAFTA's notion of investment. *Ecology Law Quarterly*, 34, 655.

Knowlton, N., Brainard, R. E., Fisher, R., Moews, M., Plaisance, L., & Caley, M. J. (2010). Coral reef biodiversity. *Life in the world's oceans: diversity distribution and abundance*, 65-74.

Kotzé, L. J. (2012). Arguing global environmental constitutionalism. *Transnational Environmental Law*, 1(1), 199–233. 10.1017/S2047102511000094

Little, P. (2013). Making Law Matter: Environmental Protection & Legal Institutions in Brazil by Lesley K. McAllister. *Luso-Brazilian Review*, 50(2), 145–146. 10.1353/lbr.2013.0042

Malve, H. (2016). Exploring the ocean for new drug developments: Marine pharmacology. *Journal of Pharmacy & Bioallied Sciences*, 8(2), 83. 10.4103/0975-7406.17170027134458

Mega, V. P. (2018). *Eco-responsible cities and the global ocean: geostrategic shifts and the sustainability trilemma*. Springer.

Megdal, S. B., Gerlak, A. K., Varady, R. G., & Huang, L. Y. (2015). Groundwater governance in the United States: Common priorities and challenges. *Groundwater*, 53(5), 677–684. 10.1111/gwat.1229425406503

Mooney, H., Larigauderie, A., Cesario, M., Elmquist, T., Hoegh-Guldberg, O., Lavorel, S., Mace, G. M., Palmer, M., Scholes, R., & Yahara, T. (2009). Biodiversity, climate change, and ecosystem services. *Current Opinion in Environmental Sustainability*, 1(1), 46–54. 10.1016/j.cosust.2009.07.006

Musolino, E. (2018). A. Vercelli, Crisis and Sustainability. The Delusion of Free Markets, 2017. *Rassegna Italiana di Sociologia*, 59(2), 408–409.

Nonet, P., Selznick, P., & Kagan, R. A. (2017). *Law and society in transition: Toward responsive law*. Routledge. 10.4324/9780203787540

O'Donnell, E., Poelina, A., Pelizzon, A., & Clark, C. (2020). Stop burying the Lede: The essential role of indigenous law (s) in creating rights of nature. *Transnational Environmental Law*, 9(3), 403–427. 10.1017/S2047102520000242

Pauly, D., & Zeller, D. (2017). Comments on FAOs state of world fisheries and aquaculture (SOFIA 2016). *Marine Policy*, 77, 176–181. 10.1016/j.marpol.2017.01.006

Picou, J. S., & Martin, C. G. (2007). *Long-term community impacts of the Exxon Valdez oil spill: patterns of social disruption and psychological stress seventeen years after the disaster*. Department of Sociology, Anthropology and Social Work, University of South Alabama.

Prakash, S., & Verma, A. K. (2022). Anthropogenic activities and Biodiversity threats. *International Journal of Biological Innovations, IJBI*, 4(1), 94–103. 10.46505/IJBI.2022.4110

Raven, J., Caldeira, K., Elderfield, H., Hoegh-Guldberg, O., Liss, P., Riebesell, U., & Watson, A. (2005). *Ocean acidification due to increasing atmospheric carbon dioxide*. The Royal Society.

Rühs, N., & Jones, A. (2016). The implementation of earth jurisprudence through substantive constitutional rights of nature. *Sustainability (Basel)*, 8(2), 174. 10.3390/su8020174

Sabine, C. L., Heimann, M., Artaxo, P., Bakker, D. C., Chen, C. T. A., Field, C. B., ... Valentini, R. (2004). Current status and past trends of the global carbon cycle. *Scope-scientific committee on problems of the environment international council of scientific unions, 62*, 17-44.

Shrader-Frechette, K. (2002). *Environmental justice: Creating equality, reclaiming democracy*. Oxford University Press. 10.1093/0195152034.001.0001

Sigee, D. C. (2005). *Freshwater microbiology: biodiversity and dynamic interactions of microorganisms in the aquatic environment*. John Wiley & Sons.

Terry, A., Lewis, K., & Bullimore, B. (2019). The impact of the Marine and Coastal Access Act (2009) on Welsh inshore fisheries and marine management. *Marine Policy*, 99, 359–368. 10.1016/j.marpol.2018.11.003

Tymoshenko, Y., Kyslenko, D., Kuzmichova-Kyslenko, E., Leonenko, I., & Servetsky, I. (2022). Features of the Pre-trial Investigation of Air Pollution. Academic Press.

Valladares, C., & Boelens, R. (2017). Extractivism and the rights of nature: Governmentality, 'convenient communities' and epistemic pacts in Ecuador. *Environmental Politics*, 26(6), 1015–1034. 10.1080/09644016.2017.1338384

## KEY TERMS AND DEFINITIONS

**Anthropogenic Impact:** Human-induced pressures on marine ecosystems, such as overfishing, pollution, habitat destruction, and climate change, threatening biodiversity and ecosystem function.

**Biodiversity:** The variety of life forms at genetic, species, and ecosystem levels, including diversity within species, between species, and of ecosystems.

**Ecosystem Services:** Benefits provided to humans by marine ecosystems, such as provisioning (e.g., food, water), regulating (e.g., climate regulation, waste decomposition), cultural (e.g., recreation, tourism), and supporting services (e.g., nutrient cycling).

**Marine Ecosystem:** A complex network of interactions among marine organisms and their physical environment.

# Section 2
# Fisheries

# Chapter 4
# First Insight Into the Deployment of Artificial Reefs to Support Local Fisheries in São Tome and Príncipe


**Joao Pessoa**
*Marapa, Sao Tome and Principe*

**Frederic Airaud**
*Programa Tato, Sao Tome and Principe*

**Graciano Costa**
*Marapa, Sao Tome and Principe*

**Antunes Pina**
*Programa Tato, Sao Tome and Principe*

**Manuel Jorge Carvalho do Rio**
*Marapa, Sao Tome and Principe*

**Venceslau Soares**
*Programa Tato, Sao Tome and Principe*

**Elisio Espirito Santo**
*Marapa, Sao Tome and Principe*

**Sara Vieira**
https://orcid.org/0000-0001-6113-1862
*Programa Tato, Sao Tome and Principe*

**Betania Ferreira-Airaud**
*Programa Tato, Sao Tome and Principe*

**Francisco Leitão**
https://orcid.org/0000-0003-4983-9782
*Center of Marine Science, University of Algarve, Portugal*


## ABSTRACT


*The small-scale fisheries in São Tomé and Príncipe play a critical socioeconomic role, supporting around 30,000 people and contributing to the nation's food security.*



DOI: 10.4018/979-8-3693-2436-3.ch004

*Local fishers, limited by a lack of resources and reliance on traditional techniques, face significant challenges. These have been aggravated by the increasing use of non-selective and ecologically harmful fishing methods, which have emerged as a response to fish scarcity. Here the authors detail the implementation of an artificial reef system in 2020, designed to enhance local fisheries and promote socioeconomic development. The project was led by NGOs MARAPA in partnership with Programa Tatô and involved a wide range of stakeholders. The artificial reefs, made of locally sourced inert materials and designed to foster marine life conducive to fishing, were strategically placed in the northern region on sandy bottoms to minimize environmental impact. This initiative not only aims to replenish fish resources but also serves as a tool for capacity building and co-management, promoting better decision-making processes.*

## 1. INTRODUCTION

The fisheries sector is pivotal to the livelihoods of many families in São Tomé and Príncipe. As of 2011, artisanal fishing involved 15% of the active population, encompassing fishermen and associated occupations, supporting around 30,000 people either directly or indirectly (Zacarias et al. 2022). Most of fishing activities are concentrated along a narrow coastal strip, largely due to resource limitations that prevent expansion into broader coastal areas. A large share of the fishing fleet consists of manually operated wooden boats, including canoes and sailing boats with minimal motorization. These boats are typically used in inshore waters close to fishing villages, making fishing a labour-intensive activity with fishers spending many hours at sea to reach fishing grounds and return with their catches (Maia et al. 2018, Zacarias et al. 2022).

The prevalent use of non-selective and destructive fishing methods, such as explosives and small mesh nets, has been increasing due to fish scarcity. These practices pose significant concerns for local government and environmental agencies, as fisheries are crucial for providing 70% of the animal protein (averaging 23.6 kg per inhabitant annually) consumed by the population (Nuno et al. 2021).

Moreover, the primary occupation for the vast majority of fishermen and fish sellers is fishing, with a minor percentage engaged in secondary activities such as agriculture. Housing conditions, means of transportation, and working hours reveal a community highly dependent on the sea, yet facing increasing challenges in sustaining its way of life due to diminishing fish stocks and outdated fishing practices (INE 2020).

Perceptions of the state and management of fishery resources among fishers indicate a widespread belief in the decline of marine resources. This belief, while subjective, underscores the urgent need for changes in conservation and resource management practices. Notably, destructive fishing practices are identified as a primary cause of resource depletion, with specific criticism aimed at poor selective netting methods that damage the marine ecosystem due to technological limitations (Nuno et al. 2021).

The organization level among fishermen is relatively low, with a significant majority reporting no affiliation with any professional associations, which exacerbates challenges in advocating for and implementing sustainable practices (De Labra et al. 2023).

In response to these challenges, an artificial reef system was proposed in 2020 to enhance local fisheries and promote socioeconomic development. Implemented by local NGOs and developed in collaboration with various stakeholders—including fishing village representatives, environmental and fishery agencies, and local and national authorities—the project aimed to provide a sustainable solution to fisheries enhancement in complement to other ongoing initiatives to protect marine biodiversity and create marine protected areas. The artificial reefs, constructed from inert materials like concrete, were designed to mitigate ecological impact while promoting marine biodiversity conducive to fishing activities. Deployed in the northern region on sandy bottoms, these structures are intended to create habitats for marine invertebrates that serve as a food source for commercially valuable fish species, thereby supporting the local fishing sector.

## 2. PROJECT CONTEXT AND IMPLEMENTATION STRATEGY

The project was initiated in 2020 to address the declining fish populations and the socioeconomic challenges faced by local artisanal fishers. Implemented by the local NGOs MARAPA in partnership withPrograma Tatô, the initiative aimed to enhance local fisheries through the strategic deployment of artificial reefs. The project integrated extensive stakeholder engagement to ensure community support and ecological sensitivity.

MARAPA (Mar, Ambiente e PescaArtesanal), established in 1999 following the closure of the Artisanal Fishing Support Project, focuses on sustainable development of artisanal fisheries and protection of marine environment in São Tomé and Príncipe. MARAPAleads the artificial reef project, drawing on its deep connections with local fishing communities to ensure the project's goals align with sustainable practices and community needs. Programa Tatô, officially formed as an NGO in 2018, focuses on the conservation of sea turtle habitats and sustainable environmental practices. In the project, Programa Tatô is responsible for ecological monitoring, ensuring that the deployment of artificial reefs adheres to the highest environmental standards and contributes positively to enhance marine biodiversity. Together, these organizations use their expertise in community engagement, resource management, and ecological monitoring to support the successful implementation of the artificial reef project.

## 3. METHODOLOGY AND PRELIMINARY RESULTS

### 3.1. Artificial Reef Committee Formation

The formation of the Artificial Reef Committee was a pivotal first step in ensuring the success of the artificial reef project, marked by meticulous planning and strategic stakeholder integration. This multi-disciplinary committee included representatives from local fishing communities, environmental agencies, fisheries departments, local authorities, the coastal guard, captaincy, and non-governmental organizations such as MARAPA and Programa Tatô. This diverse composition was intentionally chosen to bring together a broad range of expertise and local knowledge, ensuring a holistic approach to project management.

Representatives from the fishing communities provided essential on-the-ground perspectives to align the project with local needs and customs. Environmental agencies and fisheries departments contributed with their scientific and regulatory expertise, ensuring the achievement of environmental standards and conservation goals. The inclusion of local authorities, the coastal guard, and the captaincy ensured with the compliance of maritime laws and coordinated logistic issues effectively.

The committee's responsibilities played a critical role in site selection, utilizing ecological assessments and community feedback to choose locations that aimed to balance environmental sustainability with socio-economic benefits. This ensured that the selected sites would enhance fish stocks without disrupting sensitive ecosystems or local maritime activities.

To ensure the project's sustainability and the proper use by the community, the committee engaged the stakeholders in every phase of the project implementation. Regular training sessions were held, where participants were trained about the ecological functions of reefs and their roles in supporting marine biodiversity and local fisheries. These sessions also informed about the materials used and the objectives that these structures can fulfil.

As the project progressed, the committee's role evolved to the management of the fishing activity on the artificial reef ensuring a continuous involvement of local fishers in order to adapt to new challenges. The committee´s members established guidelines and protocols for regular monitoring and community engagement, which included making decisions on key project stages such as reef design, ecological monitoring, socio-economic surveys, and document validation. This management framework was designed to be adaptive, incorporating ongoing scientific insights and community feedback to optimize the reefs' effectiveness and sustainability.

### 3.2. Stakeholder Engagement and Awareness

A formal public seminar was organized to officially launch the project, marking a significant step towards community integration and participation. This event served as a crucial platform for introducing the project to the community, ensuring it was well-received and understood. The seminar enabled attendees to provide feedback, ask questions, and express concerns, fostering a sense of ownership and involvement among local stakeholders. Moreover, the event emphasized the importance of continued education and open communication, establishing ongoing dialogue between the project team and all stakeholders. This approach was designed to maintain and encourage active participation and support for the project, ensuring its alignment with community needs and expectations as it progressed.

Additionally, more than 800 fishers and traders were engaged through over 12 information sessions held across six different fishing communities along the north coast. The campaign also leveraged national television, radio, and social media to broaden the outreach and ensure widespread understanding and support. Additionally, educational materials such as posters, and brochures, were produced and distributed, providing a clear understanding of the project's objectives and the benefits it aimed to bring.

### 3.3. Site Selection and Environmental and Social Assessments

Site selection was a critical factor in the success of the project, demanding ecological assessments and detailed mapping to minimize impacts on sensitive habitats. A comprehensive study of the coastal ecosystems was conducted up to a depth of 25 meters over a total area of 25.1 km$^2$, based on Beeden et al. (2014). Habitat distribution was mapped in the field using rapid habitat surveys of major benthic types conducted around the island. The marine zone <25m deep was divided into 31 transects approximately 1 km apart running perpendicular to the coast. Transects were pre-loaded onto a GPS device before heading to the field and surveyed with the help of a small boat. Depth measures were taken using a Garmin Striker Plus echo-sounder while driving slowly along the transect. A rapid habitat survey was conducted in points at 5m, 10m, 15m, 20m and 25m deep, giving a total of 155 surveys. At each point a circular area 10 m diameter area was assessed and the benthic cover of different substrate and habitat types, such as sand, rock, rhodolite, corals, seagrass, mixed sand and rhodolite, mixed sand and seagrass, was estimated visually.

The most abundant habitat was sandy bottoms, covering 33.3% of the total area analysed, followed by rhodolite dominated areas covering 18.7%, and sand mixed with seagrass forming 17.9% of the area. Seagrass habitats and areas dominated by sand and rhodolith also presented significant coverage. The least abundant were rocky habitats (2%), coral-dominated habitats (0.2%), and mangroves (0.7%).

Substrate and habitat helped in selecting the most suitable locations for reef placement. The focus was on areas that would support the project goals while ensuring environmental integrity. Impact assessments were rigorously carried out to ensure that all chosen sites were located away from vulnerable areas, such as seagrass beds and coral habitats. Preference was given to sandy bottoms, which are less likely to be ecologically disturbed by the installation of artificial reefs, thereby minimizing the potential negative impacts on the marine environment. The site selection process was fundamental to aligning the project with sustainability goals and ecological conservation standards.

*Figure 1. Map with artificial reef deployment area*

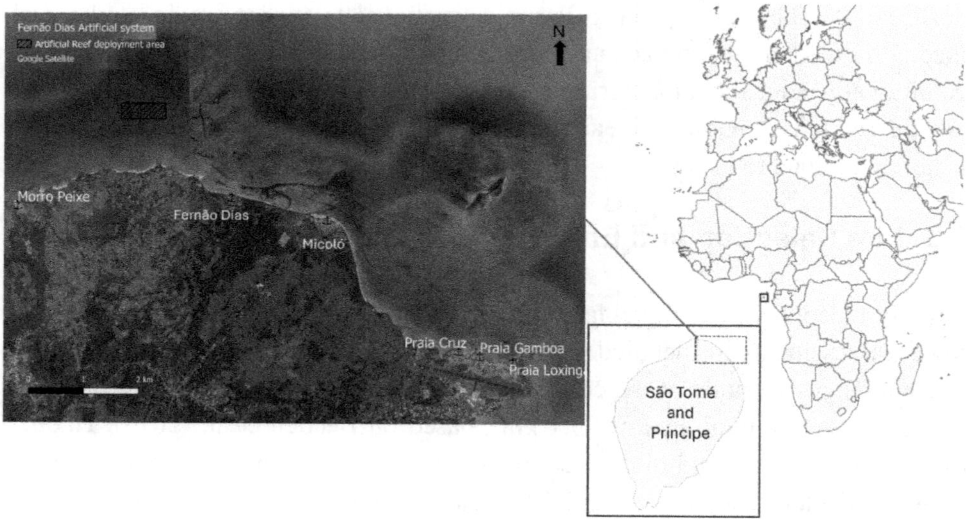


Further ecological considerations included the selection of potential locations based on environmental conditions such as current patterns, water depth, and bottom substrate. Throughout this process, a series of environmental considerations were evaluated to optimize the artificial reefs' ecological characteristics and anticipate their effects. This included analysing physical variables such as currents and wind impact on surface currents, essential for ensuring the stability of reef structures. The selected areas were also subjected to analysison sediment movement and water quality to avoid areas prone to sedimentation that could bury the reefs or affect their effectiveness.The project also accounted for marine activities, avoiding areas heavily trafficked by boats or used for commercial purposes, thereby reducing potential conflicts.

Supporting the site selection, both ecological and socioeconomic assessments were thoroughly integrated. Ecological monitoring involved setting up two sampling stations to confirm the suitability of the pre-selected sites. Local fishers from the communities of Morro Peixe and Micoló assisted in these efforts, with the first station positioned 1km north of Ponta Cruzeiro, designated for artificial reefinstallation, and the second serving as a control zone. Throughout the seven months of ecological monitoring, from June to December 2021, a total of 31 species were captured across 26 fishing outings using longline and hand line (time and number of hooks were standardized). Among these, 16 were demersal fish species, 5 pelagic fish species, 3 species of molluscs, 1 species of echinoderm, and 1 crustacean, along with 5 unidentifiable fish species known only by local names. A total of 922 individuals were captured, weighing a collective 128.97 kg. This data revealed that the control station (Station 2) generally had a higher species diversity and catch volume compared to the station near the artificial reef (Station 1), indicating variations in local marine biodiversity. This data collection was crucial in ensuring that the reefs not only provide environmental benefits but also were well received within the local communities, integrating clearly into their economic and social structures.This foundational phase of ecological monitoring, conducted before the installation of the artificial reefin March 2022, provided essential baseline data that will be compared with the subsequent monitoring campaigns to evaluate the artificial reef impact on local fish populations, diversity, and overall ecosystem health. This ongoing research is crucial for confirming the effectiveness of artificial reefs in enhancing marine biodiversity and supporting local fisheries (Claudet &Pelletier2004; Leitão et al. 2008; Leitão et al. 2009).

Simultaneously, socioeconomic monitoring was conducted through 116 interviews with fishers from Morro Peixe, Fernão Dias, Micoló, Praia Cruz, and Praia Gamboa. These interviews, held in November and December 2021, aimed to gather both quantitative and qualitative data to understand the socioeconomic context of the fishers, their fishing activities, and perceptions of the artificial reefs.

All participants were identified as commercial fishers, with the majority serving as direct fishers, and others in roles such as fishing masters or deckhands. This diversity underscores the complex social structure within these fishing communities.

In addition to fishing, more than half of the interviewees (54%) had a secondary activity, such as agriculture and construction. This indicates a multi-faceted economic reliance that extends beyond the fishing sector, highlighting the need for conservation projects to consider broader socioeconomic impacts.

Regarding fishing methods, the data revealed a near-even split between those using hand lines (45%) and those employing purse seine nets (44%), pointing to a wide range of fishing techniques used by the local communities. The majority of fishers operated small non-motorized canoes or motorized boats, suggesting a range of fishing scales from subsistence to more extensive commercial operations. The fishing trips typically lasted about 7.15 hours, with fishermen going out an average of 6.5 times per week, mostly within 10 miles of the coast. This regular and frequent use of marine areas highlights the critical role that habitat structures, such as artificial reefs, could play in sustaining fish populations and by extension, local livelihoods. Targeted species included locally significant fish like "Fulufulu" (*Auxis thazard* or *Euthynnus alletteratus*) and "Cavala" (*Decapterus punctatus* or *Decapterus macarellus*), both small pelagics, which are crucial for the local diet and economy.

*Figure 2. List of commercial local species expected to be found in São Tomé according to their affinity to artificial reef structures. Type A: benthic inhabitants' species that prefer physical contact with the reef. Type B: swim around the reef, and/or remain close to the bottom (demersal fish), without direct physical contact with the reef, but can explore the interior spaces of structures. Type C: swim in the intermediate and superficial layers of the water column, remaining some distance from the reef. Includes pelagic fish that are probably linked to them through sight and sound; Type D: tend to remain stationary on the seabed around reefs. This includes flat fish, such as sole and turbot, spider fish or other bottom fish (see: Okubo & Kakimoto, 1991)*

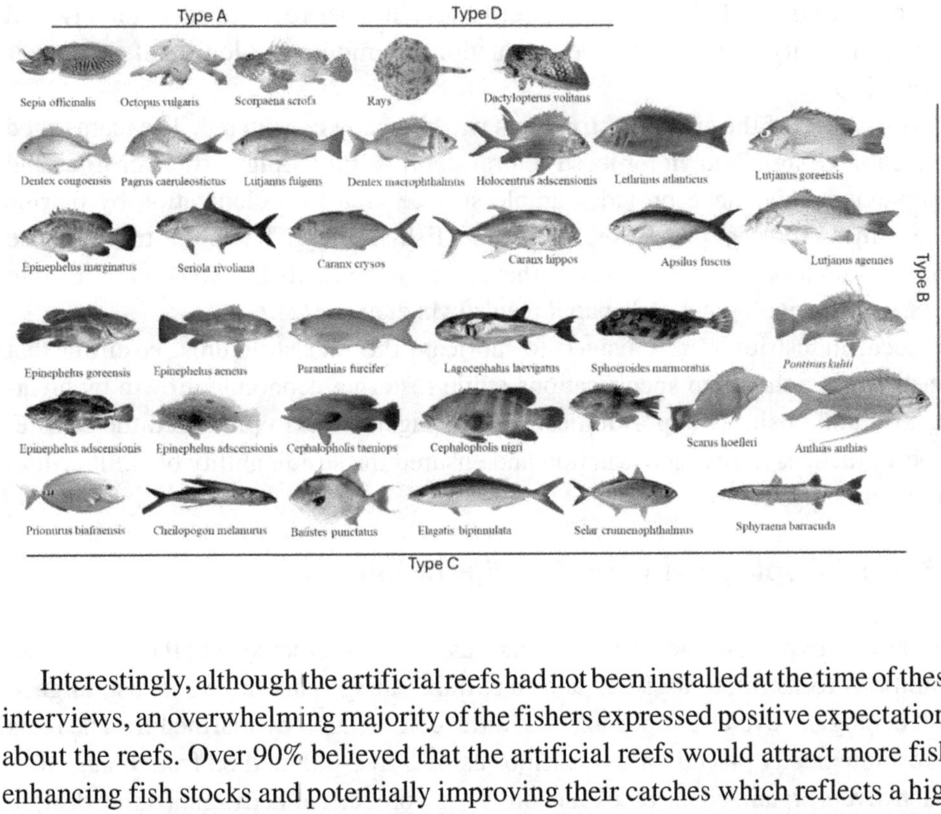


Interestingly, although the artificial reefs had not been installed at the time of these interviews, an overwhelming majority of the fishers expressed positive expectations about the reefs. Over 90% believed that the artificial reefs would attract more fish, enhancing fish stocks and potentially improving their catches which reflects a high level of community support and optimism about the integration of artificial reefs into their fishing practices.

These findings emphasize the importance of maintaining regular communication with these communities to monitor changes in fishing practices, perceptions, and the impacts of the artificial reefs once installed. This ongoing dialogue will be crucial for evaluating the effectiveness of the artificial reefs and ensuring that they continue to meet the needs and expectations of the local fishers, thereby fostering sustainable management of marine resources (Lima & Zalmon, 2022).

## 3.4. Project Logistics, Infrastructure Preparation, and Reef Deployment

The planning and execution of the artificial reef project involved particular logistics and infrastructure preparation to ensure both environmental integrity and community benefit. The project team selected materials that were not only ecologically sound but also supported marine life and thus aligned with sustainability principles. Inert materials like concrete were chosen for the construction of the reefs due to their non-polluting nature, effectiveness in other reef projects globally, and local availability which facilitated production and minimized logistical challenges (Baine 2001).

The design of the artificial structures was critically considered. The team opted for a cubic design, known for its structural stability, providing vertical profiles and complexity. This shape provides ample surface area for colonization by marine organisms and offers protection for various fish species. The cubic structure also supports modular use, which means that reefs can be easily scaled and arranged to meet specific ecological and spatial needs (Ma et al. 2022).

Local industries were engaged to fabricate the reef structures, ensuring that construction adhered to specifications while fostering economic growth by creating jobs and business opportunities. This engagement also promoted the transfer of best practices in reef construction and ensured the sustainability of skills within the community.

## 3.5. Reef Deployment and Configuration

The deployment of the artificial reefs was timed to coincide with the calmest sea conditions to facilitate logistics and maximize safety. This timing is also aligned with ecological cycles to optimize the initial colonization by marine life. The reefs were strategically placed on the seafloor using a free-fall method from boats, after which divers organized the blocks according to a predetermined organization design. This placement was essential for creating the structural complexity needed to attract diverse marine life and ensure the stability of the reefs in strong maritime currents.

*Figure 3. Concrete artificial reef cubit units deployed in São Tomé*

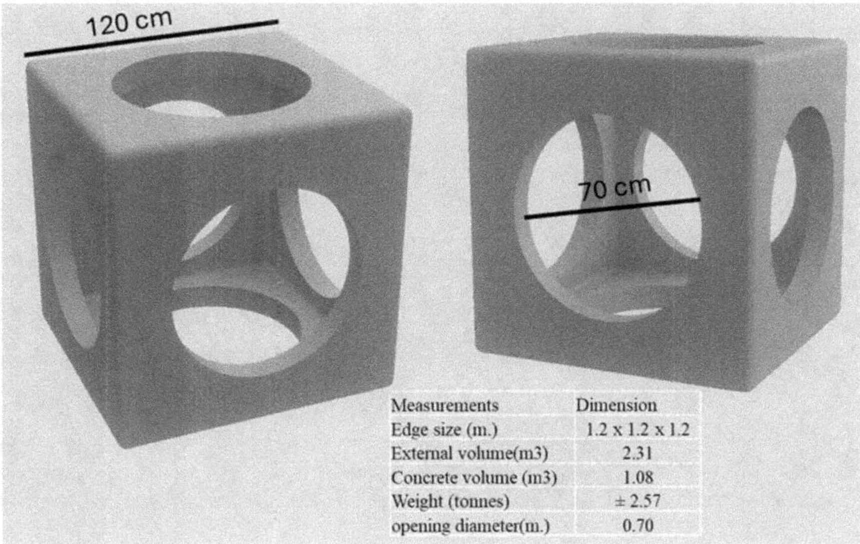


| Measurements | Dimension |
|---|---|
| Edge size (m.) | 1.2 x 1.2 x 1.2 |
| External volume(m3) | 2.31 |
| Concrete volume (m3) | 1.08 |
| Weight (tonnes) | ± 2.57 |
| opening diameter(m.) | 0.70 |

The vertical profile of these reef structures is critical in determining the overall fish species composition and biomass, with low-profile reefs providing ideal habitats for demersal species and high-profile structures more effective for pelagic species. Stability was a major consideration, involving careful calculation of weight and surface area to prevent displacement by marine currents, waves, or fishing activities, ensuring that the reefs could withstand significant marine forces and remain anchored to the seabed.

The artificial reef project strategically deployed a total of 200 concrete blocks, each measuring 1.2 meters in length, width, and height, with round wholes in every side of 70 cm diameter. These blocks were divided into 11 reef sets, each comprising 18 blocks, positioned to maximize ecological connectivity and benefit local fishing activities. The reef sets were spaced approximately 70 meters apart to ensure efficient habitat connectivity and to reduce fishing conflicts, optimizing the balance between ecological benefits and accessibility for local fishers. This arrangement was aimed to maximize ecological functionality and optimize fish production, with the units organized in groups spaced sufficiently to mitigate conflicts among fishing activities and distribute fishing effort. Each reef set was carefully placed to maintain connectivity between them, avoiding the pitfalls of isolated reef effects and ensuring a unified, effective reef system.

*Figure 4. Underwater perspective of the organization of the reef sets of 18 units each*

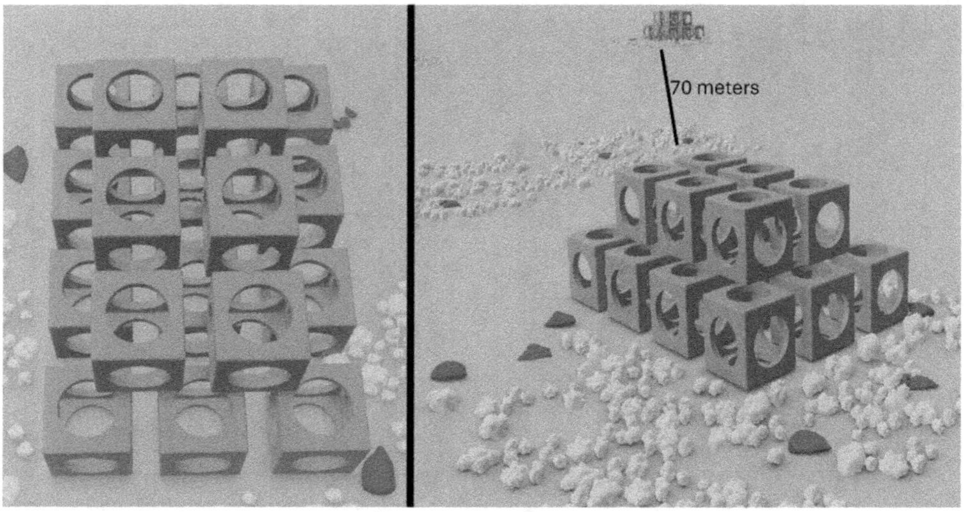


Furthermore, the materials and construction approach were designed for durability, required to maintain structural integrity in harsh marine conditions for at least 20 to 30 years, following international standards. The entire reef system covered approximately 0.03 hectares, a significant area that will foster marine colonization and contributes to the local ecosystem's diversity and productivity.

## 4. COMPLIANCE WITH INTERNATIONAL AND ENVIRONMENTAL GUIDELINES

The project adhered to both national and international environmental laws to ensure that the deployment of the artificial reefs provided ecological benefits without causing detrimental impacts. This commitment to regulatory compliance involved following established guidelines and regulations specifically related to the use of artificial reefs. These regulations covered aspects such as the appropriate materials for construction and broader environmental protection measures to prevent any negative impacts on marine ecosystems.

Moreover, the project placed a strong emphasis on sustainable practices throughout its execution. The team prioritized the use of environmentally friendly materials and techniques that would not only support the health of the marine environment but also align with international standards for ecological preservation. A key aspect of this commitment was to follow international guidelines for the deployment of artificial reefs in marine waters (OSPAR 2013). These standards are designed to ensure that projects effectively contribute to marine conservation while minimizing their environmental footprint. Through strict compliance with these regulatory frameworks, the project aimed to set a benchmark for sustainability and responsible environmental stewardship in marine development initiatives.

*Figure 5. Photos illustrating the different stages of the artificial reef implementation process in São Tome and Principe*

## 5. FINAL REMARKS

The strategic deployment of artificial reefs in São Tomé Island addressed both the ecological needs and socioeconomic challenges facing the artisanal fishing communities in São Tomé and Príncipe. The project's success centres on its comprehensive methodology, which integrates extensive stakeholder engagement and meticulous site selection to ensure the sustainability and ecological efficacy of the artificial reefs. Considering extensive stakeholder engagement and meticulous site selection, the project demonstrated significant community integration and skilled professionals. The information sessions in the local communities and media coverage, with radio and television programs, also contributed to widespread awareness and support the project goals.

The data collected from ecological monitoring revealed potential impacts on local marine biodiversity. The difference in species diversity and fish catches between the two monitoring stations highlights the artificial reefs' potential to diversify local marine environments. Although the station near the future artificial reef's potential location showed lower diversity and fish biomass compared to the control station, these findings are not uncommon in the early stages of reef establishment. Over time, artificial reefs typically see an increase in biological diversity and biomass as the structures become more integrated into the marine ecosystem (Baine, 2001; Leitão, 2013).

The socioeconomic assessments provided insightful data on the fishing communities' practices and their perceptions of the reefs. Most fishers expressed a positive outlook on the potential of artificial reefs to enhance fish stocks, which is promising for future community support and the successful integration of these structures into local fishing practices. However, continuous engagement and monitoring are crucial to ensure that these benefits are realized and that the reefs do not lead to unintended negative impacts such as overfishing (Lima & Zalmon, 2022).

*Figure 6. Schematic representation for the activities expected in São Tomé Island artificial reef system. Activities with spearfishing and static fishing gears (gill net, longline and hand line fishing) can operate over the reef while seine fishing should be avoided over the reef.*

The project presents significant advantages, including redistributing fishing efforts to relieve pressure on nearshore habitats, fishers' wellbeing and income. The deployment of artificial reefs could catalyse new recreational and commercial activities, such as diving and spearfishing, while fostering community participation in fisheries management decisions (Dupont 2008).

However, the project also faced challenges, such as reef transportation and the absence of professional diving companies for reef placement. Regulatory gaps concerning species biology and fishing seasons, along with weak fishing activity surveillance capabilities, underscored the need for improved governance and logistical support to optimize the project's impact.

In conclusion, the ongoing success of the artificial reef project in São Tomé and Príncipe depends on continuous stakeholder engagement, adaptive management strategies, and rigorous monitoring to mitigate threats and capitalize opportunities. By addressing regulatory gaps, enhancing logistical support, and fostering community involvement, the project can serve as a transformative model for sustainable fisheries management and marine conservation initiatives globally. Ongoing collaboration between local communities, NGOs, regulatory bodies, and scientific institutions will be essential for realizing the full potential of artificial reefs in enhancing marine biodiversity and supporting resilient coastal livelihoods.

## ACKNOWLEDGMENT

We would like to thank João Paulo Gouveia by the 3D artificial reefs designs.

## REFERENCES

Baine, M. (2001). Artificial reefs: A review of their design, application, management and performance. *Ocean and Coastal Management*, 44(3-4), 241–259. 10.1016/S0964-5691(01)00048-5

Beeden, R. J., Turner, M. A., Dryden, J., Merida, F., Goudkamp, K., Malone, C., Marshall, P. A., Birtles, A., & Maynard, J. A. (2014). Rapid survey protocol that provides dynamic information on reef condition to managers of the Great Barrier Reef. *Environmental Monitoring and Assessment*, 186(12), 8527–8540. 10.1007/s10661-014-4022-025179944

Claudet, J., & Pelletier, D. (2004). Marine protected areas and artificial reefs: A review of the interactions between management and scientific studies. *Aquatic Living Resources*, 17(2), 129–138. 10.1051/alr:2004017

De Labra, G., Vilela López, B., Prieto Porriños, G., Blanc, P. P., Vasta, A., & Anibal, O. (2023). *A cadeia de valor dos pelágicos costeiros em São Tomé e Príncipe: Relatório de síntese*. FAO.

Dupont, J. M. (2008). Artificial Reefs as restoration tools: A case study on West Florida Shelf. *Coastal Management*, 36(5), 495–507. 10.1080/08920750802395558

INE. (2020). *Perfil da Pobreza com base no Inquérito de Orçamento Familiar de 2017*. Cidade de São Tomé, República Democrática de São Tomé e Príncipe.

Leitão, F. (2013). Artificial reefs: From ecological processes to fishing enhancement tools. *Brazilian Journal of Oceanography*, 61(1), 77–81. 10.1590/S1679-87592013000100009

Leitão, F., Santos, M. N., Erzini, K., & Monteiro, C. C. (2008). Fish assemblages and rapid colonisation after enlargement of an artificial reef off the Algarve coast (Southern Portugal). *Marine Ecology (Berlin)*, 29(4), 435–448. 10.1111/j.1439-0485.2008.00253.x

Leitão, F., Santos, M. N., & Monteiro, C. C. (2009). Diplodus spp. assemblages on artificial reefs: Importance for near shore fisheries. *Fisheries Management and Ecology*, 16(2), 88–99. 10.1111/j.1365-2400.2008.00646.x

Lima, J. S., & Zalmon, I. R. (2022) Impacts of Artisanal Reefs on Artisanal Fishers: Southeastern Brazil. 10.4018/978-1-6684-2344-8.ch005

Ma, Q., Ding, J., Xi, Y., Song, J., Liang, S., & Zhang, R. (2022). An evaluation method for determining the optimal structure of artificial reefs based on their flow field effects. *Frontiers in Marine Science*, 9, 962821. Advance online publication. 10.3389/fmars.2022.962821

Maia, H. A., Morais, R. A., Siqueira, A. C., Hanazaki, N., Floeter, S. R., & Bender, M. G. (2018). Shifting baselines among traditional fishers in São Tomé and Príncipe islands, Gulf of Guinea. *Ocean and Coastal Management*, 154, 133–142. 10.1016/j.ocecoaman.2018.01.006

Nuno, A., Matos, L., Metcalfe, K., Godley, B. J., & Broderick, A. C. (2021). Perceived influence over marine conservation: Determinants and implications of empowerment. *Conservation Letters*, 14(3), e12790. 10.1111/conl.12790

Okubo, H., & Kakimoto, H. (1991). Changes in the community composition around artificial reefs. in Recent advances in aquatic habitat technology. Proc. Japan-U.S. Symposium on artificial habitat for fisheries, 161-168.

OSPAR Commission. (2013). OSPAR Guidelines on artificial reefs in relation to living marine resources. London: OSPAR 99/15/1-E, Annex 6.

Zacarias, W. B. M., Dai, X., Kindong, R., Sarr, O., & Moussa, A. H. (2022). Analysis of Fishery Resource Management Practices in São Tomé and Príncipe: Perception of the Dynamics of Catches from 1950–2020. Recommendations and Strategies for Future Research. *Sustainability (Basel)*, 14(20), 13367. 10.3390/su142013367

# Chapter 5
# Same Ocean, Different Currents: A Review of Socioeconomic Variables in Small Pelagic Fishery Assessment

**Cristina Pallero**
https://orcid.org/0000-0002-5443-1506
*Institute of Marine Science of Andalusia, Spain*

**Silvia Fernández del Reguero**
https://orcid.org/0000-0002-4694-1346
*Institute of Marine Science of Andalusia, Spain*

**José L. Oviedo**
https://orcid.org/0000-0003-2043-5020
*Institute of Marine Science of Andalusia, Spain*

## ABSTRACT

*Fisheries management has evolved and addressed diverse disciplines including the application of mathematical models, economic analysis, and sector behavior considerations. Models use interrelated variables that require a balanced assessment, while recent literature stresses the importance of integrating biological/ecological with social/economic aspects. This study conducts a review of 35 studies focusing on the state-of-the-art of two key currents: management strategy evaluation and bioeconomic analysis. The analysis was done for small pelagic fisheries due to their global significance. The results show the importance of socio-economic variables and emphasizes the need to address the behavioral and economic facets specific to fisheries which have been insufficiently explored. The emphasis on prioritizing*

DOI: 10.4018/979-8-3693-2436-3.ch005

*analytical reality over theoretical models highlights the importance of basing policy decisions on empirical evidence and practical considerations. The study identifies areas in need of further research and includes a critical discussion of the main approaches to a common resource.*

## 1. INTRODUCTION

Sustainable environmental management must concurrently address diverse ecological and societal goals. It navigates uncertainties from natural variability and incomplete system knowledge that may lead to disparate model forecasts, alongside evolving ecosystems (Uusitalo et al., 2022). A realistic model of decision- in fisheries would provide a valuable tool for predicting the decisions responding to different factors, such as management measures (e.g., catch or gear restrictions and marine protected areas), fluctuations in fish prices and changes in fish abundance (Wise et al., 2012).

The global adoption of harvest strategies that incorporate biological reference points and robust decision rules to determine the Total Allowable Catch (TAC) is becoming increasingly prevalent in fisheries management. These strategies comprise three key components: monitoring and data collection (stock, size and distribution), assessment routines (growth rates, reproductive success and mortality rates) and decision rules (Harvest Control Rule-HCR-) that involves setting clear guidelines to adjust the TAC based on changes in the stock. These guidelines are designed to inform decisions regarding the sustainable harvesting of fish while minimizing negative impacts on the stock (Smith et al., 2008, 2014; Punt et al., 2016).

A further significant issue is how to incorporate the uncertainties into fisheries assessments to inform management decisions regarding potential impacts (Anderson & Seijo, 2010; d'Eon-Eggertson et al., 2015; Punt et al., 2016). These uncertainties are defined as incomplete knowledge of the state or processes of nature (Caddy and Mahon, 1995). The list of sources of uncertainty in the management of a fishery includes biases both in the calculation of reference points and in the assessment of the state of the stock with respect to these reference points, as well as those associated to observations, processes, models, estimations, applications and institutions (Rosemberg and Restrepo, 1994; Francis and Shotton, 1997). In this context, models are essential for projecting how regulatory decisions within the fishing industry affect measures in three dimensions: ecological, economic and social (Dudley, 2008; Sigurdardottir and Shruben, 2014).

Regarding the economic and social dimensions, managers and scientists recognize the importance of understanding fishers' behavior amidst evolving management systems and the need todevelop decision-making models. The increasing social–

ecological approach for fisheries management would force us to look more closely into the interactions between the natural and human components of the system (Makino 2018; Jang et al., 2019). Upon an initial review, the literature on fisheries management has contributed to the discussion regarding the introduction of economics into this field by defining annual quotas or studying the impacts of alternative policy scenarios/objectives (Da Rocha et al., 2010; Da Rocha and Gutiérrez, 2011; Kvamsdal et al., 2016; Emery et al., 2017; Voss et al., 2017). However, the current definition of diverse strategies, particularly HCRs, lacks consideration of economic factors such as prices and costs (Rosa et al., 2022). Moreover, conflicting perspectives between scientists and fishers' groups impede agreement on corrective actions (Mulazzani and Malorgio, 2013). Despite the efforts of management authorities, several fish stocks continue to be harvested at unsustainable levels. In this context, a new fishery economics literature using more complex population dynamics has contributed to approximate academic research and policymaking (Nielsen et al., 2018).

When addressing simulations for the evaluation of stocks and considering socio-economic factors of fishing activity, there are two prominent approaches: Management Strategy Evaluation (MSE) and bioeconomic models. In fact, the relevance of these approaches is included as one of the main tasks in the ICES Science Plan (2019) "develop and conduct management strategy evaluations, address uncertainty, and improve the transparency, robustness, efficiency, and repeatability of stock assessment", as well as in several technical approaches of the working groups that develop recommendations.

The MSE approach is a way to test the effectiveness of including environmental variables to improve management under uncertainty (Licandeo et al., 2023). This approach uses simulation to compare the performances of candidate management strategies in relation to a set of management objectives and priorities defined by stakeholders (Punt et al., 2016). There are a multitude of theoretical developments of the models, different applications and proposals for operational tools. Its use is common in large pelagic and demersal fisheries, and it is also being applied to other fisheries. However, its application is still in its early stages in small pelagic or data-poor fisheries, where it faces significant challenges.

By other hand, integrating socio-economics into catch advice via bioeconomic modelling is increasingly advocated for effective fishery management (Beddington et al., 2007; Dichmont et al., 2010; Voss et al., 2017). Bioeconomic models consider, by definition, both the biological and economic dimensions of fisheries to simulate different scenarios and to obtain an evaluation of risks associated with different levels of resource exploitation (Cochet and Gilly, 1990; Kjaersgaard 2007; Silvestri and Maynou, 2009) and the effect of management strategies (Prellezo et al., 2012), providing a common ground for discussion among the different actors in fisheries systems. While this literature contributes to the debate on introducing economics into fisheries management, defining HCRs through bioeconomic modeling optimization remains uncertain (Rosa et al., 2022). As a result, in real-world practice, economics is typically not embedded in HCRs, but is only an ad-hoc output of management performance evaluations (Benson and Stephenson, 2018). Integrating socio-economics into catch advice using bioeconomic modelling is increasingly advocated as key in developing more effective fishery management (Beddington et al., 2007; Dichmont et al., 2010; Voss et al., 2017; Rosa et al., 2022).

In addition to the challenges inherent to fisheries management, short-life cycle species such as small pelagic (often data-poor resource) show high complexity. Small pelagic show huge fluctuations in annual recruitment due to the sensitivity to environmental conditions in large cyclic fluctuations in abundance and, occasionally, stock collapse (Freon et al. 2005; Ruiz et al. 2006, 2014; Barange et al., 2009; de Moor et al., 2011; Sánchez et al. 2014; Aoki et al. 2018; ICES 2023). By definition, small pelagic fisheries (SPFs) include small to intermediate-sized pelagic species such as sardines, anchovies and herrings that are a primary food source for several marine predators (Cury et al., 2000; Overholtz et al., 2000; Furness, 2003; Alder et al., 2008; Butler et al., 2010; Szoboszlai et al., 2015). These SPFs represent the world's largest species group, comprising approximately 30 per cent of the global marine capture production and contributing to 20 per cent of the global catch values of all marine fisheries (Pikitch et al. 2014; FAO 2016). The SPFs fulfil an important ecological role in the marine food web (processed into fish meal and fish oil), and are indispensable for food security in many developing countries (Alder and Pauly 2008; Tacon and Metian 2009; Yatsu 2019; Siple et al., 2021). Moreover, these species exhibit high fluctuations in abundance with economic and social implications (Hamilton et al. 2006; Tacon and Metian 2009; Mullon et al. 2009; Kim 2010). There is currently limited information available regarding how external factors such as climate change affect the stocks of small pelagic fisheries. Additionally, harvesting costs have increased due to the growth in fuel oil prices, and demand has growth on specific markets (Mullon et al. 2009). The fishing industries targeting short-lived species are prone to overcapitalization because industry investment increases rapidly in response to periods of high stock abundance, while disinvestment does not occur to the same degree during periods of low biomass (Fréon et al., 2008). This makes the industry vulnerable to external economic factors such as changes in fish prices and fuel costs (Hoshino et al., 2012).

The objective of this study is to make a critical review of the literature on how socio-economic variables are considered for the management of small pelagic fisheries with the two main approaches selected as representative of current trends: MSE and bioeconomic analysis. This review focuses on published scientific papers as it does not intend to exhaustively analyze everything published on the subject in all platforms and formats (scientific reports, deliverables, advice among others). Our aim is to present an objective analysis of the main focuses and their implications in a comprehensive manner.

## 2. METHODOLOGY

Our research focuses on published scientific papers examining fisheries management with socio-economic variables and indicators. The methodological procedure was inspired by the proposal of Liberati et al. (2009) for systematic reviews and meta-analyses of studies (PRISMA method). However, due to the specificity of the scopus, the universe of information is more biased than other systematic reviews. Accordingly, we identified keywords as those utilizing the Management Strategy Evaluation (MSE) and a bioeconomic approach. First, we defined two keyword searches: "management strategy evaluation" + "fishery/ies" + "small pelagic" OR "bioeconomic" + "fishery/ies " + "small pelagic". Second, we selected the Web of Science and ScienceDirect databases as the main platforms for the search and considering publications until November 2023. The selection of papers followed a structured procedure shown in Table 1.

*Table 1. Methodological Procedure for Selection of Works*

<table>
<tr><th></th><th>Step</th><th>Keywords</th><th>Total results</th><th>Selection by criteria</th></tr>
<tr><td rowspan="2">WEB OF SCIENCE</td><td>1st SCREENING</td><td>"management strategy evaluation" + "fisheries" + "small pelagic"</td><td>18</td><td>6</td></tr>
<tr><td>2nd SCREENING</td><td>"bioeconomic" + "fisheries" + "small pelagic"</td><td>23</td><td>11</td></tr>
<tr><td rowspan="2">SCIENCE DIRECT</td><td>3th SCREENING</td><td>"management strategy evaluation" + "fisheries" + "small pelagic"</td><td>53</td><td>6</td></tr>
<tr><td>4th SCREENING</td><td>"bioeconomic" + "fisheries" + "small pelagic"</td><td>68</td><td>10</td></tr>
<tr><td colspan="4">MANUAL SEARCH</td><td>2</td></tr>
<tr><td colspan="4">TOTAL NUMBER OF PAPERS SELECTED FOR REVIEW</td><td>35</td></tr>
</table>

From the total results obtained in both screenings, we applied exclusion criteria: i) theoretical or non-applied discussion papers; ii) application to other non-SP species; iii) approaches that, despite including keywords, showed different methodology or approach from the MSE and/or bioeconomic model; iv) academic dissertations. We also excluded exploratory or descriptive works of the resource from a bio/ecological (e.g. linked to population dynamics or genetics) or legislative (e.g., fishing agreements) perspective, as well as publications dealing exclusively with behavioral science unrelated to SPF dynamics. We also identified and included manually 2 papers considered relevant and detected in the reading process of the papers identified with the keyword search (see table 1).

Next, we present the main features of the trends observed in the works found, such as the species regions (stock locations) studied. We considered the families Engraulidae, Scombridae, Clupeidae and species of other families (which include the species *Boops boops*, *Trachurus trachurus or Cololabis saira*) with similar characteristics (size, gregarious, purse seine fishery).

The first general analysis includes descriptive data about: i) target species/s; ii) location; iii) conceptual approach and/or uncertainty addressed; and iv) socio-economic approach (YES/NO). The second more specific analysis identifies socio-economic indicators and variables. For comparative diagnosis and coherent classification of selected studies, we define four socio-economic items for five categories: i) Profit focus; ii) Net Present Value (NPV); iii) Income or Total Rent; iv) Behavioral; v) Other approaches. That is, those aspects that address the behavior and economic performance of the activity from any approach, whether qualitative, quantitative and/or projection of trends and scenarios. Subsequently, we performed a qualitative analysis and the results obtained were interpreted in order to define the most decisive approaches, their aspects and the possible gaps.

## 3. RESULTS AND DISCUSSION

### 3.1. General Descriptors of Sample Studies

In the 35 selected papers, we discovered a wide range of methodologies and procedural approaches, despite the specificity of the search criteria. Most of the papers analyzed (62.8%) studied one or more species from the same family (e.g., different species of sardine), while 25.7% of the papers analyze species from different families, and only 4 papers (11.4%) are presented as multi specific but do not specify the species included in the work (Figure 1).

*Figure 1. Target Small Pelagic Species in Selected Studies*

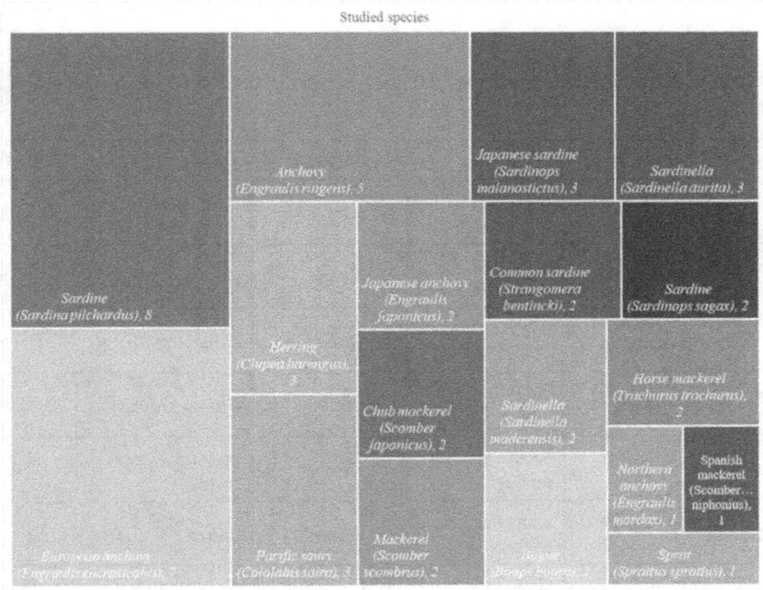


Regarding study locations, we observe regional concentrations. Nine studies iIn the southwestern sector of Europe and fivein the Mediterranean concentrate 38.9% of the total studies. In North-East Asia, 7 studies (19.4% of the total) were conducted. The remainder of the research is spread across various regions: northern Europe (3 studies), southwest Africa (4 studies), southwest South America (3 studies), northwest North America (2 studies), with one study each in Southeast Asia and the US Northeast Large Marine Region. One study does not specify its exact location. The concentration of selected studies dealing with similar approaches and target species in certain regions and the lack of studies in other regions where purse seine fisheries are present suggest specific trends or methodological influences. In order to obtain a global perspective, the selected approaches should be broadened.

Regarding the focus of the papers, 14 papers use MSE, 19 papers use bioeconomic analyses and two papers follow both approaches. Based on the methods and uncertainties addressed, 21 papers addressed socioeconomic issues in the development of the work (Table 2, note that the full reference is available in the cited references section).

We found a clear difference among the papers that include socioeconomic variables, depending on the objective of the study. For example, 12 out of the 14 MSE papers did not address the socio-economic approach. In two of the self-described bioeconomic studies, socio-economic variables have not been developed. Finally, two papers that claimed to follow both the MSE and bioeconomic approaches did include socioeconomic variables and parameters.

*Table 2. Focus, Approaches and use of Socio-Economic Variables in the Reviewed Papers*

| FOCUS | AUTHORS[1] | APPROACH | | | SOCIOECON* |
|---|---|---|---|---|---|
| MSE | Hurtado-Ferro et al. 2010 | operating models (OM) | catch rules | environmental uncertainty | NO |
| MSE | de Moor et al. 2011 | operating models (OM) | simulation | risk | NO |
| MSE | Arteaga et al.2014 | operating models (OM) | abundance | variability | NO |
| MSE | Smith et al. 2015 | ecosystem model | simulation | fishing pressure | NO |
| MSE | Punt et al.2016 | ecosystem model | simulation | performance metrics | NO |
| MSE | Clarke and Egan, 2017 | management strategy | stock trends | impact | NO |
| MSE | Sánchez et al. 2019 | simulation | MSE | uncertainties | YES |
| MSE | Chen et al. 2018 | management procedures | simulation | biological dynamic | NO |
| MSE | Yatsu 2019 | biological dynamic | simulation | catch | NO |
| MSE | Siple et al. 2021 | comparative | MSE | uncertainties | NO |
| MSE | Handjar et al. 2022 | biological dynamic | status | fishery | NO |
| MSE | Kim et al. 2022 | simulation | ecosystem approach | risk | YES |
| MSE | Licandeo et al. 2023 | population dynamic | HCR | simulation | NO |
| MSE | Chen et al. 2023 | population dynamic | uncertainties | parameters | NO |
| BIOECO | Sandberg et al. 1998 | TAC | uncertainty | population dynamic | YES |
| BIOECO | Cho et al. 2005 | economic value | ecosystem approach | societal choices | YES |
| BIOECO | Domínguez-Torreiro and Surís-Regueiro 2007 | game theory | cooperative exploitation | simulation | YES |
| BIOECO | Silvestri and Maynou 2009 | simulation | management conditions | bio and eco conditions | YES |
| BIOECO | Da-Rocha et al. 2011 | uncertainties | population dynamics | fleet productivity | YES |

continued on following page

*Table 2. Continued*

| FOCUS | AUTHORS[1] | APPROACH | | | SOCIOECON* |
|---|---|---|---|---|---|
| BIOECO | Wise et al. 2012 | behavior | cognitive maps | simulation | YES |
| BIOECO | Mulazzani and Malorgio 2013 | game theory | population dynamics | interactions | YES |
| BIOECO | Mulazzani et al. 2013 | interactions | simulation | functions linking | YES |
| BIOECO | Maouel et al. 2014 | indicators | simulation | management actions | YES |
| BIOECO | Torralba and Besada 2015 | stochastic optimization | simulation | economic performance | YES |
| BIOECO | Ruiz et al. 2017 | population dynamic | simulation | TAC | YES |
| BIOECO | Agmour et al. 2018 | carrying capacity | simulation | catch | NO |
| BIOECO | Ba et al. 2019 | marine protected area | simulation | management scenarios | YES |
| BIOECO | Jang et al. 2019 | management actions | population dynamics | socioeconomic outcomes | YES |
| BIOECO | Díaz and Flores 2021 | simulation | management scenarios | bioeconomic | YES |
| BIOECO | Ba et al. 2022 | simulation | management scenarios | subsidies | YES |
| BIOECO | Ni et al. 2022 | population dynamic | simulation | indicator | NO |
| BIOECO | Rosa et al. 2022 | HCE | biological dynamic | simulation | YES |
| BIOECO | Purwanto et al. 2023 | tradeoffs | social dimension | optimization | YES |
| MSE-BIOECO | Hoshino et al. 2012 | harvest | MEY | adaptative management | YES |
| MSE-BIOECO | Anda-Montáñez et al. 2017 | simulation | uncertainty | biological dynamic | YES |

[1]The complete reference is detailed in section 6 "References"

*SOCIOECON: Studies addressing socioeconomic indicators and/or variables

Regarding the approaches selected, we note issues that add complexity. According to Punt et al. (2013), MSE is defined as the simulation of the entire management cycle, including data collection, evaluation, application of HCR, implementation of regulations and the consequences of removals on the dynamics of the study population. However, other authors, such as Ives and Scandol (2013), clarify that the MSE "modelling methodology" is also known as "Harvest Strategy Evaluation", "Operating Management Procedures" or "Management Procedures Evaluation". The above references are examples of the controversy and problems of the execution and progress in the practical implementation of the MSE approach. Besides, the societal uncertainties can influence the identification of management objectives, performance measures and management strategies considered in participatory MSE processes and their implementation (Fulton et al., 2014; Punt, et al., 2016). The definition of the model is neither unanimous nor homogeneous, which implies added complexity to its development and practical implementation.

## 3.2. Comparative Analysis of Socio-Economic Variables and Indicators

After selecting the 21 papers that address socio-economic variables, we carried out a comparative analysis of the indicators and parameters considered. Even though they are works defined as MSE and/or bioeconomic models to analyze small pelagic fisheries, there is a wide range of approaches used in these papers. This state-of-the-art review should be interpreted as a path through publications that respond to a series of common elements, but which are developed in different ways. The analysis and diagnosis are presented in thematic blocks (without prejudice to the particular distinctive elements) to facilitate reading and comparison.

### I) Profit Focus

The studies that focus on profit present discrete socio-economic indicators that are often insufficient to provide an integrating dimension between the analysis of the resource (operational models or estimates of population dynamics) and the trade-offs that define fishing pressure. However, the models presented (Table 3) (with greater or lesser development of socio-economic variables) do not detract from the interest, justifying the limitations to have a complete perception of the economic-environmental variables that constitute a whole in fisheries management.

*Table 3. Profit Focus References*

| Reference | Socio-economical approach and contributions |
|---|---|
| **Focus I.** Profit focus | |
| Hoshino et al. (2012) | The profit function where they include the price; the cost at time t; and the yield of the dominant fleet summed over the fishing seasons. The difference in this work is that the price is calculated using the observed maxima and minima and an adjustment parameter. |
| Mulazzani and Malorgio (2013) | They compare the profits of different fisheries to which they assign the profits and unit costs for a fishery. Consider the price (exogenous in this work), the catchability coefficients; intrinsic growth rates and carrying capacities; fishing efforts; the cost per unit of fishing effort and supplementary revenue proportional to effort. |

continued on following page

*Table 3. Continued*

| Reference | Socio-economical approach and contributions |
|---|---|
| Jang et al. (2019) | Analyzes the possibility of a stock collapse given the presence of multi-decade cyclical fluctuations and trade-offs. |
| Diaz and Flores (2021) | They calculate the benefit of fishmeal production from the fishery including: price per ton of fishmeal; cost of manufacturing; production surplus of the fishing area; efficiency factor in the production of fishmeal (tons of fishmeal per ton of yield, with an approximate value of 21% for anchovy); cost per ton of yield obtained by the artisanal fleet; effort of artisanal fishing; cost per unit of applied industrial effort and industrial effort (total trips) |
| Purwanto et al. (2023) | The proposal for profits, the authors specifically include the labour variable with the number of vessels and associated personnel (crew members per standard vessel), including the average price per unit of biomass previously calculated. |

MSE-BIOECONOMIC APPROACH
BIOECONOMIC APPROACH

## II) Net-Present-Value (NPV)

The net present value is defined as the difference between the present value fishery benefit and the value of the present value of the fishery cost. Different studies have relied on NPV to estimate fishing effort and pressure on the stock from different perspectives and levels of detail (Table 4).

*Table 4. Net-Present-Value (NPV) Focus References*

| Reference | Socio-economical approach and contributions |
|---|---|
| **Focus II.** Net-Present-Value (NPV) | |
| Anda-Montáñez et al. (2017) | This work considers the NPV and includes Total revenues at time *t*, which depends on the price per ton of Pacific sardine (and other species) and catch of Pacific sardine at time *t* (and other species); also, Total Cost at time *t* (including travel cost of a fishing day with the discount rate of 4% for this study). |
| Sánchez et al. (2019) | The economic model includes: (i) a price function to model the prices for the anchovy stock (estimated given the data on landings and prices from the auction market for 2000-2006 using a semi-log price model to estimating the price function), and fixed price for the rest of the species; (ii) a Schaefer type production function (Schaefer 1954). Three fleets were considered Costs parameters (fixed and variable ones) were assumed constant and different for each fleet. Profits were computed given the total costs but excluding capital costs. The fleet gross revenue was defined as the sum of the anchovy revenues and other fisheries gross revenues where the landing fee is a percentage of the gross revenue imputed to anchovy and other species. About shared cost, they included fuel, bait, ice, food, and social security costs paid by the vessel. In addition, the cost of fuel was considered as one of the key uncertainties to fixed costs, given the importance for the activity and the fluctuation observed. |

continued on following page

*Table 4. Continued*

| Reference | Socio-economical approach and contributions |
|---|---|
| Rosa et al. (2022) | The authors consider regime shifts in marine populations and estimate a bioeconomic model with critical depensation. The authors studied how differences in initial harvest/biomass trajectories resulting from applying the HCR vs. the actual bioeconomic optimal solution affect the evaluation criteria (e.g. biomass recovery and NPV). The annual net revenues from the fishery considered the total catch and the mean annual sardine price, and the cost function (they used data on variable fishing cost). While they started by assuming a cost function that explicitly depended on the fish stock, this variable was not statistically significant. |

MSE-BIOECONOMIC APPROACH
BIOECONOMIC APPROACH
MSE APPROACH

One of the great challenges when simulating future scenarios is precisely in properly defining the indicators that are included for a TAC and in evaluating how these measures are executed in practice, analyzing possible gaps in a feedback analysis cycle of information, changes in the sector or behavior and what are the drivers that generate it to progressively redefine the scenarios and measures such as TAC. Sánchez et al. (2019) demonstrated the utility of variable costs in determining the minimum Total Allowable Catch (TAC) necessary to sustain the fleet in the short term. In the most adverse scenario, this minimum TAC would prevent losses exceeding fixed costs. However, this short-term stability is not indicative of long-term viability, as continued operation under such conditions would ultimately lead to fleet disappearance. Therefore, the economic break-even point becomes the pertinent parameter for consideration. This economic model proposed lacks realism in its integration of the socio-economic dimensions of the fleet, particularly in its exclusion of capital costs. As we have mentioned, there are social reasons why the activity must be maintained despite low or even circumstantially negative profits.

## III) Gross and Net Value Added

First of all, in order to address this thematic block of analysis, it is worth briefly expanding on some concepts. Considering ecological and economic sub models up to fleet income, developing the indicators of Gross and Net Value Added (GVA and NVA) (according to Ba et al., 2019). GVA is the difference between the value of landings and intermediate consumption inputs (all goods and services used during the production process) while NVA is equal to GVA less depreciation and subsidies. It is also equal to the sum of the net remuneration of production factors (private profit and labor remuneration) plus the net income of the State. GVA is an indicator of consumption creation by the fishery (contribution to the national Gross Domestic Product) (Ba et al., 2019). Studies focused on these concepts with respect to socioeconomic variables are shown in Table 5.

*Table 5. Income or Total Rent Focus References*

| Reference | Socio-economical approach and contributions |
|---|---|
| **Focus III.** Income or Total Rent | |
| Sandberg et al. (1998) | The authors examined the cost function of society including total variable costs, fishing mortality and β coefficient. Conversely, the vessel owners´ (private) cost function is considering the total revenue from the fishery, the share of the total revenue received by the crew, the fishing mortality and coefficient β. Finally, prices per kg catch, consider the minimum price, and adjusting coefficients. To sum up, the two different cost functions were used in order to evaluate economic benefits at the fleet level both from the perspective of the vessel owners and from the perspective of the society. However, the authors assumed that the fixed cost are not affected by TAC. They have not included a possible empirical relationship between TAC and investment, therefore their approach excluded fixed costs from the analysis. This implies that the economic benefit they calculate is equal to the contribution margin. Economic benefits realized now will have a higher value than economic benefits realized later, since net economic benefits from catches this year can be invested and give a rate of return. To adjust for this Sandberg et al. (1998) reduced the net revenue from future years by a factor that reflects the rate of return in the market, i.e. the discount rate. |
| Ruiz et al. (2017) | The biology and economy sub models are coupled in this paper. The economic sub model is based on a system where employees and ship owners share 50% of the GVA net of Social Security contributions (called *Montemayor*). Catches and fishing effort provide the linkage between biological and economic sub models. Catches result from the combination of the stock size, as derived from the biological sub model, and fishing effort. The model operates under the behavioral assumption that fishers maximize profits via revenue maximization within this management framework. A survey on the vessels' cost structures and an input market analysis of were carried out in order to estimate the parameters and specifications of the model's cost and revenue structure. For this study, this price (€/Kg) at the first sale was sampled every year from the normal distribution in agreement with the evolution of import prices and the verbal information gathered from the fleet. A survey on the vessels' cost structures and an input market analysis of were carried out in order to estimate the parameters and specifications of the model's cost and revenue structure. This analysis also determines how the price of fuel oil affects the cost and revenue structure of the model, to finally analyze the GVA. |
| Ba et al. (2019) | The authors determined fishing mortality by the dynamics of fishing effort, which is itself a function of past economic performance, technological change and management measures. The exogenous parameters of the model (price, costs, recruitment, growth and natural morality) are either determined before the start of the simulation or redefined during simulation. Fleets spatiotemporal dynamics is determined by the availability of the resource in time and space and endogenous economic results. For economic module of the bioeconomic model, they considered: fixed cost (they are independent of the activity rate of the fleet but proportional to the fishing capacity as constants inputs, depreciation, maintenance of equipment and insurance); and variable costs (proportional to the active fleet but are also be related to landed quantities). In this model, the labor remuneration is obtained from the balance to be shared, obtained by the difference between turnover and variable costs, and that is multiplied by the relative share going to the crew. The profit is the remuneration of private investors (vessel owners). It is equal to turnover minus the variable costs (common costs shared between crew and owner), fixed costs and labor remuneration. The cumulative profit per year and fleet is obtained by summing the profit on each annual time interval. The economic rent (per fleet per year) corresponds to the private profit corrected for the net transfers between the state and the fishery (taxes less subsidies). |

continued on following page

*Table 5. Continued*

| Reference | Socio-economical approach and contributions |
|---|---|
| Ba et al., 2022 | They focused on the effect of subsidies on the profitability and fishing effort of a fishery. They included the monthly fuel subsidy per fleet, the values of the monthly fuel consumption "all taxes included" per fishing unit, the monthly "tax free" fuel consumption per fishing unit and the active fleet for each type of fishing units, and the number of fishing units (fishing capacity) by type of fishing units and for year. Payment exemptions that affect profitability are also considered (annual engine tax exemption, per fishing unit and the annual exemption of fishing gear taxes, by fishing unit). Based on the considerations of subsidies and exemptions, the total economic rent of the fishery is calculated by including them and addressing the present value, benefits and labour remunerations. |

MSE-BIOECONOMIC APPROACH
BIOECONOMIC APPROACH
MSE APPROACH

In Sandberg et al. (1998) we observed an important limitation since the authors assume in their hypotheses that there is no relationship between investment and TAC. The motivation of the investments made by the owners is generated by the expectations of performance improvement and by the possibilities of being able to develop the activity with an improvement of the effort derived from the investment. Not taking this into account means dissociating two key components for understanding the profitability and evolution of the fishing fleet and, in addition, incomplete cost accounting and cost linkage

We have also observed the importance of including the support role of administrations, even when evaluating the economic dimension. Fisheries subsidies, widely used in developed countries (Sumaila et al., 2019), are the subject of much criticism. International organizations have issued recommendations to drastically reduce state support for capacity building and fishing efforts, also contributing to illegal, unreported and unregulated fishing (Ba et al., 2022). Nevertheless, artisanal fisheries sometimes reach subsistence roles or provide jobs for the most vulnerable communities, and thus should be supported by programs, which foster the sustainability of their activity (Ba et al., 2022). Finally, purse seiners with different power and lengths are not the same as identical vessels in very different socio-economic contexts (e.g., subsistence or traditional monospecific fisheries vs. multispecies pivotal fisheries or in a scenario of greater social resilience). The approach before the selection of models and methods is fundamental for the assessment to be initiated based on the inherent characteristics of the fishery in question.

## IV) Behavioral

Fishing behavior is interpreted here as the analysis of the variables that condition and/or explain the performance of the fleet. Depending on the author (Table 6) the determinant variables will be different.

*Table 6. Behavioral Focus References*

| Reference | Socio-economic approach and contributions |
|---|---|
| **Focus IV.** Behavioral | |
| Domínguez-Torreiro and Surís-Regueiro (2007) | A methodology was developed for the optimization of fisheries profit in a cooperative management scenario vs a competition scenario (dynamic optimization problem for maximizing behavior in a competitive scenario). Functional form for the fishing industry technology follows Cobb–Douglas type. Optimization proposed for Domínguez-Torreiro and Surís-Regueiro (2007) including the price per ton fished in the country, the cost per standard unit of effort (fishing days in the average vessel), the discount factor in each country analyzed and parameter that appears from the negotiation process. |
| Silvestri and Maynou (2009) | The authors studied the effects of different fuel prices and an alternative number of days at sea. As stated above, fishers buy fuel at a subsidized price, so the fiscal burden cannot be further reduced. In this proposal, the authors including:<br>- Cost structure of the fleet (economic parameters): Capital in €/vessel; Annual costs in €/vessel; Commercial costs in %; Owner's share in %; Daily fuel consumption in liters/vessel; Daily ice expenses in €/day; and Demolition, sale, transfer €/GRT<br>- Cost structure (technical structure of the fleets): GRT/vessel; HP/vessel; Crew/vessel; Daily fishing hours; Annual fishing days; and Number of vessels.<br>- Cost structure of the fleet (market parameters): Sardine price in €/kg; and Anchovy price in €/kg<br>- Other economic factors: Opportunity cost in %; Financial cost in %; and Fuel. |
| Da-Rocha et al. (2011) | They contributed with their study of stochastic modelling to quantify the effects of fishing periods on the stock and relate the calculation of the TAC to biological constants and environmental factors, expected resource values and institutional restrictions. Catch performance and the level of fishing effort will depend on whether or not restrictions on the number of fishing days exist. In this way, they can define the three different scenarios: centralized management without fishing periods (the total catches are equal to the TAC); centralized management with fishing periods (if fishing periods are set, the regulator calculates the effort which will be necessary to reach the desired TAC); and decentralized management (status quo) (given that both countries share the regulatory objective, the total catch is set as before). |
| Weise et al. 2012 | The parameters are evaluated through the selectivity of purse seiners and their relationship with the market prices of the catches and their sizes. The authors consider the vessel's selectivity-at-age, the age in years, fleet capacity and landing limit. Regarding the price, the authors calculated the total weight of fish landed. To validate, three scenarios were simulated: to evaluate the model fit to observed data, focusing on the behavior of the purse seine skippers; the value of all fish species by multiplying the concept "Price" by a factor of two (equivalent to a change in the market regulation implying an overall increase in fish first sale price); and simulates the overall decrease in the abundance of all fish species (e.g. due to overfishing and habitat destruction), accomplished by modifying the initial number-at-age of each species. |

continued on following page

*Table 6. Continued*

| Reference | Socio-economic approach and contributions |
|---|---|
| Maouel et al. 2014 | The bio-economic model MEFISTO (Mediterranean Fisheries Simulation Tool) is a multispecies, multi fleet and multi gear simulation model. It is a dynamic model consisting of three modules or boxes: a stock box, a market box and a fisherman box. In this study only one main specie is included (sardine). The market box converts the fish catches of each species generated by the stock model into revenues, via constant price functions and then revenues are imported into the fisherman's box. The revenues of a vessel correspond to the income deriving from the sale of the target species. The economic parameters were estimated from field interviews (1998-2007) carried among vessel owners: volume of catches, exvessel sale price, detailed variable costs, detailed fixed costs, amortization, capital, engine power, length and tonnage, labour costs, duration of day trips and annual fishing days. Other factors influencing price were disregarded because they were considered not important for this species in the study area (e.g. change in price due to fish size or due to fish imports). |
| Torralba and Besada (2015) | The authors used a dynamic optimization model in continuous time with an infinite time horizon in which the evolution of the natural stock is defined as a stochastic differential equation that includes prize per ton of fish, cost per unit of fishing effort and fishing effort. They assumed certain functional relationships between the variables involved in the problem and estimated one structure for the dynamics of the sardine biomass and another one for the technological function of production (i.e. the number of sardines within a period is the difference between the gross growth of the stock and the catches over the previous period). The values obtained estimate that the fundamental input is the fishing effort. The authors estimated that it is probably because the technology used by the purse seine fleet is so efficient at detecting schools of fish that it achieves captures in every set. |
| Kim et al. (2022) | Data used for indicator estimation of risk scores for socio-economic indicators were: profit conditions in fishery management survey; profit and loss relation ratio in fishery management survey; catch and production by fishery and species in fishery production trend survey; catch by year, zone number, fishery, and species; fishing effort by year, zone number, fishery, and species; the growth rate of the economically active population by age in economically active population survey; and fishery status and activity in fishery management survey. Changes in risk as a result of the implementation of the management measures were monitored so that management objectives, strategies, and tactics can be reestablished or an evaluation or forecasting can be re-conducted through MSE and management tactics evaluation (MTE). The methodological development of the aforementioned indicators is not published in this work. |

MSE-BIOECONOMIC APPROACH
BIOECONOMIC APPROACH
MSE APPROACH

Silvestri and Maynou (2009) mainly regarded management aspects based on effort control whose regulation allows pressure on the overexploited stocks to be reduced to guarantee their conservation and jobs, safeguarding the investment made in the sector. On the basis of the bioeconomic results obtained for Silvestri and Maynou

(2009), the proposal for management of SPFs is articulated around a combination of measures: i) applying a targeted support policy in defense of smaller vessels, and in particular, introducing a compensatory policy; ii) guaranteeing greater profits for the sector by means of cost reduction or/and value added policies (promotion, labelling, etc.); iii) promoting sector restructuring (commerce, aquaculture, tourism, etc.); iv) placing fishing effort in terms of maximum number of days at sea at between 205 and 215 days per year.

According to Kim et al. (2022), lowering the risk index of the socioeconomic benefit objective requires a reduction in the number of fishing vessels, an improvement of fishing gear to reduce operating costs, and measures to improve employment.

Regarding the relation of active fishing days, on the one hand, maintaining the stock at higher levels increases the catches per unit effort and the profitability of the fleet; on the other, a restriction on the number of fishing days generates lower catches and reduces profits. The results obtained by Da-Rocha et al. (2011) show that setting restrictions on the maximum number of fishing days that the fleet can use to reach the TAC reduces the risk of the resource being overexploited. In a framework of uncertainty, the regulator has to set the TAC on the basis of the inference it makes on the state of the resource at the decision-making stage which, in general, does not correspond with the real state of the resource. Therefore, the regulator sometimes overestimates and sometimes underestimates the resource condition. Limiting fishing may lower economic returns but could be necessary if the boom and bust cycle incurs significant socioeconomic costs. For instance, the fluctuation in small pelagic fish stocks has led to problems like overinvestment by fishers, the establishment of assistance programs exacerbating overfishing, and the overflow of fishing capacity into other fisheries (Jang et al., 2019).

Several works in which models were developed to analyze and predict the behavior of fishers fail to consider this individual decision-making process, examining, instead, the behavior of the fleet as a whole and relying on data only indirectly related to the behavior of fishers, such as the level and placement of fishing effort (Valcic, 2009). Fisher´s decision-making process is based on their knowledge about the fishery (how, where and when to fish) gained by experience and cultural transmission. Therefore, one way of modelling the behavior of a fishing fleet would be to take into consideration the behavior of fishers in response to a wide range of situations (Wise et al., 2012). The diversity of fleets within a country poses challenges in comparing their capacity and associated costs, leading to varying operational behaviors. In Waise et al. (2012) the authors show how data collected on board fishing vessels, a source of data currently available for many fleets and fisheries around the world, can be combined with Rule Based Fuzzy Cognitive Maps, a modelling technique that is able to deal with both quantitative and qualitative data. Game Theory can be also applied to understand purse seine fleet dynamics. Dominguez-Torreiro and

Surís-Regueiro (2007) suggest coastal states evaluate the feasibility of cooperative resource management. Economic agents' involvement is determinant for the legitimacy and enforcement of agreements. Cooperation among agents and authorities should extend to the allocation of fishing quotas post-agreement.

## V) Other Socio-Economic Approaches

Given the diversity of methodological approaches found in the selected studies, there are papers that do not specifically address any of the approaches described above (despite the heterogeneity shown within the thematic blocks themselves). Because of their contribution and relevance, the following papers should be highlighted (Table 7).

*Table 7. Other Socio-Economic Approaches References*

| Reference | Socio-economical approach and contributions |
|---|---|
| **Focus V.** Other socio-economic approach | |
| Cho et al. (2005) | The authors calculated the added value of cetacean sightings concerning the herring stock, which was decisive for cetaceans. The aggregate value was defined by trips (i.e., services) provided by whales to users, the price (i.e., the cost of participation) of a trip, the quality of the whale-watching experience as perceived by the trippers socioeconomic and others variables that affect the demand for whale-watching. Emphasis has shifted from commercial exploitation (consumptive use) to public fascination (non-consumptive use). The opportunity cost of marginal reductions in TAC to "reserve" amounts for the ecosystem is measured by reduced profits to herring fishers that are made possible by a larger herring stock. |
| Mulazzani et al. (2013) | They studied the governance of species that have international economic relevance and that are characterized by shared stocks between countries and integrated markets. The Institutional Analysis and Development (IAD) framework is frequently used to investigate the governance of marine commons. The framework can facilitate the analysis of complex social–ecological systems by developing an appropriate theoretical language and focusing on several aspects characterized by cause–effect relationships. These aspects are the following: the exogenous variables, or attributes, of the context (biophysical attributes, community's attributes, institutional setting); the action arena (the social place where participants, considering potential outcomes, action–outcome linkages, cost and benefits associated with actions and outcomes, make decisions); the interactions generated and the outcomes produced. Outcomes feedback onto the participants and the context and may transform both over time. At a lower level, and in the short run, participants can simply change their strategies and routine decisions; at a higher level, and in the long run, if interactions and outcomes are deemed unsatisfying or unfair, some will raise questions about trying to change what was initially considered exogenous, namely the rules and the institutional setting. |

MSE-BIOECONOMIC APPROACH
BIOECONOMIC APPROACH
MSE APPROACH

One of the proposals for bioeconomic models with a very particular approach is the work of Cho et al. (2005). Using non-market valuation methods, the authors estimated the economic value of reserves in the ocean. This approach broadens the understanding of the value of herring beyond the values associated with specific fisheries. The non-consumptive use value of marine mammals is so labeled because observers are able to enjoy whales and other mammals by viewing them but without consuming them to the exclusion of others. From studies of whale watching, we do know that non-consumptive uses of marine mammals are important to many people. There are well-established methods for estimating these benefits. They are based on the "resource rent' of incremental change in discounted profit which would follow a change in Ecosystem Reserves. These values are the opportunity cost of ecosystem reserves in terms of reduced net profit for fishers. The concept of opportunity cost is a reciprocal one; the opportunity cost of increased TAC includes the reduction in ecosystem benefits. Of direct commercial exploitation of whales and indirect exploitation associated with conflicting types of resource development (Samples 1986).

In addition, although bioeconomic models are useful, they face several challenges and limitations. It´s important to note that socio-economic data do not always accurately reflect the needs and attitudes of fishing communities. To improve the efficacy of these models, the socio-economic data used should be more representative of fishing communities. Mulazzani and Malorgio (2013) demonstrate how bioeconomic variables and policies shape the system's equilibrium. Factors like prices, costs, catchability, and biological parameters influence fleet and fish stock dominance. In cases such as those included in the publication of Mulazzani and Malorgio (2013), like marginal regions or for unemployed fishers, authorities may opt for open access policies to boost employment despite fleet overcapitalization. However, these exogenous changes can lead to challenges in achieving equilibrium over time. Steady-state models merely outline bioeconomic trends rather than solving mathematical problems precisely. Reference price determination for value and profit calculation lacks consistency and transparency across analyzed works. Diverse criteria and selected values contribute to this issue.

## 3.3. Transversal Diagnosis on Socio-Economic Dimension and Fisheries Management

The typology provides a systematic analysis and assessment of how participation operates in practice and what institutional mechanisms may address the risks associated with stakeholder interactions (Ballesteros and Dickey-Collas, 2023). The complexities of management can be addressed by directly involving stakeholders in the management process and applying the adaptive management cycle (Dichmont et al. 2013). Several authors highlight the need for the participation of decision makers and stakeholders (particularly fishers) in all stages of the development of the plan to properly address the management and fishing needs of the concerned fishery, but there is little return from processes that have involved consultation and active participation in the entire procedure (including simulation) leading to management measures (see good practices in the anchovy case in Uriarte et al., 2023).

Likewise, stock assessment data often lack sufficient details to precisely outline key population processes, with uncertainties across these sources typically intertwined. Notably, it's essential to clarify that correlations among uncertainties do not imply causation. Heightened perceived uncertainty may prompt evaluators to allocate higher importance scores to specific variables (Leach et al., 2014). To derive results directly applicable to management, emphasis should be placed on modelling biological/operational realism and estimating the uncertainty associated with these processes (Butterworth et al., 2010). However, it is also worth mentioning that including too many sources of uncertainty is likely to lead to results that are difficult to interpret (Punt et al., 2016). Furthermore, there are costs to collecting ecosystem information (e.g. Essington et al., 2018) and incorporating it in an MSE or bioeconomic model. It is not about undervaluing simulations either, but it is not convenient in simulations based on historical data in that they assume that an event that has not happened will never happen and that possibly rare events in the past will always repeat themselves in the same way (Schruben and Singham 2010).

Finally, while guidelines can be established for the implementation of common fisheries policies or stocks, the casuistry and realities are so varied that it is unrealistic to consider MSE beyond a procedural approach. Additionally, the International Council for the Exploration of the Sea (ICES) has taken consistent steps towards integrating socio-economic analysis with policy and management in its recently adopted Strategic Plan (ICES, 2021). In this context, ICES has initiated a set of initiatives to facilitate cooperation between marine and social scientists. For example, the Strategic Initiative on Human Dimensions (SIHD) and the Human Activities, Pressures, and Impacts Steering Group (HAPISG). The progress in scientific knowledge and experience with methodologies and modelling has evolved notably.

## 4. CONCLUSION

The review allows for critical reflection on the main currents, including Management Strategies Evaluation (MSE) and Bioeconomic models. It is concluded that the accuracy and reliability of the assessment tool's assumptions will impact the capability to provide effective management recommendations. A significant amount of data does not always guarantee reliable information for decision-making being rich in data but often poor in information. Even some of the terminology surrounding data-limited methods can be ambiguous, leading to interpretations.

Therefore, it is imperative to comprehend the dynamics of the species and the fleet within a given fishery, which constitutes the management unit, before selecting and applying a specific tool or methodology. However, for the purposes of the accounting unit, improvements must be made so that the economic (or bioeconomic) models consider the business scale (ship or boat owners). In this way, a truly operational management unit can be built, where the projected scenarios match reality in all its dimensions.

MSE and bioeconomic methods use error models to enhance accuracy and discard improbable scenarios, incorporating ecological, economic, and socio-administrative factors. This includes considering feasibility, acceptance levels, and user dependence, affecting both measure implementation efficiency and model data. Sector behavior is crucial in evaluating management strategies, though subjective evaluations like preferences and behavioral simulations are often absent from bioeconomic models. A social-ecological approach to fisheries management would imply a greater focus on the interactions between the natural and human components of ecosystems (Makino 2018). Setting short-term objectives in fisheries management, though deemed risky, may be warranted to understand how economic interests influence decision-making, while adaptive management prioritizes continuous and iterative learning and decision adaptation over fixed knowledge states (Hurtado-Ferro et al., 2010; Hoshino et al., 2012).

Currently, there is a gap between scientific recommendations and operational management in the sector. Sectoral approaches and methods are prevalent, but they are not adequately integrated with the resource's biology and the sector's economic and social aspects. It is observed that the main legislating and managing institutions of fisheries, such as the European Community, are currently focused on consolidating and adapting to new management scenarios, rather than creating new models and tools (software). This includes incorporating adjustments and uncertainties resulting from new developments in the Fisheries Policy and upcoming measures. Therefore, modelling should not be seen as a priority, but rather as a tool to form the basis for making decisions. We detected a disproportionate prioritization of modelling and simulation over an integrated and comprehensive understanding of the reality of a given stock.

Finally, the confluence between scientists and decision-makers is still slow to emerge. After all, there are different currents within the scientific community, using different models and complementary approaches. The same is true of the standards used to define policies and strategies that can be transferred to regulations. Many of the consulted authors agree that a paradigm shift is necessary for the evaluation and management of fisheries. This shift involves effectively integrating the socio-economic and ecological dimensions within a shared, dynamic world facing ambitious challenges that demand greater resilience. Understanding the behavioral dynamics is essential for devising effective management strategies that account for the complexities of human interactions with marine ecosystems.

## REFERENCES


Agmour, Bentounsi, Achtaich, & Fountayeni. (2018). Carrying capacity influence on the incomes of seiners exploiting marine species in the Atlantic coast of Morocco. *Mathematical Biosciences*, 305, 10-17. 10.1016/j.mbs.2018.08.012

Alder, J., Campbell, B., Karpouzi, V., Kashner, K., & Pauly, D. (2008). Forage Fish: From Ecosystems to Markets. *Annual Review of Environment and Resources*, 33(1), 33, 153–166. 10.1146/annurev.environ.33.020807.143204

Alder, J., & Pauly, D. (2008). A comparative assessment of biodiversity, fisheries and aquaculture in 53 countries' exclusive economic zones. *Fisheries Centre Research Reports*, 16, 90.

Anderson, L. G., & Seijo, J. C. (2010). *Bioeconomics of Fisheries Management.* Wiley-Blackwell.

Aoki. (2018). *Fish population dynamics, monitoring, and management—sustainable fisheries in the eternal ocean.* Springer Nature.

Arteaga, M., Ernst, B., Vásquez, S., & Gatica, C. (2014). Bases conceptuales para la aplicación de una evaluación de estrategias de manejo (EEM) en sardina común (Strangomera bentincki) y anchoveta (Engraulis ringens) en la zona centro-sur de Chile. *Latin American Journal of Aquatic Research*, 42(3), 3, 445–467. 10.3856/vol42-issue3-fulltext-6

Ba, Chaboud, Brehmer, & Schmidt. (2022). Are subsidies still relevant in West African artisanal small pelagic fishery? Insights from long run bioeconomic scenarios. *Marine Policy*, 146, 105294. 10.1016/j.marpol.2022.105294

Ba, A., Chaboud, C., Schmidt, J., Diouf, M., Fall, M., Dème, M., & Brehmer, P. (2019). The potential impact of marine protected areas on the Senegalese sardinella fishery. *Ocean and Coastal Management*, 169, 239–246. 10.1016/j.ocecoaman.2018.12.020

Ballesteros & Dickey-Collas. (2023). Managing participation across boundaries: A typology for stakeholder engagement in the International Council for the Exploration of the Sea. *Marine Policy*, 147, 105389. 10.1016/j.marpol.2022.105389

Barange, M. A., Bernal, M., Cergole, M. C., Cubillos, L. A., Cunningham, C. L., Daskalov, M., & De Oliveira, G. (2009). Current trend in the assessment and management of stocks. In *Climate Change and Small Pelagic Fish.* Cambirdge University Press.

Beddington, Agnew, & Clark. (2007). Current problems in the management of marine fisheries. *Science*, 316(5832), 1713–1716. 10.1126/science.1137362

Benson & Stephenson. (2018). Options for integrating ecological, economic, and social objectives in evaluation and management of fisheries. *Fish and Fisheries*, 19(1), 40–56. 10.1111/faf.12235

Butler, C. M., Rudershausen, P. J., & Buckel, J. A. (2010). Atlantic Bluefin Tuna (Thunnus thynnus) feeding ecology and potential ecosystem effects during winter in North Carolina. *U.S. National Marine Fisheries Service. Fish Bulletin*, 108, 56–69.

Butterworth, D. S., Johnston, S. J., & Brandão, A. (2010). Pretesting the likely efficacy of suggested management approaches to data-poor fisheries. *Marine and Coastal Fisheries*, 2(1), 131–145. 10.1577/C08-038.1

Caddy, J. F., & Mahon, R. (1995). *Reference points for fisheries management. FAO Fisheries Technical Paper. 347*. FAO.

Chen, Zhang, Sun, Xu, Xue, Ren, & Chen. (2018). The impact of natural mortality variations on the performance of management procedures for Spanish mackerel (Scomberomorus niphonius) in the Yellow Sea, China. *Acta Oceanologica Sinica*, 37(8), 21–30. 10.1007/s13131-018-1234-0

Chen, Liu, & Yin. (2023). Evaluation of Yellow Sea anchovy (Engraulis japonicus) fishery for management strategies using a data-limited management toolkit. *Heliyon*, 9, e18058. 10.1016/j.heliyon.2023.e18058

Cho, J. H., Gates, J. M., Logan, P., Kitts, A., & Soboil, M. (2005). The Economic Values of Atlantic Herring in the Northeast Shelf Large Marine Ecosystem. *Large Marine Ecosystems*, 13, 215–228. 10.1016/S1570-0461(05)80034-5

Clarke, M., & Egan, A. (2017). Good luck or good governance? The recovery of Celtic Sea herring. *Marine Policy*, 78, 78, 163–170. 10.1016/j.marpol.2016.10.025

Cochet, Y., & Gilly, B. (1990). *Computer programmes for fisheries bioeconomic analysis. BEAM 3: Bioeconomic analytical simulation model, for tropical shrimp fisheries with fixed or random recruitment. Fisheries Technical Papers, no: 310.2*. FAO.

Cury, P., Bakun, A., Crawford, R.J.M., Jarre, A., Quinones, R.A., Shannon, L.J., & Verheye, H.M. (2000). Small pelagics in upwelling systems: Patterns of interaction and structural changes in "wasp-waist" ecosystems. *Ices Journal of Marine Science*, 57(3), 603–618. 10.1006/jmsc.2000.0712 d'Eon-Eggertson, F., Dulvy, N. K., & Peterman, R. (2015). Reliable Identification of Declining Populations in an Uncertain World. *Conservation Letters*, 8(2), 86–96. 10.1111/conl.12123

Da Rocha, J.M., Cerviño, S., & Gutiérrez, M.J. (2010). An endogenous bioeconomic optimization algorithm to evaluate recovery plans: an application to southern hake. *ICES Journal of Marine Science*, 67(9), 1957–1962. 10.1093/icesjms/fsq116

Da Rocha, J.M., & Gutiérrez, M.J. (2011). Lessons from the long-term management plan for northern hake: could the economic assessment have accepted it? *ICES Journal of Marine Science*, 68(9), 1937–1941. 10.1093/icesjms/fsr105

De Anda-Montañez, J. A., Salas, S., & Galindo-Cortes, G. (2017). Dealing with dynamics and uncertainty of small pelagic fisheries: bioeconomic analysis of manager's responses to alternative management strategies. *Revista de Biología Marina y Oceanografía*, 52(1), 51-65. https://www.redalyc.org/articulo.oa?id=47950143004

De Moor, C. L., Butterworth, D. S., & De Oliveira, J. A. A. (2011). Is the management procedure approach equipped to handle short-lived pelagic species with their boom and bust dynamics? The case of the South African fishery for sardine and anchovy. *ICES Journal of Marine Science*, 68(10), 2075–2085. 10.1093/icesjms/fsr165

Díaz-Vega, J.P., & Flores-Arévalo, Y. (2021). Evaluating marine reserves as a management policy in the central-southern anchovy (Engraulis ringens) fishery of Chile. *Latin American Journal of Aquatic Research,* 49(2), 212-226. 10.3856/vol49-issue2-fulltext-2614

Dichmont, C.M., & Brown, I.W. (2010). A case study in successful management of a data-poor fishery using simple decision rules: the Queensland spanner crab fishery. *Marine and Coastal Fisheries: Dynamics, Management, and Ecosystem Science*, 2, 1–13. 10.1577/C08-034.1

Dichmont, C. M., Ellis, N., Bustamante, R. H., Deng, R., Tickell, S., Pascual, R., Lozano-Montes, H., & Griffiths, S. (2013). Evaluating marine spatial closures with conflicting fisheries and conservation objectives. *Journal of Applied Ecology*, 50(4), 1060–1070. 10.1111/1365-2664.12110

Domínguez-Torreiro, M., & Surís-Regueiro, J.C. (2007). Cooperation and non-cooperation in the Ibero-atlantic sardine shared stock fishery. *Fisheries Research,* 83, 1-10. 10.1016/j.fishres.2006.07.012

Dudley, R.G. (2008). A basis for understanding fishery management dynamics. *System Dynamics Review*, 24(1), 1-29. 10.1002/sdr.392

Emery, T. J., Gardner, C., Hartmann, K., & Cartwright, I. (2017). Incorporating economics into fisheries management frameworks in Australia. *Marine Policy*, 77, 77, 136–143. 10.1016/j.marpol.2016.12.018

Essington, T. E., Sanchirico, J. N., & Baskett, M. L. (2018). Economic value of ecological information in ecosystem-based natural resource management depends on exploitation history. *Proceedings of the National Academy of Sciences of the United States of America*, 115(7), 1658–1663. 10.1073/pnas.171685811529378966

FAO (Food and Agriculture Organization of the United Nations). (2016). *The State of World Fisheries and Aquaculture 2016. Contributing to Food Security and Nutrition for All.* Food and Agriculture Organization of the United Nations.

Francis, R. I. C. C., & Shotton, R. (1997). Risk in fisheries management: A review. *Canadian Journal of Fisheries and Aquatic Sciences*, 54(8), 1699–1715. 10.1139/f97-100


Francis, R. I. C. C., & Shotton, R. (1997). Risk in fisheries management: A review. *Canadian Journal of Fisheries and Aquatic Sciences*, 54(8), 1699–1715. 10.1139/f97-100

Fréon, P., Bouchon, M., Mullon, C., García, C., & Ñiquen, M. (2008). Interdecadal variability of anchoveta abundance and overcapacity of the fishery in Peru. *Progress in Oceanography*, 79(2), 401–412. 10.1016/j.pocean.2008.10.011

Freon, P., Cury, P., Shannon, L., & Roy, C. (2005). Sustainable exploitation of small pelagic fish stocks challenged by environmental and ecosystem changes: A review. *Bulletin of Marine Science*, 76(2), 385–462.

Fulton, E. A., Link, J. S., Kaplan, I. C., Savina-Rolland, M., Johnson, P., Ainsworth, C., Horne, P., Gorton, R., Gamble, R. J., Smith, A. D. M., & Smith, D. C. (2011). Lessons in modelling and management of marine ecosystems: The Atlantis experience. *Fish and Fisheries*, 12(2), 12, 171–188. 10.1111/j.1467-2979.2011.00412.x

Fulton, E.A., Smith, A.D.M., Smith, D.C., & Johnson, P. (2014). An Integrated Approach Is Needed for Ecosystem Based Fisheries Management: Insights from Ecosystem-Level Management Strategy Evaluation. *PLoS One*, 9(1), e84242. 10.1371/journal.pone.0084242

Furness, R. W. (2003). Impacts of fisheries on seabird communities. *Scientia Marina,* 67(S2), 33–45. 10.3989/scimar.2003.67s233

Hamilton, L., Otterstad, O., & Ogmundardottir, H. (2006). *Rise and fall of the herring towns: impacts of climate and human teleconnections. Climate Change and the Economics of the World's Fisheries.* Edward Elgar Publishing.

Handjar, H., Guendouzi, Y., Niar, A., Kara, H., & Benhalima, M. (2022). Stock status of the bogue Boops boops (Walbaum, 1792) in Algerian West Coast (Southwestern Mediterranean Sea). *Regional Studies in Marine Science*, 56, 102727. 10.1016/j.rsma.2022.102727

Hoshino, E., Milner-Gulland, E. J., & Hillary, R. M. (2012). Bioeconomic adaptive management procedures for short-lived species: A case study of Pacific saury (Cololabis saira) and Japanese common squid (Todarodes pacificus). *Fisheries Research*, 121, 17–30. 10.1016/j.fishres.2012.01.007

Hurtado-Ferro, F., Hiramatsu, K., & Shirakihara, K. (2010). Allowing for environmental effects in a management strategy evaluation for Japanese sardine. *ICES Journal of Marine Science*, 67(9), 2012–2017. 10.1093/icesjms/fsq126

ICES. (2023). Working Group on Southern Horse Mackerel, Anchovy and Sardine (WGHANSA). ICES Scientific Reports. Report. 10.17895/ices.pub.23507922.v1

ICES 2015. *Report of the Working Group on Southern Horse Mackerel, Anchovy and Sardine* (WGHANSA) ICES CM 2015/ACOM:16

ICES 2021. Working group on economics (WGECON; outputs from 2020 meeting. *ICES Scientific Report* (3(4)). 10.17895/ices.pub.7652

Ives, M.C., & Scandol, J.P. (2013). BIOMAS A bioeconomic modelling and assessment system for fisheries management strategy evaluation. *Ecological Modelling*, 249, 42-49. 10.1016/j.ecolmodel.2012.07.006

Jang, H. G., Yamazaki, S., & Hoshino, E. (2019). Profit and equity trade-offs in the management of small pelagic fisheries: The case of the Japanese sardine fishery. *The Australian Journal of Agricultural and Resource Economics*, 63(3), 549–574. 10.1111/1467-8489.12317

Kim, H., Kang, H., & Zhang, C.I. (2022). Ecosystem-based fisheries risk assessment and forecasting considering a spatio-temporal component in Korean waters. *Ocean and Coastal Management*, 230, 106356. 10.1016/j.ocecoaman.2022.106356

Kim, S. (2010). Fisheries development in northeastern Asia in conjunction with changes in climate and social systems. *Marine Policy*, 34(4), 803–809. 10.1016/j.marpol.2010.01.028

Kjærsgaard, J. (2007). Multiple criteria decision making in fisheries management: experiences and implications. *Acta Agric. Scand. Sect. C - Food Econ.*, 4, 49–64. 10.1080/16507540701362831

Kvamsdal, S.F., Eide, A., Ekerhovd, N.A., Enberg, K., Gudmundsdottir, A., Hoel, A.H., Mills, K.E., Mueter, F.J., Ravn-Jonsen, L., Sandal, L.K., Stiansen, J.E., & Vestergaard, N. (2016). Harvest control rules in modern fisheries management. *Elementa: Science of the Anthropocene*, 4, 000114. 10.12952/journal.elementa.000114

Lam, M.E., Pitcher, T.J., Surma, S., Scott, J., Kaiser, M., White, A.S.J., Pakhomov, E.A., & Ward, L.M. (2019). Value- and ecosystem-based management approach: the Pacific herring fishery conflict. *Marine Ecology Progress Series*, 617-618, 341-364. 10.3354/meps12972

Leach, A., Levontin, P., Holt, J., Kell, L., & Mumford, J. (2014). Identification and prioritization of uncertainties for management of eastern Atlantic bluefin tuna (Thunnus thynnus). *Marine Policy*, 48, 48, 84–92. 10.1016/j.marpol.2014.03.010

Liberati, A., Altman, D.G., Tetzlaff, J., Mulrow, C., Gøtzsche, P.C., Ioannidis, J.P.A., Clarke, M., Devereaux, P.J., Kleijnen, J., & Moher, D. (2010). The PRISMA statement for reporting systematic reviews and meta-analyses of studies that evaluate health care interventions: explanation and elaboration, *Journal of Clinical Epidemiology*, 62, 1-34. 10.1016/j.jclinepi.2009.06.006

Licandeo, R., de la Puente, S., Christensen, V., Hilborn, R., & Walters, C. (2023). A delay-differential model for representing small pelagic fish stock dynamics and its application for assessing alternative management strategies under environmental uncertainty. *Fish and Fisheries*, 24(4), 24, 544–566. 10.1111/faf.12743

Mackinson, S., Sumaila, U. R., & Pitcher, T. J. (1997). Bioeconomics and catchability: Fish and fishers behaviour during stock collapse. *Fisheries Research*, 31(1-2), 11–17. 10.1016/S0165-7836(97)00020-9

Makino, M. (2018). Rebuilding and full utilization of alternating pelagic species around Japan: A social-ecological approach. *FAO Fisheries and Aquaculture Technical Paper*, 630(2), 61–73.

Maouel, D., Maynou, F., & Bedrani, S. (2014). Bioeconomic Analysis of Small Pelagic Fishery in Central Algeria. *Turkish Journal of Fisheries and Aquatic Sciences*, 14(4), 14, 897–904. 10.4194/1303-2712-v14_4_08

Mulazzani, L., Curtin, R., Andrés, M., & Malorgio, G. (2013). Multilevel governance and management of shared stocks with integrated markets: The European anchovy case. *Marine Policy*, 38, 407–416. 10.1016/j.marpol.2012.06.020

Mulazzani, L., & Malorgio, G. (2013). Bioeconomic equilibrium and supply regime for a multi-species multi-fleet fishery: An analysis in the Mediterranean Sea. *New Medit*, 3.

Mullon, C., Mittaine, J. F., Thébaud, O., Péron, G., Merino, G., & Barenge, M. (2009). Modeling the global fishmeal and fish oil markets. *Natural Resource Modeling*, 22(4), 564–609. 10.1111/j.1939-7445.2009.00053.x

Ni, Y., Steinshamn, S.I., & Kvamsdal, C.F. (2022). Negative shocks in an age-structured bioeconomic model and how to deal with them. *Economic Analysis and Policy*, 76, 15-30. 10.1016/j.eap.2022.07.009

Nielsen, J.R., Thunberg, E., Holland, D.S., Schmidt, J.O., Fulton, E.A., Bastardie, F., & Bethke, E. (2018). Integrated ecological–economic fisheries models—evaluation, review and challenges for implementation. *Fish and Fisheries,* 19(1), 1–29. 10.1111/faf.12232

Overholtz, W. J., Link, J. S., & Suslowicz, L. E. (2000). Consumption of important pelagic fish and squid by predatory fish in the northeastern USA shelf ecosystem with some fishery comparisons. *ICES Journal of Marine Science*, 57(4), 1147–1159. 10.1006/jmsc.2000.0802

Pikitch, E. K., Rountos, K. J., Essington, T. E., Santora, C., Pauly, D., Watson, R., Sumaila, U. R., Boersma, P. D., Boyd, I. L., Conover, D. O., Cury, P., Heppell, S. S., Houde, E. D., Mangel, M., Plagányi, É., Sainsbury, K., Steneck, R. S., Geers, T. M., Gownaris, N., & Munch, S. B. (2014). The global contribution of forage fish to marine fisheries and ecosystems. *Fish and Fisheries*, 15(1), 43–64. 10.1111/faf.12004

Prellezo, R., Accadia, P., Andersen, J.L., Andersen, B.S., Buisman, E., Little, A., & Röckmann, C. (2012). A review of EU bioeconomic models for fisheries: the value of a diversity of models. *Marine Policy,* 36(2), 423–431. 10.1016/j.marpol.2011.08.003

Punt, A. E., Butterworth, D. S., de Moor, C. L., De Oliveira, J. A. A., & Haddon, M. (2016). Management strategy evaluation: Best practices. *Fish and Fisheries*, 17(2), 303–334. 10.1111/faf.12104

Punt, A. E., Trinnie, F., Walker, T. I., McGarvey, R., Feenstra, J., Linnane, A., & Hartmann, K. (2013). The performance of a management procedure for rock lobsters,Jasus edwardsii, off western Victoria, Australia inthe face on non-stationary dynamics. *Fisheries Research*, 137, 116–128. 10.1016/j.fishres.2012.09.017

Purwanto, F.E.C., White, A., & Mardiani, A.R. (2023). Multiple-goal bioeconomic programming to address conflicting management objectives in Indonesian small pelagic fisheries. *Marine Policy*, 150, 105519. 10.1016/j.marpol.2023.105519

Rosa, R., Costa, T., & Mota, P.R. (2022). Incorporating economics into fishery policies: Developing integrated ecological-economics harvest control rules. *Ecological Economics*, 169, 107418. 10.1016/j.ecolecon.2022.107418

Rosenberg, A. A., & Restrepo, V. R. (1994). Uncertainty and risk evaluation in stock assessment advice for U.S. marine fisheries. *Canadian Journal of Fisheries and Aquatic Sciences*, 51(12), 2715–2720. 10.1139/f94-271

Ruiz, J., Garcia-Isarch, E., Emma Huertas, I., Prieto, L., Juárez, A., Muñoz, J. L., Sánchez-Lamadrid, A., Rodríguez-Gálvez, S., Naranjo, J. M., & Baldó, F. (2006). Meteorological and oceanographic factors influencing Engraulis encrasicolus early life stages and catches in the Gulf of Cádiz. *Deep-sea Research. Part II, Topical Studies in Oceanography*, 53(11–13), 1363–1376. 10.1016/j.dsr2.2006.04.007

Ruiz, J., Rincon, M. M., Castilla, D., Ramos, F., & del Hoyo, J. J. G. (2017). Biological and economic vulnerabilities of fixed TACs in small pelagics: An analysis of the European anchovy (Engraulis encrasicolus) in the Gulf of Cádiz. *Marine Policy*, 78, 171–180. 10.1016/j.marpol.2017.01.022

Sánchez, S., Ibaibarriaga, L., Uriarte, A., Prellezo, R., Andrés, M., Abaunza, P., Jardim, E., Lehuta, S., Pawlowski, L., & Roel, B. (2019). Challenges of management strategy evaluation for small pelagic fish: the Bay of Biscay anchovy case study. *Marine Ecology Progress Series*, 617-618, 245-263. 10.3354/meps12602

Sánchez, S., Uriarte, A., Andrés, M., & Ibaibarriaga, L. (2014). Developments of the MSE methodology for the Bay of Biscay anchovy. Final Report for the European Commission, Directorate General for Maritime Affairs and Fisheries. Presented to STECF EWG 14-03.

Sandberg, P., Bogstad, B., & Røttingen, I. (1998). Bioeconomic advice on TAC–the state of the art in the Norwegian fishery management. *Fisheries Research*, 37(1-3), 37, 259–274. 10.1016/S0165-7836(98)00142-8

Schruben, L., & Singham, D. (2010). *Simulating multivariate time series using flocking. Proceedings of the 2010 Winter Simulation Conference*10.1109/WSC.2010.5679085

Sigurdardottir, S., & Shruben, L.W. (2014). A new approach to simulating fisheries data for policy making. *Natural Resource Modeling,* 27(3), 411-428. 10.1111/nrm.12041

Silvestri, S., & Maynou, F. (2009). Application of a bioeconomic model for supporting the management process of the small pelagic fishery in the Veneto region, northern Adriatic Sea, Italy. *Scientia Marina*, 73(3), 563–572. 10.3989/scimar.2009.73n3563

Siple, M.C., Koehn, L.E., Johnson, K.F., Punt, A.E., Canales, T.M., Carpi, P., & de Moor, C.L. (2021). Considerations for management strategy evaluation for small pelagic fishes, *Fish and Fisheries*, 22, 1167–1186. 10.1111/faf.12579

Smith, A., Hurtado, F., Klaer, N., & Fulton, E. (2015). *Review and update of harvest strategy settings for the Commonwealth Small Pelagic Fishery Single species and ecosystem considerations.*

Smith, A. D., Smith, D. C., Haddon, M., Knuckey, I. A., Sainsbury, K. J., & Sloan, S. R. (2014). Implementing harveststrategies in Australia: 5 years on. *ICES Journal Marine Science Journal Conservation*, 71, 195–203.

Smith, M. D., Zhang, J., & Coleman, F. C. (2008). Econometric modeling of fisheries with complex life histories: Avoiding biological management failures. *Journal of Environmental Economics and Management*, 55(3), 265–280. 10.1016/j.jeem.2007.11.003

Sumaila, , Ebrahim, N., Schuhbauer, A., Skerritt, D., Li, Y., Kim, H. S., Mallory, T. G., Lam, V. W. L., & Pauly, D. (2019). Sumaila U.R., Ebrahim N., Schuhbauer A., Skerritt D., Li Y., Kim H.S., Mallory T.G., Lam V.W.L. and Pauly D. (2019). Updated estimates and analysis of global fisheries subsidies. *Marine Policy*, 109, 103695. 10.1016/j.marpol.2019.103695

Szoboszlai, A. I., Thayer, J. A., Wood, S. A., Sydeman, W. J., & Koehn, L. E. (2015). Forage species in predator diets: Synthesis of data from the California Current. *Ecological Informatics*, 29, 29, 45–56. 10.1016/j.ecoinf.2015.07.003

Tacon, A.G.J., & Metian, M. (2009). Global overview on the use of fish meal and fish oil in industrially compounded aquafeeds: Trends and future prospects. *Aquaculture*, 285, 146-158. 10.1016/j.aquaculture.2008.08.015

Torralba, J., & Besada, M. (2015). A stochastic model for the Iberoatlantic sardine fishery. Global warming and economic effects. *Ocean and Coastal Management*, 114, 114, 175–184. 10.1016/j.ocecoaman.2015.06.023

Uriarte, A., Ibaibarriaga, L., Sánchez-maroño, S., Abaunza, P., Andrés, M., Duhamel, E., Jardim, E., Pawlowski, L., Prellezo, R., & Roel, B.A. (2023). Lessons learnt on the management of short-lived fish from the Bay of Biscay anchovy case study: Satisfying fishery needs and sustainability under recruitment uncertainty. *Marine Policy*, 150, 105512. 10.1016/j.marpol.2023.105512

Uusitalo, L., Blenckner, T., Puntila-Dodd, R., Skyttä, A., Jernberg, S., Voss, R., Müller-Karullis, B., Tomczak, M.T., Möllmann, C., & Pertonen, H. (2022). Integrating diverse model results into decision support for good environmental status and blue growth. *Science of the Total Environment*, 806, 150450. 10.1016/j.scitotenv.2021.150450

Valcic, B. (2009). Spatial policy and the behavior of fishermen. *Marine Policy*, 33(2), 215-222. 10.1016/j.marpol.2008.06.001

Voss, R., Quaas, M.F., Stoeven, M., Schmidt, J.O., Tomczak, M., & Möllmann, C. (2017). Ecological-economic fisheries management advice – quantification of potential benefits for the case of the eastern Baltic cod fishery. *Frontiers in Marine Science*, 4, 209. 10.3389/fmars.2017.00209

Wise, L., Murta, A.G., Carvalho, J.P., & Mesquita, M. (2012). Qualitative modelling of fishermen's behaviour in a pelagic fishery. *Ecological Modelling*, 228, 112-122. 10.1016/j.ecolmodel.2011.12.008

Yatsu, A. (2019). Review of population dynamics and management of small pelagic fishes around the Japanese Archipelago. *Fisheries Science*, 85(4), 611–639. 10.1007/s12562-019-01305-3

# Chapter 6
# Enhancing Ecosystem Services:
## The Role of Artificial Reefs

**Ana Maria Madiedo**
https://orcid.org/0009-0009-5601-3779
*University of Algarve, Portugal*

**Jorge Ramos**
https://orcid.org/0000-0002-5115-7919
*CinTurs, University of Algarve, Portugal*

**Francisco Leitão**
https://orcid.org/0000-0003-4983-9782
*Center of Marine Science, University of Algarve, Portugal*

## ABSTRACT

*Artificial reefs (AR) are human made structures deployed in the seabed to provide multiple economic, social, and ecological benefits. AR have emerged as a highly effective strategy to provide ecosystem services, such as habitat protection and restoration. Despite their acknowledged benefits, there remains a limited understanding of how these reefs are currently being used by the local communities. Ecosystem services (EC) are tangible and intangible benefits that the ecosystems provide to humans. The present study aims to identify the EC provided by the AR located on the south coast of Portugal. To achieve this, a literature review was done to set a baseline and find out what has been published in this research area.*

DOI: 10.4018/979-8-3693-2436-3.ch006

## 1. INTRODUCTION

Over time, there has been a growing awareness of the importance of nature and how it can provide multiple benefits to human beings. Specifically, regarding marine ecosystems, humanity has been using these resources since the beginning of time, whether for obtaining food, shelter, engaging in recreational activities, pollution control, and others (Barbier, 2017). These multiple benefits had been designated "Ecosystem services" and currently their high value to humanity is widely recognized (Buonocore, 2021).

Due to the high demand for resources and intensity of activities done in the different types of marine ecosystems, it has been evidenced increasing environmental impacts such as habitat loss, changes in biodiversity, physicochemical alterations, and more (Barbier, 2017). This significant habitat degradation has led to the development of tools and new methodologies that allow to mitigate the impacts, restore and protect natural resources, and enhance their productivity. These efforts aim to ensure the continued provision of those named ecosystem services (Danovaro et al., 2021).

One example of a tool implemented are the artificial reefs, which are structures designed and deployed on the seabed with the aim to simulate the natural conditions of a natural reef and promote the increase of fish density as well as the presence of other marine organisms (Pickering, 1999). These types of structures had been the main subject of different research, revealing numerous uses and benefits that can be associated with the deployment of artificial reefs.

As a management measure, artificial reefs have been implemented in multiple coastal areas around the world. A specific example of this it the Algarve, located in the south coast of Portugal. This region is one of the main tourism destinations of the country and is well known for its beaches as well as the associated marine recreational activities, such as diving, angling, dolphin watching and others (Valle et al., 2011). Apart from tourism activities, the Algarve has some large fishing communities, specifically in Fuzeta, Olhão and Quarteira (OECD, n.d.) and the tradition related to fisheries had always been an important part of the economic activities in the region (Valadas -Monteiro, 2014).

*This chapter aims to present a direct relationship between the artificial reefs and how they can provide multiple ecosystem services. To achieve this, it was used a literature review, focusing on global studies that highlight the deployment of artificial reefs and their enhancement of various ecosystem services, including cultural, provisioning, and regulation and maintenance functions. Additionally, the study emphasizes the Algarve region due to the high number of artificial reefs that had been deployed off its coastal area. Other reason is the region's close relation to recreational and fishing activities (Santos and Monteiro, 2007; Valle et al., 2011).*

## 2. BACKGROUND

### 2.1. Artificial Reefs

#### 2.1.1. Evolution of Artificial Reefs

The concept of artificial reefs has changed and adapted over the years (Lee et al., 2018). The earlies record of artificial reefs dates back to 1600s, where it was evidence of stones used as anchors to create fishing ground for catching tuna in the Mediterranean Sea. During this period, it was also evidenced an increase in marine life in sunken ships and their surrounding areas, which made these sunken ships to been seen as artificial reef structures as well (Ramm et al., 2021; Fabi et al., 2011).

Another evidenced of the beginning of artificial reefs was found in Awaji, Japan. In this area, the fishing community perceived an increase in fish population and fishing ground in the surrounding areas of a sunken ship. After these perceived benefits, they began to adopt the concept of artificial reefs. Over time, more projects started to be developed to intentionally sink vessels and new technical designs were included in the deployed structures (Pickering, 1996). In 1953, a new project was launched to build and design concrete blocks that contribute to the attraction of specific fish species (Dean, 1983).

In 1880s the concept of artificial reef expanded to the United States. Initially, an artificial reef was perceived as any type of material of opportunity that was deployed on the seabed and could be seen as a potential element to attract fish (Lee et al., 2018). At the time, within the most common elements disposed on the seabed, specifically in sandy areas, included tires, cars and construction materials. However, over time, it was evidenced that these materials were often ineffective at attracting marine life. In many cases, the artificial reefs were severely eroded by water conditions or washed up in beaches as they were not properly secured and settled at the dumping sites (Pickering, 1996).

It was evidenced that the early artificial reefs deployed in the United States lacked structural design and integrity, resulting in slow colonization and, in some cases, minimum attraction of marine life (Feigenbaum et al., 1989). Further research in the topic continued to be done, and between 1954 and 1958, the first artificial reefs with technical specifications were designed and deployed in the Gulf of Mexico and along the southern coast of California (Lee et al., 2018).

These structures proved to be effective in attracting fish and serving as nurseries for young fish, which helped reduce expenses and increase catch rates. Consequently, the perceived benefits were primarily focused on the fishing communities (Makoto, 1985). As further research continued to be done, there were found new uses and benefits that could be perceived from the implementation of artificial reefs (Baine, 2001). New structures began to emerge, concrete and steel were the main materials used and technical specifications were applied to create artificial reefs with specific purposes. Within the new benefits perceived were ecosystem protection (Pickering, 1999; Harrys, 2009), enhancement of ecological conditions and others (Falcão et al., 2007; Marin-Diaz et al., 2021). Simultaneously, there has been evidence of artificial reefs being used for recreational activities worldwide, with intentional sinking of vessels to create specific locations for SCUBA diving (Ilieva et al., 2019; Ramm, 2021).

Nowadays, artificial reefs are defined as human-made structures that are deployed in the seabed aiming to mimic the characteristics of natural reefs (Baine, 2001), and are being used as a coastal management resource tool to mitigate the impacts produced by anthropic activities such as trawling, habitat destruction, and others (Carr et al., 1997).

### 2.1.2. Artificial Reefs in Europe

In Europe, between 1980 and 1990 the use of artificial reefs significantly increased (Monteiro and Santos, 2000). In comparison to the rest of the world, Europe had deployed 57% of purpose-build reefs, where most of them were deployed as a protection tool to inhibit trawling activities (Sánchez-Jerez and Ramos-Esplá 2000 ; Ramm, 2021). The Mediterranean Sea is the area of Europe with the highest concentration of artificial reefs, primary deployed as a fishery management tool. The OSPAR (i.e., Oslo and Paris Conventions) Maritime Area, is also a region with high concentration of artificial reefs, followed by the Baltic and Black seas, where multiple structures had been deployed as a conservation and restoration tool to reduce the impact of anthropic activities in the ecosystems (Fabi et al., 2011).

France was the pioneer regarding the implementation of artificial reefs in Europe. In 1968 were deployed automobile parts as a pilot program to study the behavior of these structures as a potential artificial reef. Later, in 1980s was developed the first project using concrete modules and were deployed offshore the French Mediterranean coast (Relini et al., 2008).

In Greece, the first project aimed at enhancing fisheries through artificial reefs began in 2000. Similar initiatives were observed in Italy and Spain during the 1970s, focusing on creating nursery areas for fish and serving as management tools to support small-scale fisheries. In contrast, Malta deployed its first ship in the 1980s to create an attractive area for recreational diving (Fabi et al., 2011).

In the OSPAR Maritime Area, the implementation of artificial reefs has been relatively recent. Denmark deployed the first artificial reefs between 2005 and 2008 as a restoration tool. The Netherlands established its first experimental reef in 1992 to establish a baseline for future implementations. In Norway the first artificial reef was deployed in 2002, as a study case to assess its impact on local fauna and flora. In the United Kingdom, artificial reefs have been used since 1984 for ecosystem research and to understand their effects on tourism activities (*Op. cit.*).

### 2.1.3 Artificial Reefs in Portugal

Specifically in Portugal, from 1990 onwards, were progressively deployed seven artificial reefs along the Algarve south coast (Figure 1), by the Institute of Marine Research (now IPMA). These were implemented as a management measure, taking into consideration the intensive fishing in the area, the presence of highly productive estuarine systems and the lack of natural reefs that provided refuge for the juveniles (Monteiro and Santos, 2000). One of the most important estuarine-lagoon systems is Ria Formosa, which works as a nursery for multiple fish species which migrate to the open coastal waters, to then be potentially caught by fishermen.

*Figure 1. The Algarve is southernmost region in Portugal mainland and offshore is where most of cubic artificial reef modules are located. Source: From Santos et al. (2011).*

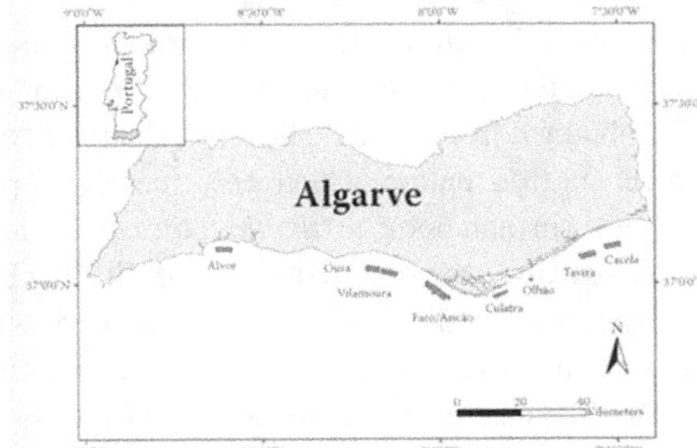


The first artificial reefs were deployed off Faro and off Olhão, which were made of concrete modules and with two distinct purposes. The ones allocated closer to the shore, where there is the most abundance of fish, were deployed to create shelter for fingerlings that migrate from Ria Formosa lagoon. The other was deployed further offshore into deeper waters, with the aim of attracting fish and their activity in the area (Santos and Monteiro, 1996). About 14 years after the deployment of these artificial reefs, it was found an increase of species availability and diversity, creating habitat to support diverse benthic and pelagic species. At the same time, the artificial reefs contributed to an increase in catch rates from 20 up to 40% (Santos and Monteiro, 2007).

In Nazaré, on the west coast of Portugal, in 2010 it was deployed an artificial reef to benefit the fishing activities, as well as to encourage the resilience of some marine organisms and to promote tourism as well as educational activities in the area. After its implementation, it was found that the concentration of phytoplankton was the same as the control site, which indicated the structures were being colonized and assimilated into the ecosystem. Moreover, there was an increase of macroinvertebrates, indicating that the artificial reefs contributed to enhance biological diversity. Concerning fishing communities, smaller vessels were the ones that used the area around the artificial reefs the most and obtained benefits from it. Prior to the reef's deployment, the most predominant species were sandy bottom related (e.g., flatfishes). However, after the reef deployment, there was an increase in species diversity, related to the refuge provided by the artificial reefs (Ramos et al., 2021).

## 2.2 Ecosystem Services

### 2.2.1 The Evolution of the Ecosystem Services Concept

The concept of ecosystem services has evolved over time and has experienced multiple classifications and adaptations before the formalization of the concept. Westman (1977), was one of the first authors to introduce the term "services" within the context of benefits obtained from nature. His research aimed to assign value to the benefits people received from nature, which were considered priceless.

Ehrlich & Ehrlich (1981) in their book "*Extinction: the causes and consequences of the disappearance of species*" were the first ones to define the concept of "ecosystem services". These authors emphasized the importance of different species and how they contribute to the adequate function of the ecosystems, as well as their role to support human life. The connection between these two factors gave origin to the term of ecosystem service.

Daily (1997), was another author who significantly contributed to the implementation and development of the term ecosystem services. In this book, the author defines ecosystem services as "(…) *the conditions and processes through which natural ecosystems, and the species that make them up, sustain and fulfill human life*". Additionally, the author characterizes how different ecosystems provide specific benefits to humanity and assess their respective value. Daily establishes a classification into three main functions: cleansing, recycling and renewal, and lists 13 specific services.

In late 1990s there were more studies involving world scientists that developed the application of the Ecosystem Services concept, sometimes raising some controversy (e.g., Costanza et al., 1997; Holmlund & Hammer, 1999; Duarte, 2000).

### 2.2.2 Current Adopted Terminology for Ecosystem Services

The term "Ecosystem Services" has come to a formalization and has been adopted for environmental policies. According to the Millenium Ecosystem Assessment (MEA, 2005), ecosystem services are direct and indirect benefits provided by the ecosystems to people. MEA considers four types of ecosystem services: provisioning, cultural, supporting and regulating.

- *Provisioning services* – are those benefits to people that can be extracted from nature, such as food, water, timber, gas, oil, plants and others.
- *Cultural services* – are nonmaterial benefits that contribute to spiritual enrichment, cognitive development, recreation and others.
- *Supporting services* – are the ones required for the adequate functioning of the other ecosystems services and bring long-term benefits. These included primary production, soil formation, nutrient cycling, provision of habitat and others.
- *Regulating services* – are those that moderate natural phenomena and help to make life possible for people, such as climate regulation, erosion control, water purification, storm protection and others.

In 2009 the European Environment Agency (EEA) developed the Common International Classification of Ecosystem Services (CICES) with the aim of creating a framework to classify and describe the ecosystem services, that could be used across different contexts of management and policymaking (EEA, 2009). According to the latest version (CICES 5.1), the ecosystem services were classified in three main categories: Provisioning, Regulation and Maintenance, and Cultural (**Table 1**). Additionally, it includes a hierarchy to better understand each of the services according to a section, division, group, class and class type (Haines-Young and Potschin, 2012)

*Table 1. Definition and Classification of Ecosystem Services According to CICES 5.1*

| Ecosystem Service | Definition |
|---|---|
| **Provisioning** | All nutritional, non-nutritional material and energetic outputs from living systems as well as abiotic outputs (including water). |

continued on following page

*Table 1. Continued*

| Ecosystem Service | Definition |
|---|---|
| **Regulation and Maintenance** | All the ways in which living organisms can mediate or moderate the environment that affects human health, safety or comfort, together with abiotic equivalents. |
| **Cultural** | All the non-material, and normally non-rival and non-consumptive, outputs of ecosystems (biotic and abiotic) that affect physical and mental states of people. |

Source: Adapted from Haines-Young & Potschin (2012).

## 3. ECOSYSTEM SERVICES AND ARTIFICIAL REEFS

Numerous studies and projects worldwide have explored the implementation of artificial reefs for various purposes and designs. The following section presents case studies and projects illustrating how artificial reefs have been used and their related benefits that can be associated to ecosystem services (Table 2).

*Table 2. Study Cases on Artificial Reefs and Ecosystems Services (Provisioning, Cultural, Regulation, and Maintenance)*

| Ecosystem service provided | Author | Year | Country | Artificial reef type |
|---|---|---|---|---|
| **Provisioning** | Polovina et al., | 1989 | Japan | Cylindrical concrete modules |
| | Ramos et al. | 2006a | Portugal | Concrete modules |
| | Santos et al. | 2011 | Portugal | Concrete modules |
| | Watanuki et al. | 2006 | Philippines and Senegal | Concrete modules |
| | Belhassen et al. | 2017 | Israel, | Intentional sunken vessels |
| | EI Gohary | 2013 | Mexico and Egypt | Underwater museum and Pharaonic Egyptian sculptures |
| | Fernando & Rathnayake | 2020 | Sri Lanka | Underwater museum |
| | Ilieva et al. | 2019 | Portugal | Intentional sunken vessels |
| **3 Regulation and maintenance** | Antsulevich | 1994 | The Neva Bay estuary (Gulf of Finland) | Culvert reefs, box culvert reefs and chaotic clusters made of concrete |
| | Falcão et al. | 2007 | Portugal | Concrete cubic modules |
| | Fan et al. | 2023 | China | Concrete modules |
| | Harris | 2009 | Florida, Thailand and the Caribbean | Reef balls |
| | Marin-Diaz et al. | 2021 | Dutch Wadden Sea | Biodegradable artificial reefs |
| | Zhu et al. | 2021 | China | Modules made of biomass fillers |

## 3.1 Provisioning

Related to the provisioning ecosystem services, artificial reefs had been implemented as a tool for fisheries development. This is because sometimes ARs can be deployed to create fishing grounds that contribute to the attraction of fish (Figure 2).

Artisanal fishing is subject to regulatory measures such as closed seasons and areas, restrictions on effort and gear, and minimum size limits, with the aim of safeguarding pressure on fishing resources and ensuring the livelihood of fishermen. However, these measures sometimes proved to be insufficient. The implementation of artificial reefs had also aimed to help in fisheries management, by limiting the access of only some gear in the reef area, functioning as a strategic tool to improve and benefit artisanal fishing. (Santos et al., 2011).

Ramos et al., (2006a) studied the usage of the artificial reef off Olhão. According to the results, it was evidenced that the surrounding areas of the reef were used mostly for commercial and recreational purposes, the first ones being the most predominant. These patterns could be related to an increase in catches around the area, translating into economic advantages for the fishing community.

To assess the efficacy of artificial reefs in enhancing local fisheries, Santos et al. (2011) studied the colonization rate and composition of fish assemblages. Four months after deploying the Faro/Ancão artificial reef, an increase in diversity and biomass was observed in the fish assemblage, with one-third of the biomass consisting of adult fish. This outcome highlights the positive contribution of artificial reefs, demonstrating their capacity to accelerate fish aggregation (with two to three times higher density). This highlights the efficiency of artificial reefs as a viable management tool, creating a potentially positive impact on local artisanal fisheries (if negative externalities created are negligible).

In the Philippines, artificial reefs were deployed to counteract the depletion of fisheries resources due to over-fishing. A similar case was studied in Senegal where there was some evidence of a decline in demersal fish resources. With assistance from Japan, it was developed a project with artificial reefs that were deployed at 20m offshore from the coast, using concrete blocks. After a year of its deployment, it was found an increase of group fish gathered around the reef and according to the testimony of the fishermen form the villages around, they found and increase in the fish catch as well as an increase in its diversity (Watanuki and Gonzales, 2006)

During the 1980s Japan spent the equivalent of about US$10 million per year to construct and deploy artificial reefs in benefit of the fisheries. It was found that over 50% of the fishermen in the East and West Shimamaki (Hokkaido, North Japan) used the reef regularly. At the same time, there was an increase in catching of flatfish, scorpionfish and octopus. About 30% of the population of fisherman recognized that they had perceived a benefit of catch increase since the implementation of artificial reefs in the area (Polovina and Sakai, 1989).

*Figure 2. Representation of changes in an ecosystem before the deployment of artificial reefs (AR), when the structures are settled and 5 years after their deployment. Source: First author own elaboration.*

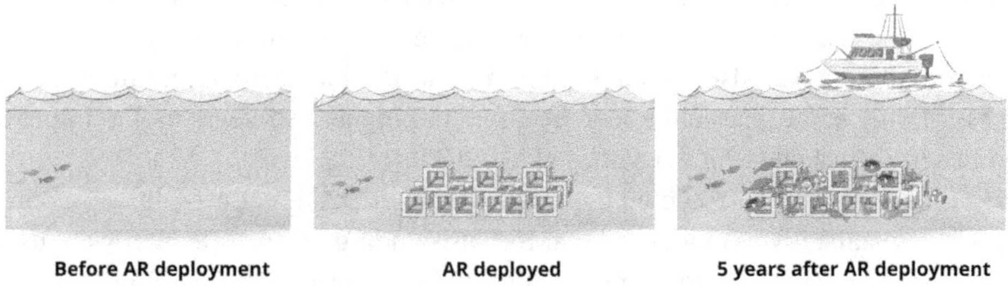


*Note. It has been evidenced that an artificial reef reaches maturity about 5 years after deployment (Harrison et al., 2020; Rallis et al., 2022)*

## 3.2 Cultural

Artificial reefs have proven to provide cultural ecosystem services related to recreational activities, as they contribute to the improvement of underwater landscape as an additional benefit that results from the increase of fish stocks. This had led to the encouragement of activities such as SCUBA diving and angling around the artificial reefs, as it is the case of the Great Barrier reef in Australia, or other reputed places to practice the activity such as Jamaica and Penghu (Chen et al., 2013). More recently Ilieva et al. (2019) described cultural services provided by intentionally submerged structures comprising cubic structures, four shipwrecks (two service vessels and two tanker warships).

A study presented by Ramos et al, in (2006b) made a comparative analysis of diver's preference for diving location. The study revealed that, on average, the sunken vessel situated off Faro ranked as the second most visited spot, following natural reefs. This reflects the role of artificial reefs in promoting ecotourism activities, as they attract divers and help to mitigate the potential mechanical damage that may occur in natural reefs due to diving activities.

Artificial coral reefs attract divers due to the increase in fish populations nearby, relieving pressure on natural reefs. In Israel, according to the study presented by Belhassen et al. (2017), the artificial reefs in Eilat exhibited higher diver density compared to the adjacent natural reefs. The research also highlighted that artificial reefs are considered more favorable for training purposes, which reflected their attractiveness to the diving community.

Among the various types of artificial reefs, artistic installations stand out, featuring sculptures crafted from eco-friendly materials that attract corals and foster the presence of diverse marine species. In Cancun, Mexico, there is an underwater museum named MUSA (Underwater Museum of Art) which contains around 400 sculptures (Figure 3). The deployment of these artworks serves to reduce pressure on natural reefs in the region, attracting divers and promoting ecotourism. A similar initiative is underway in Sharm El Sheik, Egypt, where a project is in progress to submerge Pharaonic Egyptian sculptures. The objective of this museum is to increase the tourism activities in the area as well as to benefit the overall ecological conditions of the ecosystem conditions (EI Gohary, 2013).

In Sri Lanka, an underwater museum has been established along the coastal area of Galle. This museum features a collection of statues designed and crafted by navy sailors using eco-friendly materials. The deployment of these statues serves a dual purpose: firstly, to boost diving tourism activities, and secondly, to foster the colonization of diverse marine species, thereby enhancing the overall marine life in the area. In addition to the museum, there are multiple shipwrecks that are considered cultural heritage for the local communities. This cultural heritage not only contributes to the attraction of tourists, but also inspires individuals to engage in dive trails (Fernando& Rathnayake, 2020).

*Figure 3. Sculptures of underwater museum in Cancun, Mexico. Source: Photo retrieved from Unsplash (Courtesy photo from Jamie Madden). Available from: https://unsplash.com/pt-br/fotografias/um-grande-grupo-de-corais-sob-a-agua-kkTmocR3M5U*

## 3.3 Regulation and Maintenance

Regarding the supporting ecosystem services, artificial reefs can be used as a rehabilitation tool to reduce the impact of anthropic activities. Between 2002 and 2004, Falcão et al., (2007) studied the influence of artificial reefs in the biogeochemical process in the south coast of Portugal. The study revealed that artificial reefs played an important role in enhancing sediment biomass, facilitating nutrient regeneration, and consequently, promoting increased productivity. It was verified that the high presence of several species of fish, combined with the organic matter they produced, was deposited in the sediments. It was later found that this material was reintegrated into the food chain, stimulating biological colonization.

These artificial reefs contribute essential nutrients, including nitrogen and phosphorus, which settle in the sediments. Nitrogen and phosphorus together, support the growth of phytoplankton, acting as primary producers. Artificial reefs promote biological production throughout the water column. The study observed high concentrations of nutrients (ammonium, nitrates, phosphate, and silicates) in the areas surrounding the base of the reefs, contributing to a greater presence of primary producers. Two years after deployment, the Faro-Ancão artificial reefs have shown an increase in nutrient cycling that positively impacted the entire trophic chain (*Op. cit.*).

The study conducted by Santos et al. (2011) revealed that artificial reefs strategically placed near the shore (as observed in the section of *artificial reefs in Portugal*) play a crucial role in safeguarding juvenile marine life. The deployment of these reefs led to a substantial increase in biomass colonization, with an astounding 88% of the population consisting of juveniles that directly relied on the accumulated biomass within the assemblages. Moreover, these artificial reefs served as protective shelters, providing a refuge for juveniles to escape predators and shielding them from the strong currents prevalent in the area. In terms of biomass enhancement, a notable increase in the density of *Dicentrarchus labrax* and *Diplodus sargus* was identified, constituting 75% of the total exploitable biomass.

In the Mediterranean, the *Posidonia oceanica* seagrass meadow has suffered significant damage from trawling activities. This seagrass holds significant importance, as it plays a vital role in the development of juvenile fish species that are commercially exploited, thus possessing both economic and ecological value. The removal of *Posidonia oceanica* also exacerbates erosion. To address this issue, there are some cases where artificial reefs have been employed as a physical barrier to deter trawlers from the area. This approach not only contributes to the reduction of erosion, but also safeguards the seagrass meadows which contributed to an increase in colonization (Pickering et al., 1999).

In a specific case in the southwest of Spain, prior to the implementation of artificial reefs, 48% of the seagrass was damaged, with a density of only 10 shoots/$m^2$. Post-implementation, the density increased to 60 shoots/$m^2$. A similar strategy was employed in India, where artificial reefs were deployed to stop trawling activities in locations such as Valiathra, Kerala, and Pohzihoor (*Op. cit.*).

Studies conducted in Florida, Thailand, and the Caribbean focused on investigating the erosion control ecosystem services provided by artificial reefs. In these regions, Reef Ball artificial reefs were strategically employed as restoration tools and coastal erosion protection measures, after a tsunami and hurricane event. Located offshore as breakwaters, these artificial reefs played a crucial role in stabilizing beaches by contributing to wave attenuation and wave refraction. Following the deployment of these artificial reefs, it was evidenced how they reduced the erosion of the areas of study. In the Caribbean, specifically in Gran Dominicus (Dominican Republic) there was clear evidence of accretion, leading to an increase in the shoreline by up to 13 meters in certain areas (Figure 4). Moreover, an additional 44.25 $m^3/m$ of sand was deposited without causing adverse effects on the adjacent beaches (Harris, 2009).

*Figure 4. Representation of beach erosion mitigation pre- and post-deployment of artificial reefs (upper and lower images, respectively). Source: First author own elaboration.*

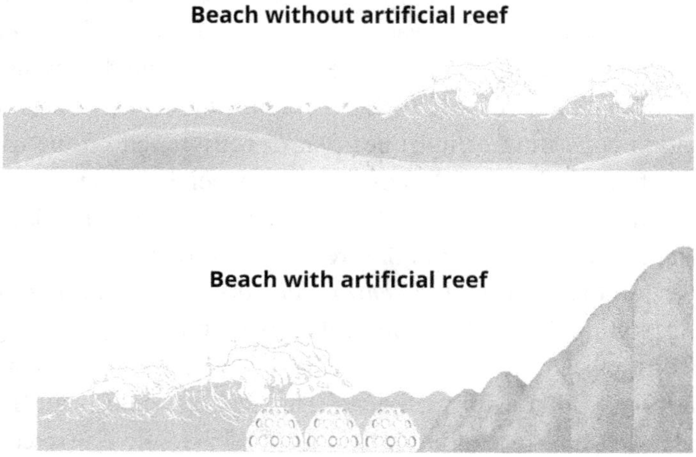


Another case study took place in the Dutch Wadden Sea, involving the installation of biodegradable artificial reefs aimed at enhancing accretion and stabilization within a tidal flat. About 3 years after its implementation, it was evidenced a reduction of 30% of the wave height of the waves coming into the area of study. Additionally, the artificial reefs played a significant role in promoting sand accretion in the targeted area (Marin-Diaz et al., 2021).

The Neva Bay estuary, an extension of the Gulf of Finland, faces consistent pressure due to ongoing industrial and domestic discharge. This continuous influx has resulted in various adverse effects, including the increase of water turbidity, diminished presence of bottom plants, and a decline in fauna and fish populations. Additionally, toxic substances have been introduced into the trophic chain. In response to these environmental impacts, a project was initiated to deploy artificial reefs aimed at fostering the growth and attraction of encrusting animals and plants with water-filtering capabilities. This filtration capacity was anticipated to contribute significantly to the enhancement of water quality (Antsulevich, 1994).

The increasing occurrences of green tides in the Bohai Sea, China, have significantly impacted tourism activities and the communities nearby. In response, a project was initiated to deploy artificial reefs as a proactive measure to control these events and mitigate their impact on the coastal area, as detailed by Fan et al. (2023). Another study conducted by Zhu et al. (2021) explored the effects of artificial reef deployment in Baiyangdian Lake, China, on phytoplankton growth. The research revealed that the strategic placement of artificial reefs created an environment that promoted nutrient transport, intersecting with currents and resulting in an increase of phytoplankton density (primary producers). Additionally, the reefs facilitated a more gradual release of nutrients from the sediments.

## 4. DISCUSSION AND FURTHER WORK

As it had been described through this chapter, artificial reefs offer multiple benefits that have significantly enhanced ecosystem services in benefit of human beings. However, further research is needed to understand the socio-economic and environmental impacts of these structures over long periods of time. This will allow to ensure that the structures implemented can be sustainable and effective over time, so its possible to assess and ensure that they don't cause any harm to marine ecosystems.

Multiple technologies and methodologies have been developed to ensure that the reefs deployed are built with adequate materials that resist constant exposure to seawater and other marine conditions. However, given the numerous unsuccessful reefs made of materials of opportunity, its important to conduct research to better understand the current conditions and impacts produced by those reefs (e.g., Blount et al., 2021). By doing this, it will be possible to develop and implement measures to improve their effectiveness, potentially making materials of opportunity, which were dumped years ago, viable for creating successful artificial reefs .

More studies should be done on the ecosystem services provided by the artificial reefs, incorporating a socio-ecological perspective. This approach will help to better understand how these structures have contributed to local communities

Regulatory frameworks should continue to be developed to provide a cohesive approach to the development, deployment, management and maintenance of artificial reefs. By doing this, artificial reefs could be seen as an accessible tool for multiple communities worldwide, potentially reducing the dumping of materials of opportunity and minimizing the implementation of unsuccessful artificial reefs. This framework should incorporate elements that promote sustainable practices in the design and construction of the artificial reefs, thereby mitigating negative environmental impacts. Furthermore, the framework should consider findings from cross-disciplinary research encompassing ecological and socio-economic elements, to ensure successful implementation and optimize reefs functionality (Bonnevie et al., 2019).

Considering the changing conditions of the marine ecosystems due to climate change, it is necessary to integrate new technologies that allow to enhance artificial reef resilience. Future designs should focus on reducing impacts produce by climate change and allowing better adaptation to changing environmental conditions. At the same time, it's important to continue exploring ways in which artificial reefs can contribute to recovery and protection of endangered species and habits. This will ensure that artificial reefs work as a tool that supports complex marine ecosystems.

In addition to the multiple ecological benefits provided by the artificial reefs, they provide significant economic benefits for local fisheries and tourism activities. In the future, the design and deployment of new artificial reefs should integrate features that not only benefit the ecological conditions but also include elements that make them appealing to tourists and suitable for recreational activities. These economic benefits will contribute to the development of coastal communities

Apart from the acknowledged benefits of artificial reefs, its necessary consider further implications of artificial reefs deployment. Potential adverse impacts caused to ecosystems and communities, should be taken into account to ensure responsible intervention of marine ecosystems. By doing this, it will be possible to mitigate possible negative impacts and promote sustainable development.

## 5. CONCLUSION

The deployment of artificial reefs has proven to be an effective measure that generates positive impacts from both ecological and socio-economic perspectives. These acknowledge benefits contribute to the protection and enhancement of the ecosystem services provided by these structures. Therefore, is necessary to continue to integrate new technologies to continue developing artificial reefs structures, that are able to adapt to various conditions and enhance ecosystem resilience. By extending the implementation of artificial reefs, it will be possible to continue working in benefit of sustainable management of marine ecosystems while supporting human needs. Some of those benefits and contributions are:

- The evidenced benefits provided by the artificial reefs to fishing communities reflects their importance as a tool that contributes to economic growth

- Artificial reefs provide multiple ecosystem services that benefit humans while preserving and enhancing the ecological conditions of the marine ecosystem
- Artificial reefs are a key element to sustainable development, that contribute to the preservation of natural resources and overall health of marine environment
- Beyond their ecological benefits, artificial reefs present a unique opportunity to combine art and other cultural aspects with methods to enhance and protect ecosystems. This can allow to increase public awareness and engagement as it's seen as more visual and cultural appealing.

## Funding

This work is funded by Portuguese national funds through FCT—Foundation for Science and Technology within the scope of the projects UIDB/04020/2020 (https://doi.org/10.54499/UIDB/04020/2020), UIDP/04020/2020 (https://doi.org/10.54499/UIDP/04020/2020), UIDB/04326/2020, UIDP/04326/2020, LA/P/0101/2020 and 2022.01783.PTDC and the Individual Calls to Scientific Employment Stimulus 2021.02180.CEECIND (https://doi.org/10.54499/2021.02180.CEECIND/CP1672/CT0003) (JR) and 2022.04803.CEECIND (FL).

## ACKNOWLEDGMENT

The authors would like to thank two anonymous reviewers whose comments substantially helped to improve the manuscript.


## REFERENCES


Antsulevich, A. E. (1994). Artificial reefs project for improvement of water quality and environmental enhancement of Neva Bay (St.-Petersburg County Region). *Bulletin of Marine Science*, 55(2-3), 1189–1192.

Baine, M. (2001). Artificial reefs: A review of their design, application, management and performance. *Ocean and Coastal Management*, 44(3-4), 241–259. 10.1016/S0964-5691(01)00048-5

Barbier, E. B. (2017). Marine ecosystem services. *Current Biology*, 27(11), R507–R510. 10.1016/j.cub.2017.03.02028586688

Belhassen, Y., Rousseau, M., Tynyakov, J., & Shashar, N. (2017). Evaluating the attractiveness and effectiveness of artificial coral reefs as a recreational ecosystem service. *Journal of Environmental Management*, 203, 448–456. 10.1016/j.jenvman.2017.08.02028837911

Blount, C., Komyakova, V., Barnes, L., Smith, M. L., Zhang, D., Reeds, K., McPhee, D., Taylor, M. D., Macbeth, W., & Needham, E. (2021). Using ecological evidence to refine approaches to deploying offshore artificial reefs for recreational fisheries. *Bulletin of Marine Science*, 97(4), 665–698. 10.5343/bms.2020.0059

Bonnevie, I. M., Hansen, H. S., & Schrøder, L. (2019). Assessing use-use interactions at sea: A theoretical framework for spatial decision support tools facilitating co-location in maritime spatial planning. *Marine Policy*, 106, 103533. 10.1016/j.marpol.2019.103533

Buonocore, E., Grande, U., Franzese, P. P., & Russo, G. F. (2021). Trends and evolution in the concept of marine ecosystem services: An overview. *Water (Basel)*, 13(15), 2060. 10.3390/w13152060

Carr, M. H., & Hixon, M. A. (1997). Artificial reefs: The importance of comparisons with natural reefs. *Fisheries (Bethesda, Md.)*, 22(4), 28–33. 10.1577/1548-8446(1997)022<0028:ARTIOC>2.0.CO;2

Chen, J. L., Chuang, C. T., Jan, R. Q., Liu, L. C., & Jan, M. S. (2013). Recreational benefits of ecosystem services on and around artificial reefs: A case study in Penghu, Taiwan. *Ocean and Coastal Management*, 85, 58–64. 10.1016/j.ocecoaman.2013.09.005

Chen, J. L., Hsu, K., & Chuang, C. T. (2020). How do fishery resources enhance the development of coastal fishing communities: Lessons learned from a community-based sea farming project in Taiwan. *Ocean and Coastal Management*, 184, 105015. 10.1016/j.ocecoaman.2019.105015

Costanza, R., d'Arge, R., De Groot, R., Farber, S., Grasso, M., Hannon, B., Limburg, K., Naeem, S., O'Neill, R. V., Paruelo, J., Raskin, R. G., Sutton, P., & Van Den Belt, M. (1997). The value of the world's ecosystem services and natural capital. *Nature*, 387(6630), 253–260. 10.1038/387253a0

Daily, G. C. (Ed.). (1997). *Nature's Services: Societal Dependence on Natural Ecosystems*. Island Press.

Danovaro, R., Aronson, J., Cimino, R., Gambi, C., Snelgrove, P. V., & Van Dover, C. (2021). Marine ecosystem restoration in a changing ocean. *Restoration Ecology*, 29(S2), e13432. 10.1111/rec.13432

Dean, L. (1983, May). Undersea oases made by man: Artificial reefs create new fishing grounds. *Oceans*, 26(May–June), 27–29.

Duarte, C. M. (2000). Marine biodiversity and ecosystem services: An elusive link. *Journal of Experimental Marine Biology and Ecology*, 250(1-2), 117–131. 10.1016/S0022-0981(00)00194-510969166

EEA. (2009). European Environment Agency. Common international classification of ecosystem services (CICES). European Environment Agency. https://cices.eu/

Ehrlich, P. R., & Ehrlich, A. H. (1981). *Extinction: the causes and consequences of the disappearance of species/Paul and Anne Ehrlich* (1st ed.). Random House.

El Gohary, G. (2013). Wet landscape: Using sculptures to form underwater landscape. *Democratic Transition and Sustainable Communities*, 6, 351.

Fabi, G., Spagnolo, A., Bellan-Santini, D., Charbonnel, E., Çiçek, B. A., García, J. J. G., Jensen, A. C., Kallianiotis, A., & Santos, M. N. D. (2011). Overview on artificial reefs in Europe. *Brazilian Journal of Oceanography*, 59(spe1), 155–166. 10.1590/S1679-87592011000500017

Falcão, M., Santos, M. N., Vicente, M., & Monteiro, C. C. (2007). Biogeochemical processes and nutrient cycling within an artificial reef off Southern Portugal. *Marine Environmental Research*, 63(5), 429–444. 10.1016/j.marenvres.2006.12.00117239434

Fan, J., Kuang, C., Liu, H., Wang, D., Liu, J., Wang, G., & Zou, Q. (2023). Impact of Artificial Islands and Reefs on Water Quality in Jinmeng Bay, China. *Water (Basel)*, 15(5), 959. 10.3390/w15050959

Feigenbaum, D., Bushing, M., Woodward, J., & Friedlander, A. (1989). Artificial reefs in Chesapeake Bay and nearby coastal waters. *Bulletin of Marine Science*, 44(2), 734–742.

Fernando, R., & Rathnayake, S. (2020). Potentials of Underwater Cultural Heritage Towards Tourism in Galle, Sri Lanka: Perspectives of Tourism Service Providers and Maritime Archaeologists.

Haines-Young, R., & Potschin, M. (2012). *Common International classification of ecosystem services*. Centre for Environmental Management, University of Nottingham.

Harris, L. E. (2009). Artificial reefs for ecosystem restoration and coastal erosion protection with aquaculture and recreational amenities. *The R Journal*, 1(1), 235–246.

Harrison, S., & Rousseau, M. (2020). Comparison of artificial and natural reef productivity in Nantucket Sound, MA, USA. *Estuaries and Coasts*, 43(8), 2092–2105. 10.1007/s12237-020-00749-6

Holmlund, C. M., & Hammer, M. (1999). Ecosystem services generated by fish populations. *Ecological Economics*, 29(2), 253–268. 10.1016/S0921-8009(99)00015-4

Ilieva, I., Jouvet, L., Seidelin, L., Best, B. D., Aldabet, S., da Silva, R., & Conde, D. A. (2019). A global database of intentionally deployed wrecks to serve as artificial reefs. *Data in Brief*, 23, 103584. 10.1016/j.dib.2018.12.02330809558

INE. (2015). Anuário Estatístico Regional do Algarve -INE. Available at: https://www.cm-albufeira.pt/sites/default/files/public/RepositorioDocumentos/plano_alinhamento_intermunicipal_algarve2020_0.pdf

Lee, M. O., Otake, S., & Kim, J. K. (2018). Transition of artificial reefs (ARs) research and its prospects. *Ocean and Coastal Management*, 154, 55–65. 10.1016/j.ocecoaman.2018.01.010

Lima, J. S., & Zalmon, I. R. (2022). Impacts of Artificial Reefs on Artisanal Fishers: Southeastern Brazil. In *Impact of Artificial Reefs on the Environment and Communities* (pp. 87–109). IGI Global. 10.4018/978-1-6684-2344-8.ch005

Marin-Diaz, B., Fivash, G. S., Nauta, J., Temmink, R. J., Hijner, N., Reijers, V. C., Cruijsen, P. P. M. J. M., Didderen, K., Heusinkveld, J. H. T., Penning, E., Maldonado-Garcia, G., van Belzen, J., de Smit, J. C., Christianen, M. J. A., van der Heide, T., van der Wal, D., Olff, H., Bouma, T. J., & Govers, L. L. (2021). On the use of large-scale biodegradable artificial reefs for intertidal foreshore stabilization. *Ecological Engineering*, 170, 106354. 10.1016/j.ecoleng.2021.106354

Millennium Ecosystem Assessment, M. E. A. (2005). *Ecosystems and Human Wellbeing: Biodiversity Synthesis*. Island Press.

Monteiro, C. C., & Santos, M. N. (2000). Portuguese artificial reefs. In *Artificial Reefs in European Seas* (pp. 249–261). Springer Netherlands. 10.1007/978-94-011-4215-1_15

Nakamura, M. (1985). Evolution of artificial fishing reef concepts in Japan. *Bulletin of Marine Science*, 37(1), 271–278.

OECD. (n.d.). Country note on national fisheries management systems – Portugal. Available at: https://www.oecd.org/portugal/34431028.pdf

Pickering, H. (1996). Artificial reefs of bulk waste materials: A scientific and legal review of the suitability of using the cement stabilised by-products of coal-fired power stations. *Marine Policy*, 20(6), 483–497. 10.1016/S0308-597X(96)00036-X

Pickering, H., Whitmarsh, D., & Jensen, A. (1999). Artificial reefs as a tool to aid rehabilitation of coastal ecosystems: Investigating the potential. *Marine Pollution Bulletin*, 37(8-12), 505–514. 10.1016/S0025-326X(98)00121-0

Polovina, J. J., & Sakai, I. (1989). Impacts of artificial reefs on fishery production in Shimamaki, Japan. *Bulletin of Marine Science*, 44(2), 997–1003.

Rallis, I., Chatzigeorgiou, G., Florido, M., Sedano, F., Procopiou, A., Chertz-Bynichaki, M., Vernadou, E., Plaiti, W., Koulouri, P., Dounas, C., Gerovasileiou, V., & Dailianis, T. (2022). Early succession patterns of benthic assemblages on artificial reefs in the oligotrophic eastern mediterranean basin. *Journal of Marine Science and Engineering*, 10(5), 620. 10.3390/jmse10050620

Ramm, L. A., Florisson, J. H., Watts, S. L., Becker, A., & Tweedley, J. R. (2021). Artificial reefs in the Anthropocene: A review of geographical and historical trends in their design, purpose, and monitoring. *Bulletin of Marine Science*, 97(4), 699–728. 10.5343/bms.2020.0046

Ramos, J., Santos, M. N., Whitmarsh, D., & Monteiro, C. C. (2006a). Patterns of use in an artificial reef system: A case study in Portugal. *Bulletin of Marine Science*, 78(1), 203–211.

Ramos, J., Santos, M. N., Whitmarsh, D., & Monteiro, C. C. (2006b). The usefulness of the analytic hierarchy process for understanding reef diving choices: A case study. *Bulletin of Marine Science*, 78(1), 213–219.

Ramos, J., Tuaty-Guerra, M., Almeida, M., Raposo, A. C., Gaudêncio, M. J., Silva, A. D., Rodrigues, N., Leandro, S. M., & Caetano, M. (2021). An artificial reef at the edge of the deep: An interdisciplinary case study. *Ocean and Coastal Management*, 210, 105729. 10.1016/j.ocecoaman.2021.105729

Relini, G., Fabi, G., dos Santos, M. N., Moreno, I., & Charbonnel, E. (2008). Fisheries and their management using artificial reefs in the northwestern Mediterranean Sea and southern Portugal. *American Fisheries Society Symposium*, 49(1), 891.

Sánchez-Jerez, P., & Ramos-Esplá, A. (2000). Changes in fish assemblages associated with the deployment of an antitrawling reef in seagrass meadows. *Transactions of the American Fisheries Society*, 129(5), 1150–1159. 10.1577/1548-8659(2000)129<1150:CIFAAW>2.0.CO;2

Santos, M. N., & Monteiro, C. C. (1997). The Olhao artificial reef system (south Portugal): Fish assemblages and fishing yield. *Fisheries Research*, 30(1-2), 33–41. 10.1016/S0165-7836(96)00567-X

Santos, M. N., & Monteiro, C. C. (2007). A fourteen-year overview of the fish assemblages and yield of the two oldest Algarve artificial reefs (southern Portugal). In *Biodiversity in Enclosed Seas and Artificial Marine Habitats:Proceedings of the 39th European Marine Biology Symposium, held in Genoa,Italy,21–24 July 2004* (pp. 225-231). Springer Netherlands.

Santos, M. N., Monteiro, C. C., & Leitão, F. (2011). The role of artificial reefs in the sustainability of artisanal fisheries. *Artificial reefs in Fisheries Management, S. Bortone, FP Brandini, G. Fabi and S. Otake, CRC Press, Taylor & Francis Group*, 221-237.

Sutton, S. G., & Bushnell, S. L. (2007). Socio-economic aspects of artificial reefs: Considerations for the Great Barrier Reef Marine Park. *Ocean and Coastal Management*, 50(10), 829–846. 10.1016/j.ocecoaman.2007.01.003

Valadas-Monteiro, P. (2014). General insights of the Portuguese maritime economy and particularly of the Algarve region: Contributing towards a strategic vision. *Journal for Maritime Research*, 11(1), 3–9.

Valle, P. O. D., Guerreiro, M., Mendes, J., & Silva, J. A. (2011). The cultural offer as a tourist product in coastal destinations: The Case of Algarve, Portugal. *Tourism and Hospitality Research*, 11(4), 233–247. 10.1177/1467358411420623

Watanuki, N., & Gonzales, B. J. (2006). The potential of artificial reefs as fisheries management tools in developing countries. *Bulletin of Marine Science*, 78(1), 9–19.

Westman, W. E. (1977). How Much Are Nature's Services Worth? Measuring the social benefits of ecosystem functioning is both controversial and illuminating. *Science*, 197(4307), 960–964. 10.1126/science.197.4307.96017784115

Zhu, H., Liu, X., Cheng, S., & Wang, J. (2021). Effects of Artificial Reefs on Phytoplankton Community Structure in Baiyangdian Lake, China. *Water (Basel)*, 13(13), 1802. 10.3390/w13131802

## KEY TERMS AND DEFINITIONS

**Angling:** It is a recreational activity where the enthusiast uses a rod and reel to fish with line and hooks (and possibly live or fake bait). It can be practiced alone or in a group and can be done on foot or on a recreational boat.

**Artificial Reef:** Structures designed with specific characteristics that contribute to the improvement of marine ecosystems

**Cultural Heritage:** Legacy from the past that can include physical (e.g., books, statues, poems) or non-physical elements (e.g., histories, customs, beliefs, knowledge), that are significant for a community or group of people

**Ecosystem Services:** Services provided to humans by nature, enhancing both the life quality and overall, well-being

**Ecotourism:** Tourism activities done in natural areas, that integrate recreational activities and at the same time contribute to the conservation of the ecosystems

**Environmental Impacts:** Changes produced to the ecosystems that produce negative effects as result of anthropic activities

**Erosion:** Process in which sediments are moved and transported away by natural forces, such as waves and wind

**Juvenile Fishes:** Young fish that is still is the process of development and haven't reach complete maturity to reproduce

**SCUBA Diving:** Type of underwater diving where oxygen tanks are used along with a regulator and other specialized equipment to breath underwater.

# Section 3
# Case Studies

# Chapter 7
# Science Communication of Coastal and Marine Issues:
## Insights From Diverse Case Studies

**Ana Matias**
https://orcid.org/0000-0003-2090-6291
*Centro de Investigação Marinha e Ambiental, Portugal*

**Bruno Pinto**
*University of Lisbon, Portugal*

**A. Rita Carrasco**
https://orcid.org/0000-0002-8980-0068
*Centro de Investigação Marinha e Ambiental, Portugal*

## ABSTRACT

*Marine and coastal science communication plays a major role in fostering community resilience and motivating collective actions for ocean sustainability, especially within initiatives such as the Ocean Decade and the UN Sustainable Development Goals. Despite growing interest and communication efforts, challenges for scientists remain. This chapter aims to present a portfolio of communication activities and critically reflect on their advantages and disadvantages. It outlines 10 Portuguese communication cases, covering beach/dune and salt marsh dynamics, research methods, aquaculture, environmental sustainability, climate change impacts, and pollution. An integrated analysis evaluates key messages, required resources, and target audiences, including a SWOT (strengths, weaknesses, opportunities, threats)*

DOI: 10.4018/979-8-3693-2436-3.ch007

*analysis and comparisons of budget, team size, and audience reach. Despite being limited to 10 specific case studies, the chapter offers a broad spectrum of communication activities, providing insights and a catalogue for marine and coastal science communication.*

## INTRODUCTION

The Earth's ocean is a fundamental planet component, encompassing 70% of its surface and playing a critical role in regulating the climate, supplying food, and sustaining diverse ecosystems (Halpern et al., 2019). The capacity to understand and predict marine and coastal threats, and effectively communicate them to society is important to implement measures of mitigation and adaptation to marine and coastal risks and thus increase the resilience of communities to global and local changes. Indeed, keeping citizens informed about the sea is important because it can motivate collective actions to address threats to coastal and marine sustainability (Matias et al., 2023). Marine issues and climate change are intertwined in various ways, as changes in the ocean can be both cause and consequence of changes related to climate change, and both topics have close relations to human activities.

Science communication brings different audiences together; it empowers individuals with knowledge, drives policy change, and inspires conservation efforts (Grorud-Colvert et al., 2010). Thus, communication on marine and environmental issues tends to be maximised within the frameworks of the Ocean Decade and the EU Mission: Restore our Ocean and waters by 2030, representing distinctive avenues for advancing this type of communication (Guan et al., 2023; Seys et al., 2022). For example, building upon the successes of science communication in addressing marine issues, governance tools have emerged as effective bridges to strengthen connections between science and end-users. These tools are central to the objectives of the Ocean Decade (Guan et al., 2023). Moreover, by enhancing understanding and raising awareness about the significance of ocean science, marine science communication plays a crucial role in achieving Goal 14 of the United Nations (UN) Sustainable Development Goals, which aims to conserve and sustainably use oceans, seas, and marine resources for sustainable development. Future conflicts arising from the use of marine resources versus marine protection will necessitate solutions that experts can explain in ways that non-specialists can easily understand.

In Portugal, there has been a renewed interest in the marine environment since the mid-1990s, and several communication efforts were developed after that. Examples include the World Exhibition Expo 98 about the oceans which implied the creation of the Lisbon Oceanarium, and the proposal for an extension of the Portuguese continental shelf submitted in 2009 to the U.N. or the establishment of the Ministry of the Sea. Despite this, research on marine science communication in Portugal is still an emerging issue (e.g., Carapuço et al., 2017; Pinto & Costa, 2018). Like in other countries, Portuguese science communication still lacks resources, recognition and professionalisation, despite its rapid growth since the mid-1990s. Moreover, public participation in research and policy in activities such as citizen science projects is growing at the national level and is viewed as an important way to further engage audiences with science (Entradas et al., 2020). Most difficulties arise from a lack of time for researchers to undertake science communication activities and from a lack of formal training in science communication (Matias et al., 2024).

From the researchers' perspective, it is noticeable that although societal impact is increasingly regarded as a key indicator of scientific quality among science policy stakeholders (Fecher et al., 2021), there is a trend within science communication research to overlook applied research in favour of basic research (Fischer et al., 2024). Furthermore, the lack of standardised methods or the adoption of less effective marine science communication approaches can result in diminished public understanding and awareness regarding ocean topics. This issue can be particularly problematic in a science-policy context where the use of the best available scientific information is mandated or binding (e.g., fisheries (Steins et al., 2023) or marine energy (Gunn et al., 2022)).

Hence, the primary objective of this chapter is to showcase a portfolio of coastal and marine science communication initiatives, serving as a catalogue of typologies for the exploration of these topics. Communication on coastal and marine issues, framed within the wider field of science communication, is examined through the analysis of ten Portuguese case studies in terms of objectives, target audiences, resources, and budgets. Additionally, valuable insights and lessons learned are derived from the comparative analysis and SWOT (strengths, weaknesses, opportunities, and threats) analysis conducted.

## FRAMING COASTAL AND MARINE SCIENCE COMMUNICATION

Science communication can be defined in various ways. According to Dijkstra et al. (2020) it describes the many ways in which the processes, outcomes, and implications of the sciences – broadly defined – can be shared or discussed with audiences. Science communication involves interaction, with the goal of interpreting scientific or technical developments or discussing issues with a scientific or technical dimension (Dijkstra et al., 2020). Currently, there are several books and reports that provide guidance, describe good practices, and offer practical examples of how scientists can engage in science communication (e.g., Bennett & Jennings,

2011; Bowater & Yeoman, 2013; Boxman & Boxman, 2017; Illingworth & Allen, 2020; Laszlo, 2006).

Different models of science communication have been followed to suit various contexts, purposes, and audiences. Historically, science communication was described as a process of information transmission, based on the assumption of "public deficiency, but scientific sufficiency", known as the 'Deficit Model'. This model asserts that science communication is guided by a perceived need for science literacy, where the public is viewed as lacking scientific knowledge until it is conveyed by the communicator, attributing public misunderstanding of science to this knowledge deficit (e.g., Nisbet & Scheufele, 2009). It is characterised as a one-way, top-down communication process.

In contrast, the 'Dialogue Model' of science communication draws on the belief that more active, open and democratic relations between science and citizens are both desirable and necessary (Irwin, 2008). Nerghes et al. (2022) summarized the three main features of the dialogue model: (1) engagement in a dialogue with the public to help explain the science, (2) listening to and consulting the public about their perceptions, concerns and needs regarding science, and (3) acknowledging that the public may have useful knowledge that can help scientific progress and policy-making. Under this model, science communication is centred around dialogic approaches that aim for a two-way conversation, allowing an exchange of perspectives between scientists and the public (Nerghes et al., 2022).

These two practices of science communication - one-way and two-way - are not mutually exclusive. In modern societies, it is natural for traditional unidirectional approaches through media such as television, radio, or newspapers to coexist and be complemented by more recent bidirectional (or dialogue) approaches, including citizen science projects and other forms of public participation. This combination has the potential to reach more people within society and hopefully promote greater engagement with science.

Effective communication of coastal and marine science by scientists follows the same general guidelines and provides evidence for best practices as described in these publications. Communicating science about coastal and marine issues can take different forms, similar to other scientific areas. One of the traditional communication formats is the press, although research indicates modest media coverage of such themes. Nonetheless, press journalists reporting on coastal and marine issues from 13 European countries consider climate change, marine pollution, and biodiversity to be the three most important topics (Pinto & Matias, 2023a, 2023b). These journalists regard compelling marine stories as those that address topics affecting people's lives (e.g., human health, coastal erosion, energy projects) and those relating to nature protection and conservation (e.g., endangered species, marine parks, pollution, accidents leading to animal fatalities). Additionally, these stories usually pique curiosity, bring fresh intelligible scientific news, or unveil the mysterious world of the deep sea.

Coastal and marine issues can also be communicated through art or using artistic techniques. In a recent literature review about the communication of marine and coastal sustainability, it was found that visual arts (57%) are the most common, followed by performing arts (33%), media arts (28%), and literary arts (20%) (Matias et al., 2023). Examples of these artistic pieces include sculptures (e.g., 'Bottled Ocean 2,117', Boswell, 2021), performances (e.g., 'Beach-dancing Day' Olsen, 2018), theatre plays (e.g. 'The Orphan Sea'; Brown, 2016), films (e.g., 'A Calling, Deeply' Frangovska, 2020), poems (e.g., 'Up/down/fragile', Nesci & Valentini, 2020), murals (e.g., 'The guardians of water', Sanchez et al., 2020). Many were done in North America, Europe and Oceania, and the most common topics were climate change, pollution and biodiversity loss. Their main objective was engagement with the public, but others such as marine conservation and restoration, management, education, and activism were also mentioned. Additionally, these authors argue that more research and new ways to assess impact are needed to better understand the effect of art on coastal and marine sustainability.

Another way to foster dialogue between science and society is through citizen science initiatives. A famous global citizen science project related to the coast is "CoastSnap" (coastsnap.com), which relies on repeating photos at the same location to track how the coast is changing over time, described in detail in Harley & Kinsela (2022). A more recent example of citizen science is the "Plankton Planet" (planktonplanet.org), which aims to mobilise citizen sailors, known as 'Planktonauts,' to provide continuous samples of plankton from all over the world to biological oceanographers (de Vargas et al., 2022).

A study by Matias et al. (2024) found that a high percentage of coastal experts engage in science communication. This professional group of researchers (coastal geoscientists and engineers) prefer direct interaction with audiences (e.g., lectures) over indirect methods (e.g., media). Their communication efforts mainly focus on coastal risk (75%) and climate change impacts (69%). The preferred audiences are the public sector (e.g., municipalities) and school students (Matias et al., 2024).

The International Marine Science Communication Conferences ("CommOcean") are another significant initiative in marine science communication. Held biennially since 2014, with the sixth edition in 2024, these conferences are promoted by the European Marine Board Communications Panel (EMBCP) in partnership with several research institutes and science communication organisations. The aim is to focus "on tools and practices for scientists and marine science communicators to develop their skills and improve dialogue with a range of target audiences including the general public and decision-makers." Themes of presentations and workshops include citizen science, science and art, engagement of local communities, ocean literacy, etc.

Regarding the interest of audiences in marine and coastal issues, a study concluded that concern about the ocean in Europe between 2010-2011 was generally lower than concern for other environmental problems and there was a schism between scientific and public perspectives on the key pressures that affect marine systems (Potts et al., 2016). They also argued that the proximity of a community to the coastal environment did not significantly change public perceptions about the sea. Another study on marine issues in ten European countries, undertaken in early 2011 and involving more than 10,000 respondents found that overfishing and marine pollution were the top priorities for marine policy (Gelcich et al., 2014). Furthermore, a Eurobarometer study conducted in 2017, which included all EU countries, found that marine pollution was the 6th most important environmental issue (mentioned by 33% of Europeans; European Commission (2017)). Nevertheless, climate change was considered the most important issue (51%) and the growing amount of waste was the third most important issue (40%), both of which include aspects related to marine threats. This interest was reinforced by a 2021 Eurobarometer study (European Commission Directorate-General for Communication, 2021), which confirmed that Europeans are highly concerned about environmental issues, with climate change being considered the most important topic in science and technology (89% of respondents).

## METHODOLOGICAL APPROACH

Ten case studies of communication about the sea, coastal areas, and the environment were selected. They were chosen to include a variety of approaches, ranging from one-way communication, such as animations and documentaries, to bidirectional interaction, such as educational hands-on kits, and coastal field visits. The topics of the described communication activities are diverse, covering aquaculture, dune dynamics, fish species, instrumentation for measuring the ocean and the impacts of sea-level rise. These recent examples are primarily centred on Portugal and are mostly based on the authors' experiences, allowing for a transparent portrayal of what is required to build these projects and offering an inside view of the challenges involved.

For each case study, the objectives, target audience, and a brief description of the activity are provided. In addition to this compilation, the human resources, costs, audience size, and type of interaction with the public were also identified. A SWOT (strengths, weaknesses, opportunities, and threats) analysis was conducted for all case studies through discussions among the authors, based on the insights gained from leading and executing these case studies. The identified strengths and weaknesses are related to the internal aspects of the communication project, whereas the highlighted opportunities and threats are external factors that can impact the success of the project.

## CASE STUDIES OF COASTAL AND MARINE COMMUNICATION

### Case 1: Kit Salt Gardens

***Objective.*** The goals are twofold, firstly, to diversify science communication efforts beyond conventional PowerPoint presentations, and secondly, to develop an engaging science communication tool specifically tailored to children, focusing on the fascinating dynamics of wetlands. The scientific messages being transferred are how wetlands are formed, how they change and how they will evolve in the future. This helps to raise awareness about the conservation of these habitats.

***Audience.*** Implemented in two primary school levels, the first one devoted to children aged 6 to 8 years old, and the second to children aged 9 to 10 years old.

***Description.*** The Kit Salt Gardens is composed of a box containing miniatures that exemplify the natural elements that compose the tidal wetlands (such as sediment, water, animal species, and typical vegetation) and auxiliary material (Figure 1) and allows hands-on a set of experimental activities related to wetlands ecogeomorphology. The characteristics of the miniatures are inspired by the Ria Formosa (Algarve, South of Portugal) wetland diversity. Students are invited to understand the physical (geomorphological) and ecological processes that regulate the formation of these environments, by constructing wetlands zones, while understanding the main threats (e.g., sediment supply, sea-level rise). For example, they will be able to play with several inundation scenarios, as a simplified way of experiencing rising sea-levels. The handling of the Kit by the students is intended to generate a critical spirit, empowering them with additional knowledge about natural sciences, and eventually inspiring them to become future scientists. The Kit is applied in a group-based context, with the supervision of a teacher. Activities included in the kit should last for about 2 hours.

*Figure 1. Kit Salt Gardens. The image illustrates the Kit after completion: at left it is possible to see the marsh zones (and correspondent plant ecological succession), water and the ruler at the end of the box, used to simulate sea-level rise scenarios; in the centre is a zoom to the plants and animals; at the right is possible to observe the plants and animals placed in the respective marsh zone.*

## Case 2: Geoscience and Dance Activity "The Sea Rolls the Sand"

***Objective.*** This activity aims to be an alternative and innovative way to engage children in ocean and coastal geology, through a combination of scientific concepts explanation and creative dancing.

***Audience.*** The activity was designed for third and fourth-grade primary school students including children with mild cognitive, neurological and physical impairment (e.g., attention deficit disorder, amblyopia, light autism, hyperactivity and dyslexia)

***Description.*** The activity combines coastal science concepts (wind, waves, currents, and sand), storytelling techniques (narrative arc), and creative dance techniques (movement, imaginative play, and sensory engagement, Figure 2). A sequence of six exercises was proposed starting with the generation of offshore ocean waves and ending with sediment transport on the beach, during storm/fair-weather conditions (Figure 2). Important associations from this activity are the connection between the atmosphere, ocean and the coast, and the insight between casual observations that the students make, i.e., their empirical knowledge of the coast, for example, breaking waves, beach width, sand grains, and the science behind it. Scientific concepts were translated into structured creative movements, within imaginary scenarios, and accompanied by sounds or music. The narrative of the exercises was plotted in a sequence of increasing complexity and excitement, with a sharp decline to relaxation, following the narrative arc. All students in the class, including children with special needs, participated jointly. Teachers assisted all sessions and had no intervention on the scientific topics or session alignment; however, occasional teacher interference occurred to assist behaviour control of the class. In one of the sessions, a teacher assigned to cognitively impaired students was also present, but no interference occurred. The science & art activity aroused emotions of enjoyment and pleasure and allowed effective communication between scientists and the schools. Details of the activities, evaluation, and analysis are published in Matias et al. (2020).

*Figure 2. Coastal environments, coastal processes and metaphors. The image illustrates waves approaching the coast from the right side (blue arrow). The direction of sediment transport (pink arrows on top of photographs) was embodied by the direction of the hand-to-hand balls (representing sediment grains) passage during storms (top right) and fair weather (bottom right). The wave-breaking position is represented in the room by two yellow vertical poles + horizontal bar, with the sub-aerial beach towards the left side (where children are resting in the middle photograph).*

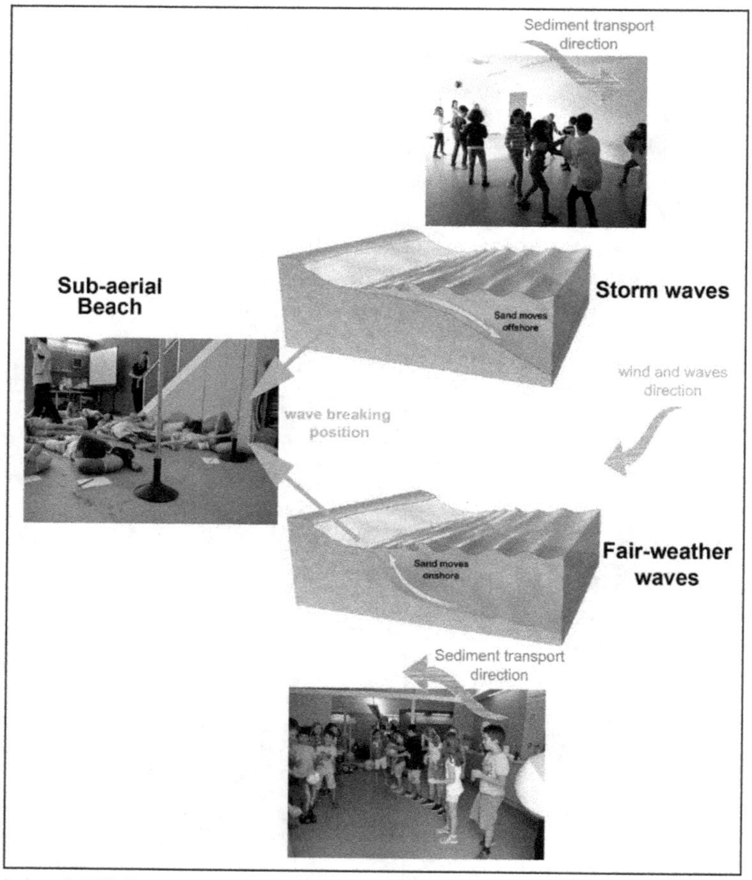


*From Matias et al. (2020)*

## Case 3: Animation Series "Coast for Kids" and "O Litoral Para Crianças"

***Objective.*** The series of animations were developed to engage children with the coast, celebrate its richness, communicate concepts of coastal geosciences, raise awareness of problems faced by coastal zones in a changing climate and show what can be done to address current and future challenges.

***Audience.*** The original animations (in English) are targeted at primary school children in the UK, and the animation adaptation is targeted at primary school students living in Portugal.

***Description.*** The original animation 'Coast for Kids' (C4K) had contributions from coastal geomorphologists, physical geographers, coastal ecologists, and human geographers from Universities in the UK, Australia, Canada, Spain, France, and Mexico (Delgado Fernandez, 2022). The series engagement strategy was to use identity and place-based elements (Figure 3). To create a sense of identity, episodes were narrated by school children aged 6-8 years old from the Liverpool City Region, UK. Place-based identification was achieved by introducing coastal elements of Liverpool Bay and different locations where the child narrators live. Engaging elements were used such as references to playing and fun activities by the coast, superheroes, the character 'Mrs. Piggy', onomatopoeia and interjection. While effective for the target audience, place-based and identity elements make the series C4K less suitable for children from other countries. Therefore, developing a version for another country, let alone for another language, implies more than translation, instead a comprehensive adaptation is necessary (Matias et al., 2022). The adaptation of C4K for Portugal - 'O Litoral para Crianças' (LPC), required several tasks: (1) adapting the script to Portuguese, including rewriting the episode about the coastal zone of Liverpool by changing it to the coastal zone of the Algarve, (2) substituting maps, (3) collecting photographs of the coast, (4) casting six children to narrate the episodes, (5) working with the volunteer families for sound recording, (6) choosing illustration avatars, (7) choosing new music, (8) creating new content to support teachers and parents, in addition to (9) the technical work of sound editing and (10) reassembling the animations. As a result of the adaptation work, LPC closely resembles the main features of C4K (Figure 3), as underlying coastal science principles apply, nonetheless, it is diverse to account for the many differences between the coastal zones, and the cultural and educational context of children in different countries.

*Figure 3. Frames from the animations "O Litoral Para Crianças" (top panel) and "Coasts for Kids" (bottom panel) to serve as examples of the original and adaptation similarities and differences. The graphical elements (typography and colours of these frames) were maintained, while natural elements (animals and plants represented in these frames) were changed according to the regional biodiversity.*

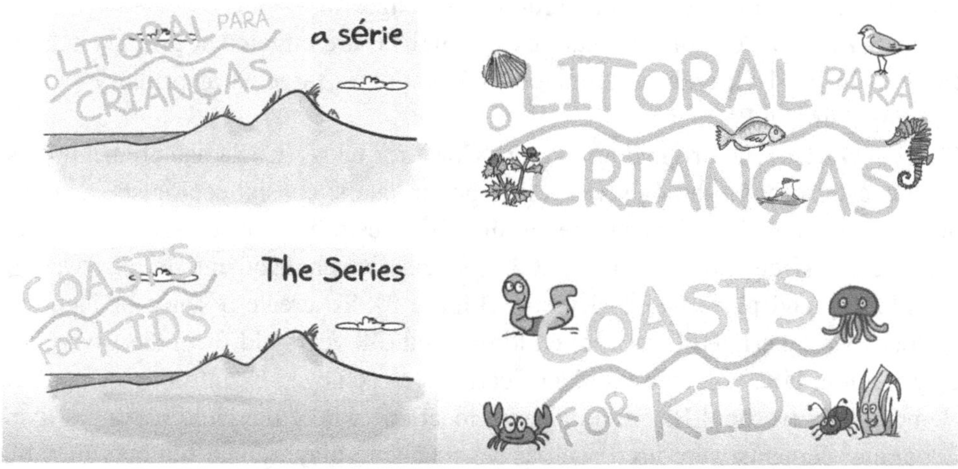


## Case 4: Comics Aquaculture

***Objective.*** The comic book "The path to the aquaculture" aims to inform about different issues concerning aquaculture, with a particular emphasis in Portugal. This includes the regimes of extensive, semi-extensive and intensive production, innovation in the industry, feeding processes, disease control, and monitoring of water quality, oxygen, temperature, etc.

***Audience.*** Portuguese teenagers and young adults are the main audiences of this book. To a lesser extent, their families are also potential audiences.

***Description.*** This comic book was proposed and financed by the Portuguese Institute for Sea and Atmosphere (IPMA), a public environmental agency. The book's main story concerns a senior couple that owns an aquaculture in the Algarve (south of Portugal, Figure 4) and the possibility of passing the torch to their young grandson - a recent graduate in Biology -, which implies changes in life and a maturity that this young adult may not have. The book was done in both Portuguese (Figure 4) and English, using paper and digital formats. The paper format is about A4 size, with 40 pages of illustrations and 52 pages in total, including a preface and a glossary about aquaculture. The script was written by a biologist and science communicator, the art was done by an artist with previous experience in science comic books, with a designer also included in the team. Additionally, several technicians of the Portuguese Institute for Sea and Atmosphere were involved in the creation of the book, assisting and revising it at different stages.

*Figure 4. Cover and page of the comic book "The Path to the Aquaculture". Published in Portuguese and English, with paper and digital formats, the main aim is to inform young audiences about the activity of aquaculture. The presented page is an example of an aquaculture company based in the south of Portugal, including its activities, installations, and staff.*

## Case 5: Salt Marsh Visits

***Objective.*** The aim is to increase the general public's knowledge of salt marshes dynamics (geomorphological and ecological processes) and ecosystem services, helping to raise awareness of the conservation of these habitats. These are focused science communication practices grounded in experimentation, observation, and heightened sensory experiences in natural environments (e.g., by seeing, touching, and smelling).

***Audience.*** Families, adults, and teenagers, with an interest in natural heritage.

***Description.*** The duration of each field visit is on average 2 hours, is conducted by boat or on foot (Figure 5), demanding the presence of a guide and a maximum of 20 participants. During the visits, the natural living resources elucidate the primary physical processes within salt marsh areas. The guide uses boards with schemes (Figure 5) to support explanations of the scientific processes occurring in the salt marshes. A dynamic question-and-answer method between the guide and participants fosters dialogue throughout the excursion. The language employed is simple and appropriate to the target audience.

*Figure 5. Field visits to the salt marshes. visits on foot (left) with university students and visits by boat (right) with geology teachers from various schools in Portugal. The image on the right shows the use of panels to support the scientific explanation.*

## Case 6: Photography Exhibition "Rescue|Search"

***Objective.*** The project aimed to explore unconventional science communication formats and create institutional partnerships by representing scientific methods of study in the ocean and on the coast.

***Audience.*** The target audience was adults and families who attend the various artistic shows on display in Faro's largest cultural events venue.

***Description.*** The project took place as part of the "Encontros do DeVIR" Festival, organised by the Algarve Performing Arts Centre, which commissioned large-format artistic photographs based on marine science (Figure 6). A science communicator liaised between the photographer and the University of the Algarve's Centre for Marine and Environmental Research, identified the scientific projects in the Ria Formosa, accompanied the photo shoots (Figure 6) and created the exhibition texts, which reflected the scientific and artistic processes. The photographer interviewed the researchers, visited the laboratories, selected the objects, and carried out the photo shoots. The exhibition was curated by the festival director and photographer, following aesthetic criteria and harmony with former photographs of the Ria Formosa, and a brochure was produced by the photographer and the science communicator. The exhibition took place between October and November 2020, in the foyer of the Faro Municipal Theatre (Figure 6). The COVID-19 pandemic caused constraints such as changes to dates, a prohibition on travelling inside the exhibition and a reduction in visitors.

*Figure 6. On the left panel, is an image of the photoshoot of a piece of equipment to monitor ocean colour handled by a researcher (Sónia Cristina), with the photographer (Vasco Célio), the assistant technician and part of the photography gear and setting. On the right panel is an image of the exhibition, with the large-format photographs (260x220 cm) in the theatre foyer, during the preparation phase.*

## Case 7: Social Media Project About Plastic Pollution in the Sea (*Plasticus maritimus*)

***Objective.*** The main objectives of this social media project based on the webpages of Facebook and Instagram (Pêgo, n.d.-a, n.d.-b) are to promote environmental education about plastic pollution and to publicize regional activities related to this issue by the biologist and beachcomber Ana Pêgo.

***Audience.*** The audience of the Facebook webpage is adults in Portugal, even though some specific posts are also bilingual and aimed at international adult audiences. The Instagram webpage is more directed to teenagers and young adults.

***Description.*** In December 2015, the Facebook webpage entitled *Plasticus maritimus* (Figure 7) was created by Ana Pêgo, often showing plastic objects collected during beachcombing activities in the Cascais region (about 25 km from Lisbon). Since then, this webpage was regularly maintained mostly with contents about plastic pollution, ocean conservation, and curiosities about marine species. It also features short video animations, beachcombing exhibitions of objects, news, presentations, live events, and other activities. Today, it has more than 25 000 followers mainly based in Portugal (77.4%) and Brazil (11,2%), in which about 78% are women. According to the author, the number of followers increased drastically with an online 7-minute video published in November 2017, and after the publication of the book *Plasticus maritimus* in Portuguese in November 2018. The Instagram page (Figure 7) featuring similar content was created in August 2018 but hacking of the page in June 2022 implied the loss of about 15000 followers. A new Instagram page was created soon after and currently has more than 3200 followers. Both social media webpages are solely maintained by Ana Pêgo and serve as a complement to her usual activities in environmental education.

*Figure 7. The main webpages of the project Plasticus maritimus in Facebook (top) and Instagram (bottom), feature a logotype and drawings directly taken from the book published under the same name and theme. Although the contents of both pages are similar, they are slightly adapted to fit adult audiences on Facebook and young audiences on Instagram.*

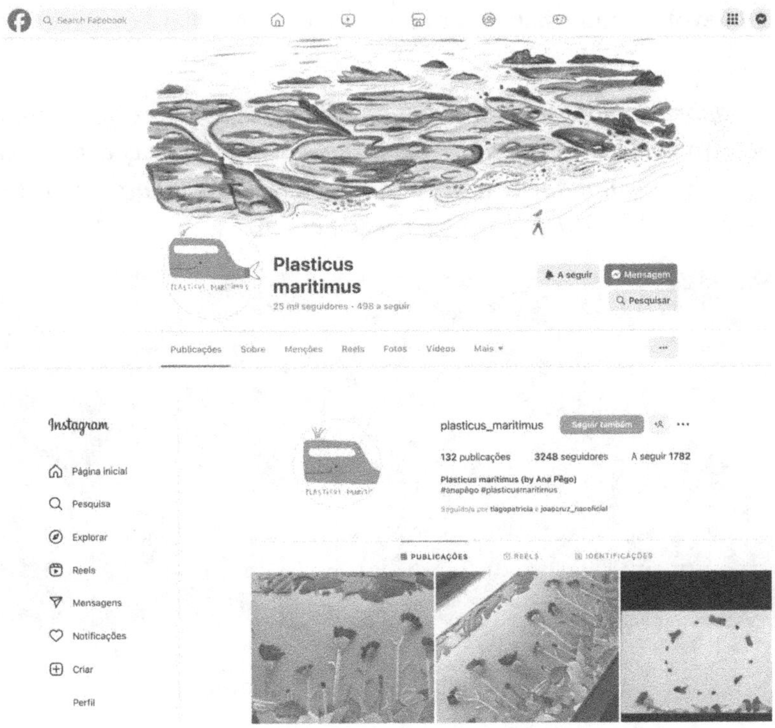


## Case 8: Public Talks About Climate Change Impacts

***Objective.*** The seminars aim to disseminate scientific knowledge on a specific topic that the presenter is actively engaged in, in a structured way.

***Audience.*** The target audience varies depending on the organisational options but commonly includes families, school students and adults with an interest in science or a particular issue.

***Description.*** As part of various science communication events, researchers are often invited to deliver lectures or talks related to their research. These events may encompass international occasions such as the European Researchers' Night, national events like the Science and Technology Week (Figure 8), or regional gatherings that highlight, raise awareness of or celebrate the existence of a specific natural park (coastal or marine). Other occasions may include Earth Day or World Oceans Day. Slide presentations are typically prepared for these occasions, but other lecture aids are also used. Scientific research tools can be showcased, or objects may be brought to elucidate natural processes (for example, using dominoes to copy wave propagation, Figure 8). Depending on the occasion and the audience, the lectures were on how waves form, how the coastline changes over time, how artificial beach nourishment evolves and what kind of effects storms can have on the coastline.

*Figure 8. Supporting material (dominoes and slides) of a lecture to sixth-grade students about waves (how they are formed, propagated until the coast, storm waves, etc.). Lecture done within the Science and Technology Week, organised locally by the science centre "Centro de Ciência Viva de Tavira".*

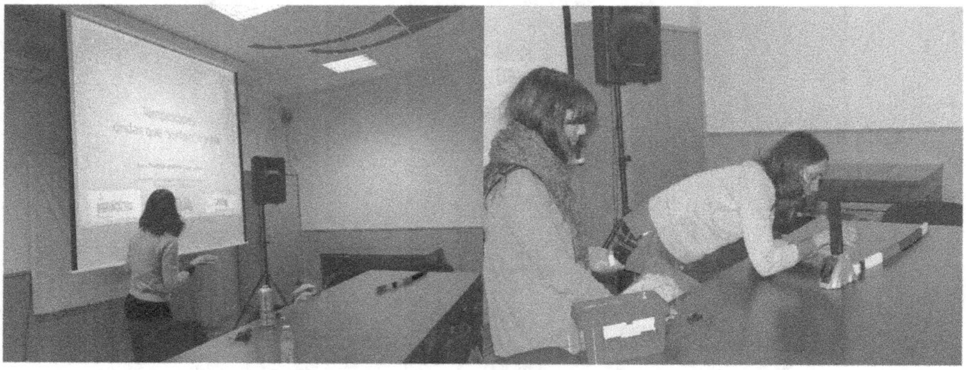

## Case 9: Sustainable Food Marine-Related Radio and Podcast Episodes

***Objectives.*** The main objective of this podcast/radio program was to communicate about the current Portuguese research on sustainable food at the national level (Figure 9). This communication is considered important to increasing awareness about these issues.

***Audience.*** The target audience was adults who listen to podcasts on the 'RTP Play' platform and/or live radio on the Portuguese public channel Antena 1. This audience is diffuse and not well characterized in Portuguese media studies. The produced episodes were heard in the live radio broadcast of Antena 1 potentially by about 100,000 listeners, according to broadcasting data of RTP in December 2022. There is currently no available data about the number of times these episodes were heard as podcasts.

***Description.*** A podcast/radio program with ten audio episodes was produced in Portuguese concerning national research on sustainable food production (Figure 9). Each episode had 9-10 minutes of duration, with a radio journalist as host. On every episode, the chosen theme was introduced, and reporting on-site (e.g. the lab or the field) was made, interviewing one or more of the researchers involved. Different issues related to food sustainability were covered, such as the production of fibres from milk waste. Diversity was also achieved with research institutions, scientists, and sites, including distinct regions such as the Azores archipelago. The Portuguese national radio channel (Antena 1) is one of the three radio channels of the Portuguese public broadcasting entity. The radio episodes were broadcast on weekdays between the 6th and 17th of February 2023 at 2:40 pm. After broadcasting, all episodes were made available on the RTP Play podcast platform.

*Figure 9. The Podcast Artwork for the podcast/radio programme "The Challenges of Sustainable Food", that features several elements mentioned in the episodes such as insects, broccoli, soil, olives, sea urchins, and mussels.*

## Case 10: TV Documentary About Sardines

***Objectives.*** The sardine is one of the most popular and consumed fish species in Portugal. Therefore, the main idea of this documentary was to present the sardine in its several dimensions. The focus was the management and conservation of stocks in Portugal, but its economic, historical and cultural importance was also approached.

***Audience.*** This documentary is destined for a national TV audience of adults and teenagers, particularly if they have an interest in natural history and culture.

***Description.*** "Sardinha portuguesa: riqueza natural" (in English "Portuguese sardine: natural richness") is the title of a 50-minute documentary about the sardine in Portugal (Figure 10), which was produced between 2021 and 2022 for the Portuguese national television (RTP1). It included the biology and ecology of the species, the monitoring of its stocks, the fishing and canning process (Figure 10), the historical relation between sardine and Roman culture, and its use as a symbol of popular festivals, and the country Portugal. Moreover, research on the production of sardines in aquaculture, genetics and traceability with this species is also shown in the documentary. This documentary was shown in RTP1 on the night of 11 July 2022 and became available online on the RTP site since then. A team of three main people conceived and developed this documentary: two science communicators at the University of Lisbon and the University of Aveiro, and one video professional. The technical team also included a video editor, a sound editor and a voice-over narrator.

*Figure 10. The image above represents a frame of the documentary in which the title is presented, and the image below was taken behind the scenes in an interview about the sardine canning industry and its importance to the Portuguese economy.*

## COMPARATIVE ANALYSIS OF CASE STUDIES

Developing ten coastal and marine communication case studies required a range of resources to reach diverse audiences of varying sizes, each employing distinct communication modes (outlined in Table 1). Budgets were less than 100 euros for the science & dance activity (Case 2), salt marsh visits (Case 5), social media (Case 7), and public talks (Case 8). Between 100 and 1,000 euros were needed to develop the educational kit (Case 1), while 1,000 to 10,000 euros were necessary to produce the animation series (Case 3), the photography exhibition (Case 6) and the radio/

podcast (Case 9). More than 10,000 euros were necessary to write, print and put on sale the comics (Case 4) and to produce the TV documentary (Case 10, Table 1).

While a single individual can effectively engage in science communication on social media (Case 7), collaborative efforts are more common in coastal and marine communication activities. Two-person teams are typical for field visits or public talks, usually with one person leading the activity, and another handling logistics and organisation. Larger projects, however, like animation series, radio/podcasts and TV documentaries, required teams of four or more individuals (Table 1), often needing highly specialised technical skills.

In the case studies involving direct contact with audiences, such as science & art activities and field visits, the audience size for each interaction was typically small (fewer than 30 people). However, in public talks, 30 to 100 people have attended. When communication is conducted through a product (educational kit and comics), it can reach larger audiences without additional effort from the team, being used, re-used, or commercialised on a scale of up to 1,000 copies.

The described case studies encompassed one-way communication methods, such as the animation series and the photography exhibition, as well as two-way communication, exemplified by the science & art activity. Public talks and field visits can be conducted as either one-way communication or as a dialogue with the audience, involving explanation and learning. This latter approach was used in the case studies described here (Cases 5 and 8).

*Table 1. Resources, Audience Size, and Interaction Type of the Described Case Studies*

| Case Studies | Budget[a] | Team[b] | Audience Size per Interaction[c] | Interaction Type[d] |
|---|---|---|---|---|
| **1. Educational kit** | €€ | II | ♥♥ | → |
| **2. Science & dance activity** | € | II | ♥ | ↔ |
| **3. Animation series** | €€€ | IIII | ♥♥♥♥ | → |
| **4. Comics** | €€€€ | III | ♥♥♥ | → |
| **5. Salt marsh visits** | € | II | ♥ | ↔ |
| **6. Photography exhibition** | €€€ | IIII | ♥♥♥♥ | → |
| **7. Social media** | € | I | ♥♥♥♥♥ | ↔ |
| **8. Seminars** | € | II | ♥♥ | ↔ |
| **9. Radio/podcast** | €€€ | IIII | ♥♥♥♥ | → |
| **10. Documentary** | €€€€ | IIII | ♥♥♥♥♥ | → |

[a]Budget intervals: € <100 euros; €€ 100 - 1,000 euros; €€€ 1,000 – 10,000 euros; €€€€ >10,000 euros;

[b]Team size intervals: I 1 person; II 2 people; III 3 people; IIII 4 or more than 4 people;

[c]Audience size intervals: ♥ < 30 people; ♥♥ 30 – 100 people; ♥♥♥ 100 – 1,000 people; ♥♥♥♥ 1,000 – 10,000 people; ♥♥♥♥♥ > 10,000 people;

[d]Interaction types: → one-way communication; ↔ two-way communication.

For each case study, strengths, weaknesses, opportunities, and threats were identified and detailed in Table 2. A notable strength of hands-on, face-to-face activities is their capacity to achieve a high level of engagement (Table 2). Examples include the educational kit (Case 1), the science & art activity (Case 2), the salt marsh visits (Case 5) and the public talks (Case 8). Another key asset in science communication is the use of storytelling, employed in the animation series (Case 3), the comics (Case 4) and the TV documentary (Case 10). The ability to revisit an exhibition or have ongoing access to a podcast are important strengths for audiences with limited time. The potential to reach large audiences is a significant strength of traditional media, such as TV documentaries (Case 10), as well as social media (Case 7).

An identified weakness of some of the conducted science communication initiatives is that they can be time-consuming, with some case studies being more demanding than others (Table 2). This is particularly true for the production of animation series (Case 3), comics (Case 4), and TV documentaries (Case 10). Ephemeral activities, limited in space and time, are also a weakness as their impacts may be restricted to that occasion; examples include field visits, public talks, and science & dance activities.

Opportunities for coastal and marine communication can be identified by leveraging existing logistics for events such as the "European Researcher´s Night", organised by a partner entity, to host public talks or promote comics. Another example is capitalising on high-budget projects with dedicated tasks for dissemination and communication. These opportunities can support activities that require substantial budgets and time but are highly engaging, reach larger audiences, and are readily available on broadcasting platforms (Case 10).

The most frequent threats to science communication activities include the potential instability of online hosting, hacking, adverse weather conditions, dependence on non-compliant organisations, and logistical challenges with schools.

## INSIGHTS AND LESSONS LEARNED

### What Were the Key Messages?

The described case studies cover various scientific topics: beach/dune dynamics (through the geoscience and dance activity, an animation series, and public talks), salt marsh dynamics (through an educational kit, and field visits), scientific research methods (through a photography exhibition and podcast), aquaculture and environmental sustainability (through a TV documentary, podcast, social media and comic book), impacts of climate change (through public talks, field visits, an educational kit, and an animation series), and pollution and ocean threats (through the social media). These topics align with the main emergent issues in marine and coastal matters - climate change, marine pollution, and biodiversity loss - identified by the European press journalists (Pinto and Matias, 2023) and the Eurobarometer (European Commission, 2017, 2021).

For the geoscience and dance activity (Case 2) and the animation series (Case 3) targeting primary school children, the message focuses on the natural processes that occur in coastal zones and how they shape the coastline geomorphology and evolution. The content serves as an educational complement (a strength of this case study, see Table 2) and is suitable for children very familiar with beaches, dunes and the sea but unaware of their functioning. A similar approach is taken with the wetlands educational kit (Case 1), especially in the instructions and activity book. Here, the main message is the very existence of wetlands, their geomorphology, and the relationship between animal and plant species and the tide.

Though dependent on the audience, the message during the salt marsh field visit (Case 5) is about collectively interpreting the landscape, highlighting its natural elements, and contextualising observations within a broader scientific framework (Figure 5). Although it can serve as an educational complement, this is not its primary goal. These are casual situations, in which participants are close to each other, where public curiosity and the possibility of dialogue are stimulated, thus the scientific content may be variable on group dynamics (strengths identified in Table 2).

Public talks are the most common communication format for geoscientists (Liverman & Jaramillo, 2011; Matias et al., 2024). While the message may be specific to a recent investigation topic, most lectures do not delve into the latest and more complex issues researchers are tackling, as these are reserved for communication among peers, who can follow and benefit from such specific knowledge advancements. The messages for teenagers and adults are more intricate than those for activities with children. Older audiences can make more connections to aspects of their people's lives related to the topic. For instance, when discussing coastal storms (Case 8), recent or particularly impactful episodes can be mentioned, or known locations can be referenced when explaining coastal evolution processes.

The photography exhibition (Case 6) and the podcast (Case 9) focused on scientific methods, instrumentation and research purposes. However, the photography exhibition message was almost the opposite; the artistic view of scientific objects intends to deliver an aesthetic perspective, without concerns about scientific results, rigour, or teaching new information about the ocean or the coast. The podcast episodes faithfully reported what was being investigated, how, and for what purpose, with the science journalist providing language translation. Nonetheless, the artistic view of scientific objects, somewhat encrypting the message, can also be seen as a weakness of this science communication activity (Table 2).

The comic book's message aims to inform about aquaculture production, innovation, commercialisation and its importance for the sustainability of fisheries and fish consumption, trying to clarify concepts and ideas about aquaculture. Sustainability messages are also addressed in the TV documentary, particularly regarding fish species in Portugal (sardines), focusing not only on their ecology and conservation but also on their significance as cultural patrimony.

## What Were the Required Resources?

Of the science communication activities described in section 3, four were conducted (and can be done) with a relatively small budget (~<100 euros): the science & dance activity, the salt marsh visits, social media, and public talks (Table 1). However, some additional costs may arise, such as a boat trip for the salt marsh visit, props for the science & dance activity (Figure 2) and supporting materials for the public talks (Figure 8), although these are optional.

On the other end of the budget spectrum of the described activities (i.e., expensive activities) are the TV documentary (> 10,000 euros; Figure 10) and the comic book (Figure 4). It is worth mentioning that none of these communication products were commercialised, as they were supported by the TV channel and a research project, respectively. High budgets are one of the main weaknesses of these projects (Table 2).

The size of the teams is also somewhat related to the budget and technical expertise required to develop the communication activity. The smallest team is needed for social media communication (Table 1); even for talks and visits, a mediator typically organises the event, books the venue, advertises the activity, and contacts the school's direction bodies or teachers. Only on specific occasions, the researcher who delivers the public talk may also be responsible for its organisation. The dependence on other organisations for logistics or event dissemination poses a threat to several communication activities (such as the educational kit, the talks, and the comic book, Table 2).

The inclusion of technical experts is mandatory for certain communication types such as TV documentaries. However, for activities such as podcasts, there are both professional, high-budget options (such as the case described here) and more affordable alternatives, as argued by Illingworth & Allen (2020). On the other hand, social networks can be created and managed with minimal resources and staff, but reaching larger audiences might require support from communications offices or significant time investment from researchers, probably more time than they are typically willing to spend. The major issue is that to reach a wide audience, content must be created regularly (several times per week) over an extended period (months to years). This is relevant because the lack of time is the main communication barrier identified by scientists in several studies (e.g., Matias et al., 2024; Rodrigues et al., 2023; The Royal Society, 2006).

*Table 2. Strengths, Weaknesses, Opportunities, and Threats (SWOT) Analysis of the Described Case Studies*

| Case Studies | Strengths | Weaknesses | Opportunities | Threats |
|---|---|---|---|---|
| **1. Educational kit** | + High level of engagement, as a face-to-face activity<br>+ Hands-on activity<br>+ Uses storytelling<br>+ Can be reused<br>+ Can become an educational resource<br>+ Flexible application venues | - It requires a substantial effort to create, with specific resources, for a limited range (tailored materials), which also implies a cost.<br>- Money is needed | + Can be reused for other purposes<br>+ Can be marketed | - Organising logistics with schools can be complicated<br>- Dissemination is dependent on an organisation that does not fully comply |
| **2. Science & dance activity** | + Stimulates creativity<br>+ High level of engagement, as a face-to-face activity<br>+ Inclusive activity<br>+ Low cost<br>+ Direct interaction that allows adjustment and adaptation to the audience<br>+ Flexible application venues | - It requires a person with a very particular profile, probably not applicable to the majority of people<br>- Not all children are receptive<br>- Ephemeral activity, limited in space and time | + Can be implemented at regional, national and international events such as "European Researcher Night" or the Portuguese "Science and Technology Week" | - Organising logistics with schools can be complicated<br>- Dissemination is dependent on an organisation that does not fully comply |

continued on following page

*Table 2. Continued*

| Case Studies | Strengths | Weaknesses | Opportunities | Threats |
|---|---|---|---|---|
| **3. Animation series** | + Children's engagement with animation is typically high<br>+ Uses storytelling<br>+ It is very flexible (making things up, going into the past or the future, creating on top of reality)<br>+ Available for viewing at any time<br>+ Can become an educational resource<br>+ It is in two languages, which allows for greater outreach | - It is relatively expensive<br>- It is labour-intensive<br>- Require national adaptations<br>- Require animation professionals | + Can be reused for other purposes | - Possible instability of online hosting<br>- Dissemination is dependent on an organisation that does not fully comply |
| **4. Comic book** | + It is very flexible (making things up, going into the past or the future, creating on top of reality)<br>+ Uses storytelling<br>+ Can become an educational resource<br>+ It can be produced in 2 formats (paper and digital)<br>+ In two languages, which allows for greater outreach | - It is relatively expensive<br>- It is labour-intensive<br>- Requires illustration professionals<br>- There can be conflicts between the illustrator and scriptwriter<br>- Can portray a current situation and therefore become out-of-date | + The book can be presented at various occasions, not only at school events but also at comic festivals, book fairs, etc.<br>+ Can be reused for other purposes | - Possible instability of online hosting<br>- Dissemination is dependent on an organisation that does not fully comply |

continued on following page

*Table 2. Continued*

| Case Studies | Strengths | Weaknesses | Opportunities | Threats |
|---|---|---|---|---|
| **5. Salt marsh field visits** | + High level of engagement, as a face-to-face activity<br>+ Enjoyment of natural spaces<br>+ Appeal to memories lived in a place<br>+ Multisensorial experience<br>+ Low cost<br>+ Allows adjustment and adaptation to the audience | - It requires a researcher expert in the scientific area and the geographical area to be visited<br>- Ephemeral activity, limited in space and time | + During the visits, the public may suggest other visits not envisaged by the researcher | - Dependent on the weather<br>- Dissemination is dependent on an organisation that does not fully comply |
| **6. Photography exhibition** | + Reaching the cultural events public<br>+ Great visual impact due to the size of the panels<br>+ Possibility of re-visit during the exhibition time | - Requires an artist (photographer)<br>- Ephemeral activity, limited in space and time<br>- The message is somewhat encrypted by the artistic view of scientific objects<br>- Given the exhibition size, other venues are limited | + Can be reused in other indoor cultural events | - Only suitable for a particular context |

continued on following page

*Table 2. Continued*

| Case Studies | Strengths | Weaknesses | Opportunities | Threats |
|---|---|---|---|---|
| **7. Social media** | + It is always available and easily accessible<br>+ Direct interaction and feedback from the audience<br>+ Possibility to reach large audiences<br>+ Few resources are required<br>+ Created content can be used in more than one social media<br>+ Unmediated activity, therefore, it does not depend on third parties to create, disseminate, analyse, change | - Requires persistence to keep social media alive<br>- Post takes time and effort to prepare<br>- Constantly looking for new things to say and be updated<br>- Constantly dealing with trolls, negationists, anti-science people | + Can allow networking with other profiles that amplify the outreach and dissemination<br>+ Useful to publicise and get other work opportunities | - Can be hacked<br>- Possible instability of online hosting<br>- Excess of online information/other profiles can make the penetration difficult<br>- Dependence on algorithms to remain relevant |
| **8. Public talks** | + Requires a few additional resources<br>+ The preparation effort can be relatively small<br>+ High level of engagement, as a face-to-face activity | - Ephemeral activity, limited in space and time | + Can be implemented at regional, national and international events such as "European Researcher Night" or the Portuguese "Science and Technology Week" | - Organising logistics dependent on an organisation that does not fully comply<br>- Dissemination is dependent on an organisation that does not fully comply |
| **9. Radio/ podcast** | + The podcast is always available and easily accessible<br>+ Possibility to reach the large audiences of the radio channel | - Requires professional sound editors<br>- Can become out-dated<br>- Interesting field or laboratory sounds are needed therefore these tasks need to be ongoing at the recording timing | + It can reach people while they do other things, such as driving, taking public transport, doing housework, sports, etc. | - Possible instability of online hosting<br>- Dissemination is dependent on an organisation that does not fully comply |

continued on following page

*Table 2. Continued*

| Case Studies | Strengths | Weaknesses | Opportunities | Threats |
|---|---|---|---|---|
| **10. Documentary** | + Being an audio-visual medium, it is very engaging<br>+ Possibility to reach the large audiences of the TV channel<br>+ Always available on the broadcasting online platform<br>+ Can become an educational resource | - Requires several audio-visual professionals<br>- It is expensive<br>- It is labour-intensive<br>- Can become out-dated<br>- It portrays only live situations on film (not past or future) | + Suits high-budget projects that intend to reach large audiences<br>+ Public Portuguese TV allow partnerships and has the means to disseminate the product | - Possible instability of online hosting<br>- Dissemination is dependent on an organisation that does not fully comply |

## Who Were the Targeted Audiences?

Generally, there is a relation between the budget and the audience size reached with each interaction. Smaller budget activities, such as science & art activities with a single school class, typically reach smaller audiences, while larger budget activities, such as TV documentaries, reach larger audiences (Table 1). However, some cases deviate from this pattern. This is the case of social media that, depending on various factors, can reach medium-sized audiences with a minimal budget. Another aspect to consider is that a one-way communication interaction may not reach the same level of engagement as a two-way communication, thus the efficiency of reaching large audiences may be counter-balanced by the effectiveness of an engaged audience. Additionally, if the target audience is children, animations, while reaching fewer people (not teenagers, nor adults), are tailored to their audience and therefore should be considered more adequate.

The only activities designed with families in mind are the field visits. The format is flexible enough to cater to teenagers, older citizens, or adults alone. An experienced researcher can adapt the language and the visit itself (e.g., duration, sites to visit, points of interest) based on public interaction, including non-verbal cues.

The photography exhibition had a relatively large team and a higher budget (Table 1). As a one-way communication format with an artistic perspective, traditional learning objectives were not fulfilled (Table 2). In a cost-benefit analysis, this project may be considered of limited effectiveness. However, the interaction between the artist and researchers during the exhibition's development included a 'science-society' component, showing scientific objects to audiences at cultural events, potentially reaching beyond traditional science communication boundaries.

The ten cases, with different types of interaction with the public (Table 1), follow both deficit and dialogue models of science communication (described e.g., in Bauer et al., 2007), aligning with the growing perspective that both are necessary (Lewenstein, 2011; Metcalfe, 2022). The vision held by English environmental scientists and science communicators is that it is desirable to simultaneously educate and fill knowledge gaps through a two-way dialogue process and empower a democratic decision-making process thus ultimately leading to a more desirable state of the environment (Riesch et al., 2016).

## FINAL REMARKS

The primary objective of this work was to present a concise portfolio featuring practical science communication cases tailored for diverse audiences, accommodating various budgets and team sizes. The analysis assesses their respective strengths, weaknesses, opportunities and threats. Given the limited published work that comparatively and comprehensively describes marine and coastal communication formats, benefits, and implementation barriers, this study was written from a scientist's perspective. It aims to provide valuable insights from our case studies to inform and enhance future communication efforts.

This chapter is limited by its reliance on our ten case studies, which possess specific characteristics that may not be generalizable to other experiences. A different group of cases might offer more comprehensive perspectives. However, such an approach would preclude the provision of "behind the scenes" insights, which are crucial for planning science communication. Despite this limitation, the case studies presented here encompass a broad spectrum of science communication activities conducted by researchers, with or without the support of other professionals. The target audiences include primary school students, teenagers, families and nationwide TV audiences. The diversity of communication formats (ranging from field visits to podcasts), team sizes (from one to more than four people) and budgets (from less than 100 euros to more than 10,000 euros) is significant. This diversity is relevant because future real-world situations may impose constraints that influence scientists' choices. Understanding the nature of these activities is useful for making informed decisions. This portfolio can serve as a catalogue of communication typologies for addressing marine and coastal issues, promoting standardisation within the field of marine and coastal science communication.

## ACKNOWLEDGMENT

This study had the support of the Fundação para a Ciência e Tecnologia (FCT), through the strategic projects UID/MAR/00350/2020 (CIMA) - DOI 10.54499/UIDP/00350/2020, and UID/MAR/04292/2020 (MARE) - DOI 10.54499/UIDB/04292/2020, and the project LA/P/0069/2020 granted to the Associate Laboratory ARNET – DOI 10.54499/LA/P/0069/2020. Bruno Pinto was supported by the contract CEECIND/03059/2017, and A. Rita Carrasco was supported by the contract https://doi.org/10.54499/CEECINST/00052/2021/CP2792/CT0007, both funded by FCT.

## REFERENCES

Bauer, M. W., Allum, N., & Miller, S. (2007). What can we learn from 25 years of PUS survey research? Liberating and expanding the agenda. *Public Understanding of Science (Bristol, England)*, 16(1), 79–95. 10.1177/0963662506071287

Bennett, D. J., & Jennings, R. C. (2011). *Successful science communication: Telling it like it is*. Cambridge University Press.

Boswell, R. (2021). Art and the senses for ocean conservation. *Journal of Marine and Island Cultures*, 10(1), 39–56. 10.21463/jmic.2021.10.1.03

Bowater, L., & Yeoman, K. (2013). *Science communication: a practical guide for scientists*. Wiley-Blackwell.

Boxman, R., & Boxman, E. (2017). *Communicating science: a practical guide for engineers and physical scientists*. World Scientific Publishing.

Brown, K. (2016). Oceanic geographies. *Performance Research*, 21(2), 118–124. 10.1080/13528165.2016.1162506

Carapuço, M. M., Taborda, R., Silveira, T. M., & Andrade, C. (2017). Upstream public engagement on coastal issues: Audience response to a science-based exhibition. *Ocean and Coastal Management*, 144, 83–89. 10.1016/j.ocecoaman.2017.04.008

de Vargas, C., Le Bescot, N., Pollina, T., Henry, N., Romac, S., Colin, S., Haëntjens, N., Carmichael, M., Berger, C., Le Guen, D., Decelle, J., Mahé, F., Poulain, J., Malpot, E., Beaumont, C., Hardy, M., Guiffant, D., Probert, I., Gruber, D. F., & Prakash, M. (2022). Plankton Planet: A frugal, cooperative measure of aquatic life at the planetary scale. *Frontiers in Marine Science*, 9(August), 1–17. 10.3389/fmars.2022.936972

Delgado Fernandez, I. (2022). Coast for Kids (C4K): A transdisciplinary science communication effort. *EGU General Assembly*, 2022, EGU22–EGU13427. 10.5194/egusphere-egu22-13427

Dijkstra, A. M., de Bakker, L., van Dam, F., & Jensen, E. A. (2020). Setting the Scene. In van Dam, F., de Bakker, L., Dijkstra, A. M., & Jensen, E. A. (Eds.), *Science Communication. An Introduction* (pp. 1–16). World Scientific Publishing. 10.1142/9789811209888_0001

Entradas, M., Junqueira, L., & Pinto, B. (2020). Portugal: The late bloom of (modern) science communication. In T. Gascoigne, B. Schiele, J. Leach, M. Riedlinger, B. V Lewenstein, L. Massarani, & P. Broks (Eds.), *Communicating science - a global perspective* (pp. 693–741). Australian National University Press. 10.22459/CS.2020

European Commission. (2017). Attitudes of European citizens towards the environment. In *Eurobarometer*. European Commission.

European Commission. (2021). *European citizens' knowledge and attitudes towards science and technology*. 10.2775/071577

European Commission Directorate-General for Communication. (2021). *Citizens' knowledge, perceptions, values and expectations of science - Report*. (Issue May). Publications Office of the European Union. 10.2775/071577

Fecher, B., Kuper, F., Sokolovska, N., Fenton, A., Hornbostel, S., & Wagner, G. G. (2021). Understanding the Societal Impact of the Social Sciences and Humanities: Remarks on Roles, Challenges, and Expectations. *Frontiers in Research Metrics and Analytics*, 6(July), 1–14. 10.3389/frma.2021.69680434278206

Fischer, L., Barata, G., Scheu, A. M., & Ziegler, R. (2024). Connecting science communication research and practice: Challenges and ways forward. *Journal of Science Communication, 23*(02). *JCOM, Journal of Science Communication*, 23(2). Advance online publication. 10.22323/2.23020501

Frangovska, A. (2020). Inter/Trans-Disciplinarity as a Platform for Seeking Universal Common Ground: From Speculation to Pragmatism Through Kristina Pulejkova's Wedding Route Art Project. *AM Journal of Art and Media Studies*, 21(21), 147–158. 10.25038/am.v0i21.366

Gelcich, S., Buckley, P., Pinnegar, J. K., Chilvers, J., Lorenzoni, I., Terry, G., Guerrero, M., Castilla, J. C., Valdebenito, A., & Duarte, C. M. (2014). Public awareness, concerns, and priorities about anthropogenic impacts on marine environments. *Proceedings of the National Academy of Sciences of the United States of America*, 111(42), 15042–15047. 10.1073/pnas.14173441125288740

Grorud-Colvert, K., Lester, S. E., Airamé, S., Neeley, E., & Gaines, S. D. (2010). Communicating marine reserve science to diverse audiences. *Proceedings of the National Academy of Sciences of the United States of America*, 107(43), 18306–18311. 10.1073/pnas.091429210720427745

Guan, S., Qu, F., & Qiao, F. (2023). United Nations Decade of Ocean Science for Sustainable Development (2021-2030): From innovation of ocean science to science-based ocean governance. *Frontiers in Marine Science*, 9(January), 1–8. 10.3389/fmars.2022.1091598

Gunn, C. M., Amerson, A. M., Adkisson, K. L., & Haxel, J. H. (2022). A Framework for Effective Science Communication and Outreach Strategies and Dissemination of Research Findings for Marine Energy Projects. *Journal of Marine Science and Engineering*, 10(2), 130. Advance online publication. 10.3390/jmse10020130

Halpern, B. S., Frazier, M., Afflerbach, J., Lowndes, J. S., Micheli, F., O'Hara, C., Scarborough, C., & Selkoe, K. A. (2019). Recent pace of change in human impact on the world's ocean. *Scientific Reports*, 9(1), 11609. 10.1038/s41598-019-47201-931406130

Harley, M. D., & Kinsela, M. A. (2022). CoastSnap: A global citizen science program to monitor changing coastlines. *Continental Shelf Research*, 245(June), 104796. 10.1016/j.csr.2022.104796

Illingworth, S., & Allen, G. (2020). *Effective science communication: a practical guide to surviving as a scientist* (2nd ed.). IOP Publishing. 10.1088/978-0-7503-2520-2

Irwin, A. (2008). Risk, science and public communication: Third-order thinking about scientific culture. In Bucchi, M., & Trench, B. (Eds.), *Handbook of Public Communication of Science and Technology*. Routledge. https://doi.org/https://doi.org/10.4324/9780203928240

Laszlo, P. (2006). *Communicating science: a practical guide*. Springer Berlin Heidelberg New York., 10.1007/3-540-31920-4

Lewenstein, B. V. (2011). Changing our ideas. *International Journal of Science Education. Part B*, 1(1), 17–21.

Liverman, D., & Jaramillo, M. (2011). Communicating environmental geoscience - An international survey. *Episodes*, 34(1), 25–31. 10.18814/epiiugs/2011/v34i1/004

Matias, A., Carrasco, A. R., Pinto, B., & Reis, J. (2023). The role of art in coastal and marine sustainability. *Cambridge Prisms: Coastal Futures, 1*(e25), 1–10. https://doi.org/https://doi.org/10.1017/cft.2023.13

Matias, A., Carrasco, A. R., Ramos, A. A., & Borges, R. (2020). Engaging children in geosciences through storytelling and creative dance. *Geoscience Communication*, 3(2), 167–177. 10.5194/gc-3-167-2020

Matias, A., Delgado-Fernandez, I., Costas, S., & Loureiro, C. (2022). Adaptation of 'Coasts for Kids' animations: an exercise of cultural diversity in coastal communication. *5th International Marine Science Communication Conference - CommOcean22*, 6–7.

Matias, A., Pinto, B., Areia, N. P., & Carrasco, A. R. (2024). How do coastal experts communicate science? Insights from an international survey. *Egusphere, EGU24-1703*. https://doi.org/https://doi.org/10.5194/egusphere-egu24-17037

Metcalfe, J. (2022). Science communication: A messy conundrum of practice, research and theory. *JCOM, Journal of Science Communication*, 21(7), 1–8. 10.22323/2.21070307

Nerghes, A., Mulder, B., & Lee, J. S. (2022). Dissemination or participation? Exploring scientists' definitions and science communication goals in the Netherlands. *PLoS ONE, 17*(12), 1–29. 10.1371/journal.pone.0277677

Nesci, O., & Valentini, L. (2020). Science, poetry, and music for landscapes of the Marche region, Italy: Communicating the conservation of natural heritage. *Geoscience Communication*, 3(2), 393–406. 10.5194/gc-3-393-2020

Nisbet, M. C., & Scheufele, D. A. (2009). What's next for science communication? Promising directions and lingering distractions. *American Journal of Botany*, 96(10), 1767–1778. 10.3732/ajb.090004121622297

Olsen, A. (2018). Dancing in wild places: Seaweed and ocean health. *Coreographic Practices*, 9(1), 97–118. 10.1386/chor.9.1.97_1

Pêgo, A. (n.d.-a). *Plasticus maritimus*. Facebook Page @plasticusmaritimus. Retrieved 10 May 2024, from https://www.facebook.com/plasticusmaritimus/

Pêgo, A. (n.d.-b). *Plasticus maritimus*. Instagram Page @plasticus_maritimus. Retrieved 10 May 2024, from https://www.instagram.com/plasticus_maritimus/

Pinto, B., & Costa, J. L. (2018). What are the practitioners' views about past marine science communication to the public? An example from Portugal. *Regional Studies in Marine Science*, 18, 27–34. 10.1016/j.rsma.2017.12.013

Pinto, B., & Matias, A. (2023a). European journalists and the sea: Contexts, motivations, and difficulties. *Public Understanding of Science (Bristol, England)*, 32(4), 459–469. 10.1177/0963662522113703636448498

Pinto, B., & Matias, A. (2023b). How European journalists cover marine issues. *JCOM, Journal of Science Communication*, 22(5), 1–11. 10.22323/2.22050802

Potts, T., Pita, C., O'Higgins, T., & Mee, L. (2016). Who cares? European attitudes towards marine and coastal environments. *Marine Policy*, 72, 59–66. 10.1016/j.marpol.2016.06.012

Riesch, H., Potter, C., & Davies, L. (2016). What Is Public Engagement, and What Is It for? A Study of Scientists' and Science Communicators' Views. *Bulletin of Science, Technology & Society*, 36(3), 179–189. 10.1177/0270467617690057

Rodrigues, J., Castro, C., Costa, E., Silva, E., & Pereira, D. I. (2023). Geoscientists' views about science communication: Predicting willingness to communicate geoscience. *Geoscience Communication*, 6(1), 15–25. 10.5194/gc-6-15-2023

Sanchez, E., Vinueza, R., Izurieta, X., & Rey, N. (2020). Use of muralism to promote awareness about aquatic ecosystems and wise water consumption in northwestern Ecuador. *Ocean and Coastal Management*, 190, 105165. 10.1016/j.ocecoaman.2020.105165

Seys, J., Cox, L., Yücel, E. Ş., Ezgeta-Balić, D., Faimali, M., Garaventa, F., García-Martínez, M. C., Gili, C., Kopke, K., Petrera, F., Simon, D., Villwock, A., Zielinski, T., Moreau, K., & Piniella, A. M. (2022). *Marine Science Communication in Europe: A Way Forward*. Zenodo. 10.5281/zenodo.6444143

Steins, N. A., Baker, M. R., Brooks, K., Mackinson, S., & Stephenson, R. L. (2023). Editorial: Co-creating knowledge with fishers: challenges and lessons for integrating fishers' knowledge contributions into marine science in well-developed scientific advisory systems. *Frontiers in Marine Science*, 10(December), 1–5. 10.3389/fmars.2023.1338271

The Royal Society. (2006). *Survey of factors affecting science communication by scientists and engineers*. Author.

## KEY TERMS AND DEFINITIONS

**Aquaculture:** Aquaculture is the breeding, raising, and harvesting of fish, shellfish, algae and other organisms in all types of water environments.

**Biodiversity:** Variety of life on Earth at all its levels, from genes to ecosystems, and can encompass the evolutionary, ecological, and cultural processes that sustain life.

**Citizen Science:** Any activity that involves the public in scientific research and thus has the potential to bring together science, policy-makers, and society as a whole in an impactful way.

**Coastal Risk:** Potential for adverse effects and damage to coastal areas and communities due to natural hazards such as storms, flooding, erosion, and sea-level rise, as well as human activities.

**Eurobarometer:** A series of public opinion surveys conducted regularly on behalf of the European Commission to monitor and analyse trends in attitudes across EU member states on a wide range of topics.

**Geomorphology:** Scientific discipline that studies the Earth's landforms and the processes that shape them, including mountains, valleys, rivers, and glaciers, through processes, such as erosion, weathering, and tectonic activity.

**Habitat:** Natural environment or specific area where particular species live, grow, and thrive, providing the necessary conditions for their survival.

**Narrative arc:** (Same as story arc) Structure and progression of a story or narrative that typically includes the introduction of characters and setting, development of conflict, climax, and resolution, creating a cohesive and engaging storytelling experience for the audience.

**Salt Marsh:** Coastal wetland that is flooded and drained by salt water brought in by the tides, supporting a unique ecosystem of salt-tolerant species.

**Sustainability:** Meeting the needs of the present generation without compromising the ability of future generations to meet their own needs, encompassing three main dimensions: the economic, social and environmental.

**Wetlands:** Area of land that is permanently or seasonally inundated with fresh, brackish, or saline water and contains a range of plant species that are uniquely adapted to the degree of inundation, the type of water that is present, as well as the soil conditions.

# Chapter 8
# Applying Ecohydrology to Promote a Nature-Based Solution:
## To Improve Quality of Life and Ecosystem Health in Luanda Bay, Angola

**Maria Alexandra Anica Teodosio**
http://orcid.org/0000-0002-0939-9885
*CCMAR, Universidade do Algarve, Portugal*

**Silvana Faria**
*CCMAR, Universidade do Algarve, Portugal & Instituto Nacional de Investigação Pesqueira e Marinha, Angola*

**Vânia Baptista**
https://orcid.org/0000-0002-1428-3334
*CCMAR, Universidade do Algarve, Portugal*

**Eric Wolanski**
*James Cook University, Australia*

## ABSTRACT

*This chapter aims to develop an ecohydrological modelling approach to propose an environmental engineering solution based on historical nature-based information to alleviate the severe water degradation problem within Luanda Bay, Angola. This solution would improve local communities' ecosystem health and quality of life. The model suggests a simple and practical solution to solving the pollution*

DOI: 10.4018/979-8-3693-2436-3.ch008

*problem by opening a cut from the bay's shallows to the sea. This will essentially reproduce an old bay opening to the sea, which was filled early in the last century. The residence time of water in the shallows of Luanda Bay would be reduced to about one week by excavating the channel over the land is straightforward. This should return the system to a more natural state and improve the water quality for recreational activities, increase the quality of marine food resources, reduce toxic algae blooms, and improve the quality of life of people in the area. Nevertheless, preventing pollutants from reaching Luanda Bay is the only long-term solution to restore a healthy environment in Luanda Bay*

## 1. INTRODUCTION

Throughout the world, coastal waters have experienced environmental degradation caused by global changes, both climate and direct anthropogenic actions. Significant observed impacts in bays are the increasing eutrophication risk, Harmful Algae Bloom (HAB), muddiness and siltation, and modifications in biodiversity due to overfishing, and invasions. These result in the loss of traditional ecosystem services (coastal fisheries, leisure) and negative socioeconomic impacts (the loss of income and employment for coastal communities). This also hinders sustainable development in coastal areas worldwide, including Angola. In Angola, Luanda Bay (Figure 1) is affected namely by water quality degradation and frequent harmful algal blooms (HABs) events (Rangel et al., 2004, 2006, 2007; Blanco et al., 2010). The Luanda Waterfront Project aimed to provide a biodiversity database and a deeper understanding of the cause underlying poor water quality. The Project found significant pollution in the upper shallows of Luanda Bay, threatening the consumption of marine resources used in subsistence fishing and leisure and recreational activities (Faria et al., 2021). This was linked to runoff from the watershed, mainly associated with intense rainfall events and exacerbated by the artificialization of the bay with human structures. The bay has accumulated sediments and pollutants from the land, a vexing problem that will continue until this inland problem is addressed. Another study by the Luanda Waterfront Project (Nogueira et al., 2021) showed a moderate risk perception by the population towards environmental risks but a high emotional attachment to Luanda Bay as a formerly healthy ecosystem, reinforcing the need to restore its historical function to society.

*Figure 1. A general Google Earth map of the study area shows Luanda Bay's location and neighbouring coastal ecosystems*

To propose a nature-based solution to restore a healthy ecosystem in Luanda Bay, it is important to first understand how it got there. That history was described by Croese (2018). Portuguese explorers founded the city of Luanda in 1576, and like waterfronts elsewhere in the world, Luanda Bay has traditionally been a place of 'flow' and trade, connecting the city and state at large to the world. For centuries, the bay represented the trading heart of the city, where goods such as rum, food, clothes, sugar, tobacco, and hardware came into Luanda from sailing boats in exchange for slaves, ivory, wax, and seeds. The large sailing boats moored near the Island of Luanda were not connected to Luanda itself on the mainland (Figure 2). Local transport between the sailing boats and the land was then using small boats. The island was vegetated, and Luanda had mangroves and seagrass near the city. The construction of a new port, with its own wharf, under the urbanization plan for the waterfront of Luanda of 1943, shifted the focus of the city's movements from the centre of the bay to the mainland (Gouveia, 1970).

*Figure 2. Historical photographs of the Luanda waterfront.* ***Top*** $1^{ST}$ ***line****: left and centre: The city of São Paulo de Luanda during the XIX century; right: Luanda in 1908.* $2^{nd}$ ***line****: right: coffee commerce in 1925; centre and left: Luanda Bay at the beginning of the XX century with natural vegetation (including mangroves).* $3^{rd}$ ***line****: left and middle: Aerial images in 1956 showing the bridge connecting Luanda Island with the mainland, the channel connecting the upper bay with the ocean, and two sand bars (https://www.persee.fr/doc/jhydr_0000-0001_1957_act_4_1_3342). Right: That channel was filled thereafter.* $4^{th}$ ***line****: Recent urban development and more landfill around the bay including over the historical channel Images of the proposed plan are available at https://www.slideshare.net/Jals/projectos-para-a-baia-de-luanda -1178921 (accessed 14 March 2024).* $5^{th}$ ***line****: Left: Aerial photograph of Luanda Coastal area, Mussulo lagoon in first plan. Middle: Chicala lagoon, with intense pollution visible. Right: postcard picture of Luanda Bay showing the waterfront with more land reclaimed for recreational activities and some planted palm trees.*

Later in the mid-20th century, Luanda experienced rapid urban expansion, due to migration from the countryside and rapid demographic growth, characterized by three significant periods – late colonial (1948–1975), postcolonial amid civil war (1975–2002) and postcolonial at peace (2002 to the present; Viegas, 2016).

In this context of rapid economic growth, fuelled by a coffee boom and increasing investments in and growth of the city during the 1950s and 1960s, the Bay of Luanda became a place of wealth and leisure—the symbol of the city's transformation to modernity and the postcard image of the jewel of the Portuguese colonial empire. During this period, Luanda Island became connected to the waterfront of Luanda through a bridge (Figure 2) (da Fonte, 2007). When Angola attained independence in 1975, and during the post-colonial civil war period, the bay's waters became increasingly polluted as public investment in urban infrastructures ceased and sewerage and drainage systems stopped working. By the war's end in 2002, raw sewage continued to flow into the bay, roads and buildings on the waterfront were frequently flooded. During the postcolonial period at peace, plans were commissioned by the Ministry of Public Works to redevelop the bay. The proposal included, among other things, plans for dredging and cleaning the bay, the extension of the waterfront road and the construction of parking lots and green spaces (Croese, 2018). Land was reclaimed using dredged sediment to increase the land area around Luanda Bay for urban development, including replacing the bridge by a landfill. This cut off the ocean and upper bay's historical water connection. The construction work on the first phase in mid-2008 and was completed by August 2012 (Lopes & Eugénio, 2012), with a 3 km promenade lined by over 2,000 newly planted palm trees, 147,000 $m^2$ of pedestrian space, about three km of cycle lanes, ten new open spaces along the beachfront, three playgrounds, three sports fields, five basketball courts and five spaces for cultural events, in addition to a new wastewater system, a six-lane road (three lanes in each direction), a fuel station and a flyover connecting the waterfront to the Luanda Island (Figure 3; Croese, 2018).

Notwithstanding these developments on the land, the quality of water in Luanda Bay was severely degraded (Nicolau, 2016). Thus, the motivation for the Luanda Waterfront Project (https://ccemarluandawaterfront.com/en/home_en/), funded by Aga Khan Network (AKN) and by the Portuguese Foundation of Science and Technology (FCT), was not just to increase the knowledge about marine biodiversity and ecohydrology of this previously unstudied area but mainly to develop solutions to increase the health of this emblematic ecosystem and the life of the residents (Teodósio et al., 2018). The approach that not based on finding an exclusively engineering solution, but instead proposes a nature-based solution (i.e., an ecohydrology solution) to restore the ecological processes of a healthy coastal ecosystem including its fisheries resources and biodiversity. It integrates the hydrological processes, the trophic web dynamics, and the historical evolution of the natural system (Wolanski et al., 2008). Nature-based Solutions (NbS) were recognized by the United Nations Environment Assembly as a key approach to tackling global issues such as the one in Luanda Bay (Sowińska-Świerkosz and García 2022; Melanidis and Hagerman, 2022). In this chapter, we will present the challenges to Luanda Bay Angola, the environmental quality of the marine ecosystem, and the quality of life of the people. Our aim is to use ecohydrology science to propose a remedial environmental engineering cum nature-based solution, based on historical information, to alleviate the severe water degradation problem of Luanda Bay. This solution would improve the health of both the local communities and the coastal ecosystem.

## 2. LUANDA BAY ENVIRONMENTAL AND QUALITY OF LIFE: CHARACTERIZATION, AND CHALLENGES

Luanda Bay is 7 km long and shore-parallel, 2.3 km wide, 30-40 m deep for 2/3 of its length, but very shallow (< 0.7 m) in the upper third. It is protected from the sea by Luanda Island, limited in the North by the harbour of Luanda City (Leitão et al., 2016). It is currently separated from the Chicala and Mussulo lagoons by the landfill that constitutes the access to Luanda Island (Figure 1). It is a rich ecosystem providing wide diversity of fauna and flora species, working as growth, recruitment, feeding and breeding area for several species of fish, crustaceans, bivalves, cephalopods, etc., as well as for livelihood and income for the local fishing communities (Masifundise Development Trust, 2013; Nicolau, 2016, Baptista et al., 2020; Faria et al., 2021; Canda, 2022).

In recent years, there have been several scientific research projects on Luanda Bay, some in collaboration with international researchers. One of these projects was the Luanda Waterfront Project, which found that the bay is used by a wide variety of organisms, such as phytoplankton (dominated by diatoms and dinoflagellates; Costa, 2022), zooplankton (Andrade, 2022), fish (including commercially important species; Canda, 2022), benthic invertebrates (such as worms, cockles, clams, mussels, cuttlefish, starfish, crabs, etc.), seabirds (two of which are classified as endemic, one from the waters of the cold Benguela Current and the other from the coastal areas of south-west Africa), marine mammals and sea turtles that use the bay in search of a safe place to breed, grow and feed (Figure 3).

*Figure 3. Some inhabitants of Luanda Bay: upper – commercial fish species, Sardinella maderensis (left) and Cephalopholis taeniops (right); middle left – bivalve, Senilia senilis; middle right – cuttlefish, Sepia spp.; down left – crab, Callinectes marginatus; seabird, Egretta garzetta.*

Previous studies had reported multiple anthropogenic pressures that are threatening Luanda Bay over the last decades (e.g., Leitão, 2016; Nicolau, 2016; Pestana et al., 2020). This bay, surrounded by the largest urban area in Angola, experiences daily discharges of solid waste (including plastic), domestic wastewaters, agricultural industrial effluents, rainwater run-off, and untreated run-off from street-washing (Leitão et al. 2016; Nicolau 2016; Pestana et al., 2020) (Figure 4). All these factors result in water with high microbial activity and low oxygen concentration (Nicolau, 2016). The bay also supports the Angola's most important commercial harbour (Port of Luanda), a refinery, a naval base, a fuel station, petroleum and cargo terminals, and other activities, that also contribute to water pollution in this ecosystem (Leitão et al. 2016; Nicolau 2016; Pestana et al., 2020) (Figure 4), as well as for the biological invasions due to the continued movement of recreational and commercial boat and ships from different parts of the country and the world (Pestana et al., 2020). There is also siltation that retains pollutants, anoxic sediments, a high concentration of faecal coliforms and heavy metals, namely Cr, Cd, Cu, Zn, Pb (Santos, 2012). The Luanda Waterfront Project also found significant pollution in the upper shallows of Luanda Bay, threatening the consumption of marine resources used in subsistence fishing (Faria et al., 2021). This pollution was evident in project results parameters such as pH, dissolved oxygen, suspended solids, nutrient concentration, bacteriological analysis, and Harmful Algal Blooms (HABs). This was due to runoff from the watershed, which was associated with intense rainfall events bringing in pollutants and sediments from the land and was exacerbated by the artificialization of the bay with human structures (Figure 4).

*Figure 4. Anthropogenic pressures in Luanda Bay: upper – strong surrounding urbanization; middle left – sewage discharge point in the bay; middle right – rainwater run-off coming from a local neighbourhood; bottom – solid waste, including plastics.*

Nowadays, pollution is an obstacle for sport, tourism, and high-end recreation (Figure 5). It also has numerous socio-environmental impacts on the population including threats to public health, social inequality, pollution of water resources/ insecurity, urban pollution, unproductivity and reduction of local income, and curtailing activities that depend on environmental quality, such as fishing, and constant flooding on rainy days. This pollution results from the misfunctioning of a sewage system dumping untreated domestic effluent daily into the bay, and poor or non-existent solid waste collection services. Despite all this, Luanda Bay still provides essential ecosystem services, though much degraded, to the residents. The bay remains a refuge for small-scale artisanal fishers, mainly from Luanda Island (Muxiluanda community), who fish there and collect marine organisms by mollusc gatherers, as a means of subsistence and income. Hundreds of people also collect molluscs and sea worms to sell on the streets of Luanda (Faria et al., 2021).

*Figure 5. Touristic and recreational activities in Luanda Bay (Angola): upper left – sailing competitions; upper right – seabirds' observation; bottom – recreational vessels at the Luanda Naval Club.*

Faria et al. (2021) described a new fishing technique in Luanda Bay, whereby the fishermen use smaller and traditional vessels, including some made of Styrofoam boards (Figure 6). Fishermen operate traditional and manually hauled fishing gears with low impact on the ecosystem, such as line/hooks, shovel, seine, gillnet, trawl, traps and cast nets (Figure 7). This activity in Luanda Bay provides livelihood for local communities, with maximum earnings of about 39 € by fishing day (Faria et al., 2021), which is larger than the average income per person (20.09 € in 2019; Faria, 2021). Thus, the quality of life of these communities is highly depending on this local fishing activity (Faria et al., 2021). While fishing in the bay serves as a source of income for the local communities, existing legislation permits this activity solely for subsistence purpose (Aquatic Biological Resources Law No. 6-A/04 of October 8 by the Ministry of Fisheries, Angola, 2004; Presidential Decree No. 41/05 of June 13 by the Ministry of Fisheries, Angola, 2005). However, despite regulations such as Presidential Decree No. 28/15 of 13 January by the Ministry of Fisheries, Angola (2015), which prohibits the capture of bivalves in closed areas of Luanda Bay, inadequate enforcement leads to uncontrolled fishing practices and challenges in resource management. The situation is exacerbated by poor legislation, fostering unreported landings and the consumption of contaminated species. Moreover, the emergence of biotoxins in the bay poses serious health risks, particularly through the contamination of bivalves and certain fish species. This contamination by HABs can lead to various health issues in humans, including neurotoxic effects such as Amnesic Shellfish Poisoning (ASP) and Paralytic Shellfish Poisoning (PSP) (Vale et al., 2009; Branco et al., 2010). Consequently, the quality of life for local communities is adversely affected (Faria et al., 2021).

*Figure 6. Fishing activity in Luanda Bay: upper – catching molluscs and sea worms with the support of the adapted vessel of Styrofoam boards; bottom – using seine net (left) and gillnet (right) to catch fish.*

The communities living around Luanda Bay are conscious of the environmental problems that they face, and they are generally willing to actively engage in efforts to protect the environment (Nogueira et al., 2021). The direct relationship with Luanda Bay is part of the life of the people around it. Beyond all the impurities, the waters of Luanda Bay carry memories from generation to generation, as does the relationship between a people and the territory in which they live. The leisure activities of the people who were born and grew up there are mostly related to the beaches, even if some of them are now considered unsuitable for bathing. A clean and protected Luanda Bay would bring much more economic activity to Luanda through sustainable tourism and fishing, it would create new jobs, development, and opportunities for the city, and, more importantly, it would restore the quality of life for its people.

## 3. METHODS FOR ECOHYDROLOGICAL MODELLING TO ANSWER CHALLENGES

Luanda Bay was divided in the mouth (30 m deep in the middle) and coastal water (also typically 30 m deep), a lower bay (25-30 m deep with shallower water along the coast), an upper bay (typically 25 m deep with shallower water along the coast), and the shallows (Figure 7). The depth in the shallows is commonly no more than 0.7 m.

*Figure 7a. The 2022 bathymetry of Luanda Bay from the bathymetric chart INT 2551. The insert is our survey of the bathymetry in the shallows on a 10 X 10 m grid. All depths are in m below Lowest Astronomical Tide (LAT). Depth meshes (resolutions of 10 m and 20 m) were provided by the Portuguese navy built based on the multibeam survey carried out in 2021; with probes represented on the nautical chart 16303 – Port of Luanda. Cuts 1 (A,B) and 2 refer to the proposed engineering works described in this chapter. Cut 1A consists of a dredged channel, 61 m wide and 2 m deep below LAT, and extending 130 m into the shallows. Cut 1B is located at the same site but it is 86 m wide, and it extends 413 m into the shallows on a straight line. Cut 2 is a dredged channel just north of the marina; it is 86 m wide, 2 m deep below LAT, and extends 600 m into the shallows.*

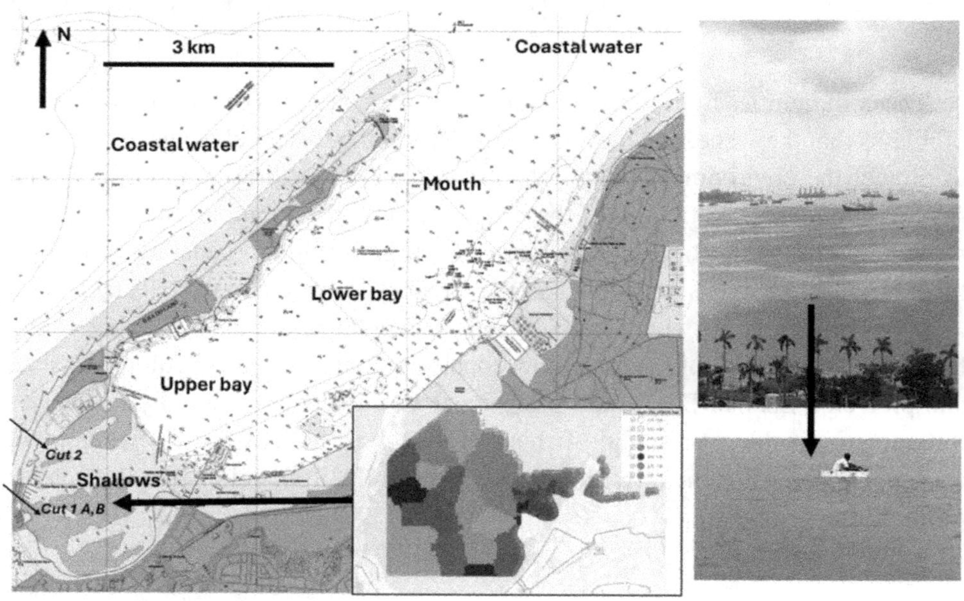


*Source: https://geomar.hidrografico.pt/*

*Figure 7b. The domain of the two models.*

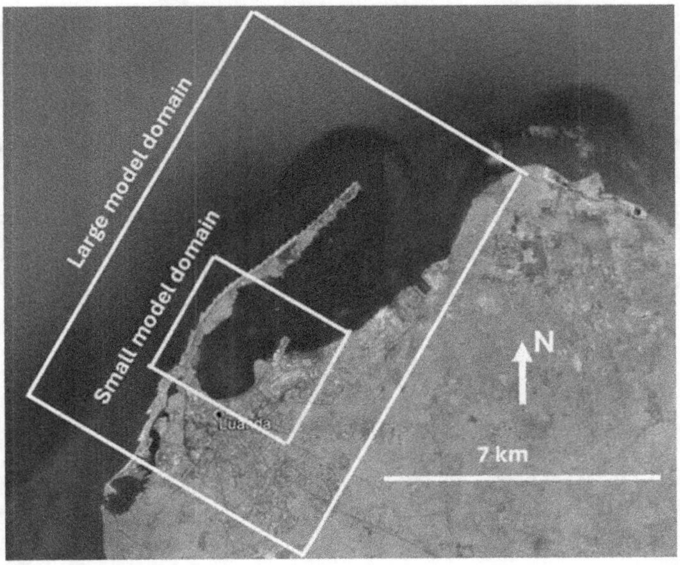


*Source: https://geomar.hidrografico.pt/*

Tidal data for Angola were obtained from the World Tidal Atlas, and the corresponding tidal currents in coastal waters were calculated from the phase differences. The mean longshore currents were obtained from satellite altimetry data (NOAA database) to be highly variable, and they often reverse direction, typically ranging from + 0.06 m $s^{-1}$ to – 0.06 m $s^{-1}$.

For modelling the hydrodynamics, we used the fully implicit, 2D hydrodynamic model of Baptista et al. (2020). To study the large-scale circulation, a large model domain was used that covered the whole of Luanda Bay and coastal waters (Figure 7b). The mesh size was 285 m. To study the flushing of the shallows and Upper Luanda Bay, a model with a smaller domain (Figure 7b) was used; it was nested inside the model with the large domain; its mesh size was 43 m. The mesh size is limited by the condition in 2D modelling that the mesh size must be larger than the depth. We focused on the dry season. The situation in the wet season could not be studied reliably because there is no data on the inflow of water and pollutants from the land. For modelling the advection-dispersion processes we used the Lagrangian model of Spagnol et al. (2002). 84,000 virtual particles were introduced in the model all over the shallows. They moved with the water currents, and they also dispersed following ambient turbulence. Following Okubo (1971), the sub-grid scale horizontal diffusion coefficient was taken to be 0.1 $m^2$ $s^{-1}$ for the small mesh. The virtual particles represent polluted water initially in the shallows.

## 4. PROJECTING RESULTS TO FUTURE SOLUTIONS TOWARDS A HEALTHY LUANDA BAY

The model revealed that the mean currents in the ocean do not strongly affect the bay (i.e., they do not intrude in the bay; Figure 8). This is due to the closing of the south end of the bay, which started in 1957 with the bridge construction and continued culminating in 2012 with more land reclaimed (Figure 2). The currents in the bay are controlled by the tides at the mouth of the bay and by the local wind over the bay. Thus, the net water circulation in the ocean and coastal water does not drive the flushing of Luanda Bay.

*Figure 8. Two snapshots of the predicted currents in Luanda Bay and coastal waters at two opposite times of the tidal cycle in calm weather. The colour bar shows the speed in m $s^{-1}$.*

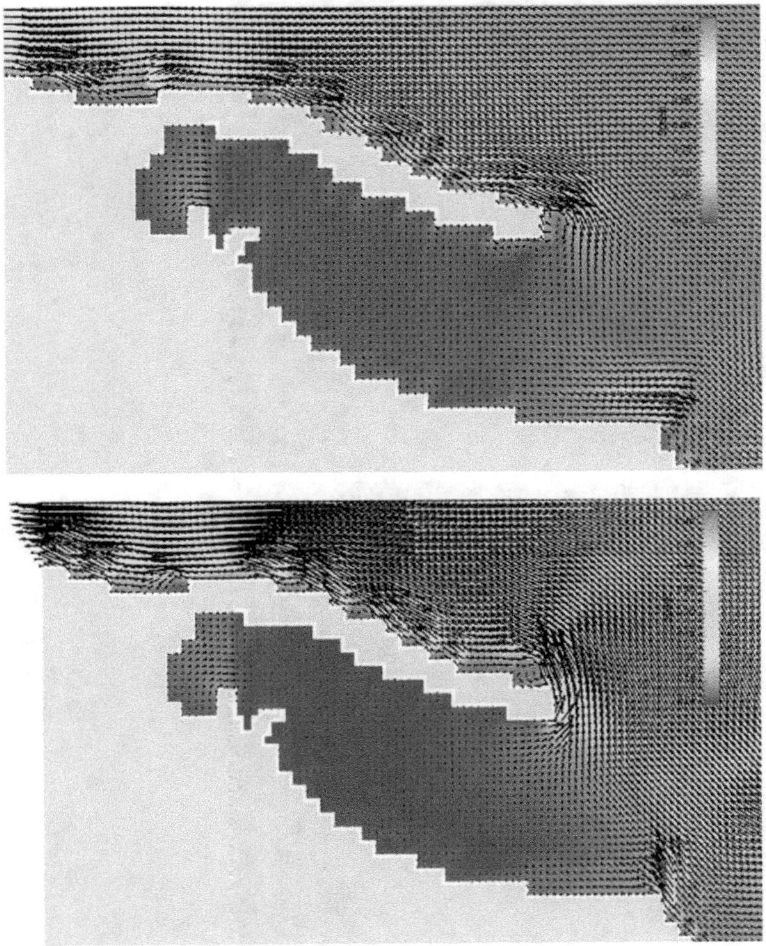

Snapshots of the currents over the shallows and Upper Luanda Bay at two different times in the tidal cycle, for a 1.4 m tide and calm weather, are shown in Figure 9. The noticeable stronger currents at the outer margin of the shallows are due to the incoming tidal wave in deep water suddenly finding shallow water. The currents are measurably stronger in coastal waters than in the bay.

*Figure 9. Snapshots of the currents over the shallows and Upper Luanda Bay and coastal waters at two different times of the tidal cycle in calm weather. The colour bar shows the speed in m* $s^{-1}$.

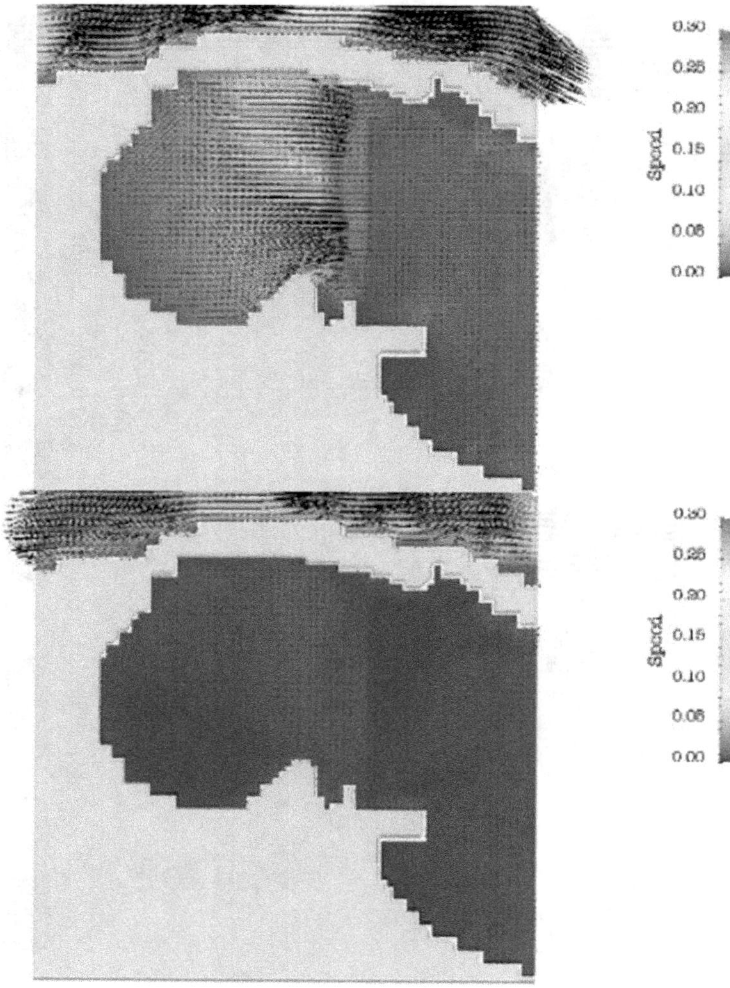


For the present condition and in calm weather, the model reveals that the shallows remain heavily polluted after 40 days, and that Upper Luanda Bay traps pollutants in the shallows (Figure 10a, b). The high number of tagged particles remaining trapped in Upper Luanda Bay implies a persistent pollution problem there. What matters for the biota is the concentration, and the concentration is the number of particles per model cell divided by the depth. The rconcentration of polluted water from the shallows after 40 days is presented in Figure 10c. The pollution of Upper Luanda

Bay is small compared to that in the shallows and the heaviest pollution occurs in the uppermost areas of the shallows.

*Figure 10. The distribution of the tagged water particles in the model (a) at time 0 when they are introduced in the model and (b) after 40 days. In (a) 1000 particles were introduced at each seeded cell. The colour bar shows the number of particles per cell. (c) The relative concentration of polluted water after 40 days of calm weather and no inflow from the land. The colour bar shows the concentration.*

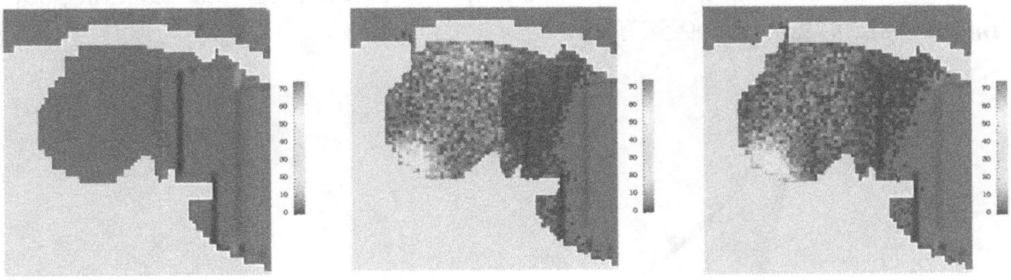

The residence time is the time it takes for 1/e (37%) of the particles to remain in the domain (Wolanski and Elliott, 2015). The residence time of polluted water over the shallows in calm weather (labelled 'As is') is 38 days (Figure 11). Thus, pollutants stay trapped for a very long time in the present situation. Our results agree with data from the Lagrangian tracers' study of Rosa (2018). These results suggest that, in the case of discharges of pollutants such as domestic effluents in this inner zone of the bay, the effluent plume is trapped in the shallows. The wind is predicted to play a minor role in the flushing of the shallows. The wind at Luanda is only 'strong' and sustained for long duration in the wet season from December to April (Rosa, 2018); but even then, a 'sustained' wind speed of 19 km $h^{-1}$ is not a strong wind that could drive strong currents. The model indeed suggests that such a wind has a negligible influence on the flushing of the shallows.

*Figure 11. Time-series plot of the number of particles seeded in the shallows remaining in the shallows. Where the line intersects the 37% line is the Residence Time (RT) in the oceanographic convention. 'As is' is the present condition (RT = 37.5 days), Cuts (1A, 1B and 2) refer to the situation following channel dredging at the sites shown in Figure 7. Cut 1A consists of a dredged channel, 61 m wide and 2 m deep below LAT, and extending 130 m into the shallows (RT = 250 hours,10.41 days). Cut 1B, is located at the same site but it is 86 m wide, and it extends 413 m into the shallows on a straight line (RT = 9.17 days). Cut 2 is a dredged channel just north of the marina, it is 86 m wide, 2 m deep below LAT, and extending 600 m into the shallows (RT = 9.58 days).*

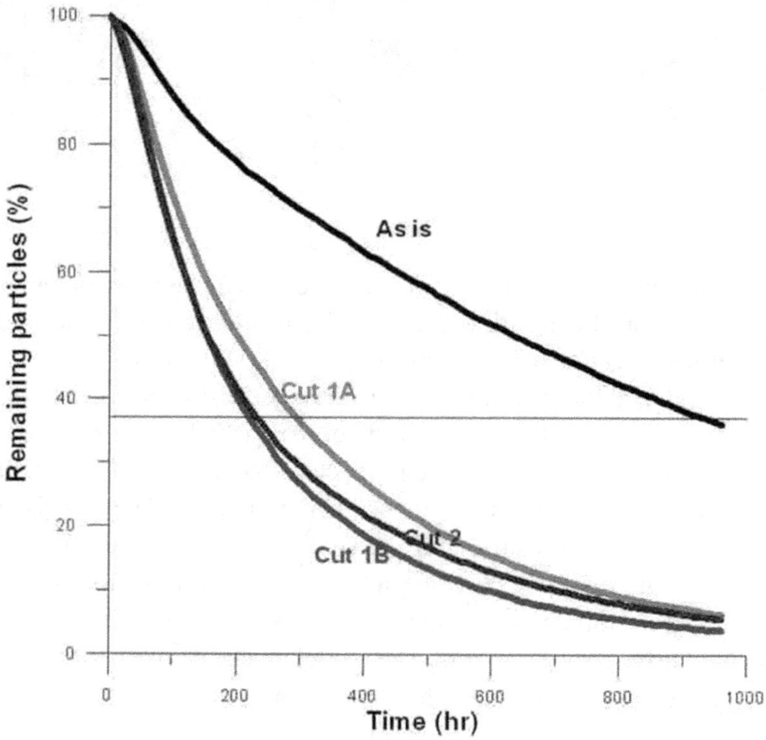


It is possible to use environmental engineering technology to reduce the residence time of polluted water in the shallows. The proposed environmental engineering measure is to dredge a channel to the open seas. It is not a magic solution as it then spreads the polluted water to coastal water where, however, flushing of the coastal water is swift due to dynamic currents in that area, thus presumably the environmental impact is small. Moreover, the new opening will also allow the tidal inflow of coastal water with its planktonic propagules from nearby lagoons (Tavares et al.,

2022) to enhance the colonization of historically relevant species, such as seagrasses, mangroves, and accompanying biodiversity.

Three likely cuts were considered (Figure 7a). Cut 1A consists of a dredged channel, 61 m wide and 2 m deep below LAT, and extending 130 m into the shallows. Cut 1B is located at the same site but it is 86 m wide, and it extends 413 m into the shallows on a straight line. Cut 2 is a dredged channel just north of the marina; it is 86 m wide, 2 m deep below LAT, and extends 600 m into the shallows. The reason for a long channel in the shallows is that the site is far away from the middle of the shallows so the channel must be longer to drain and export water at falling tide from the middle of the shallows.

The residence time of the water in the shallows for cut 1A is 8.7 days, and that for cuts 1B and 2 is measurably smaller (Figure 11). The concentration of polluted water on day 40 for Cut 1A and Cut 2 is shown in Figure 12. The cut has flushed out to the sea much, but not all, of the polluted water from the shallows. Figure 12 also reveals that the concentration in Upper Luanda Bay is always smaller than in the shallows.

*Figure 12. The distribution of the concentration of polluted water originating from the shallows on day 40 for (a) case 1A and (b) case 2. The colour bar shows the concentration.*

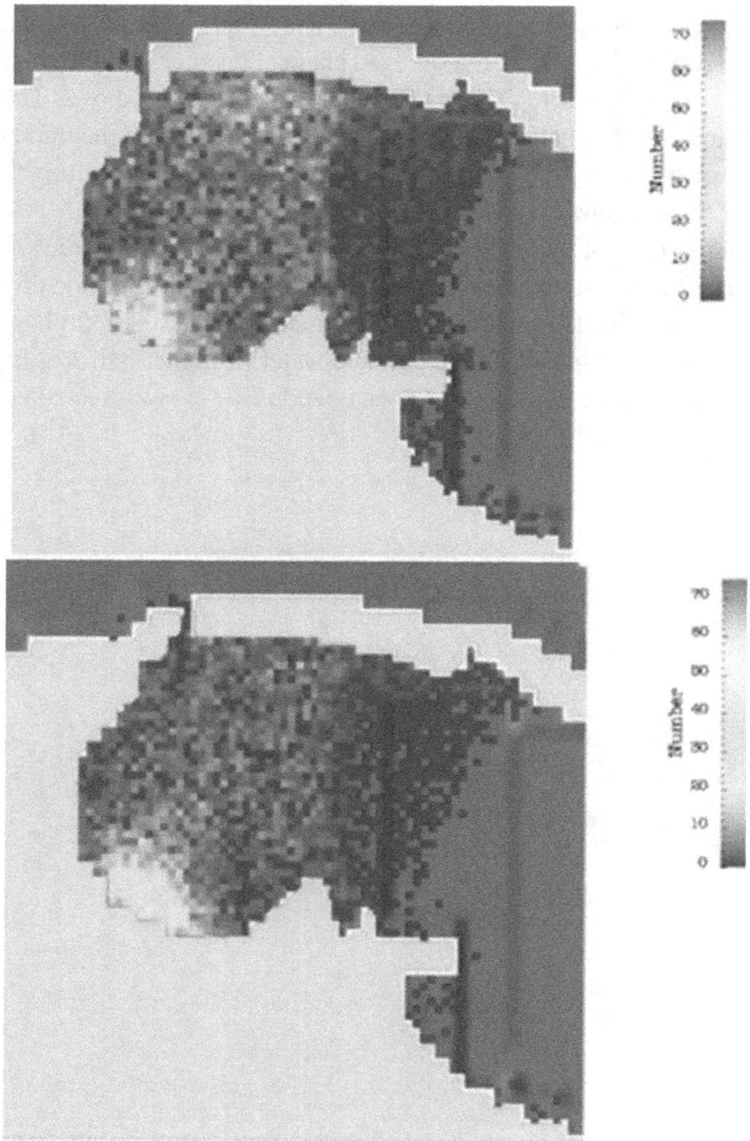

## 5. CONCLUSION

Land use in the catchment upstream of estuaries, bay and along the shores, land claim in coastal wetlands, pollution, dams/barriers, and dredging, among many other causes, have profoundly degraded many coastal areas worldwide and Luanda Bay is one of them. There are attempts worldwide in many maritime states to restore the ecosystem health of the degraded estuaries and bays. Most commonly, the experience shows that the solution combining physical intervention with wise engineering and eco-engineering (i.e., nature-based solutions) are linked to their restoration with different level of success and time of recovery (Wolanski et al 2008; Elliott et al. 2016, with world-wide examples). Those actions promote recolonization by plants and animals and acknowledge the role of the microbial populations and the materials that they exude (Wolanski & Elliott 2024, review). These responses of nature allow that recovery times are usually short (examples from 6 months are found in salt marsh ecosystems (Chicharo et al. 2008).

Luanda Bay ecosystem health depends on the rate at which water is flushed. The longer the residence time in the system, the greater the likely water quality problems. Ecological integrity also depends on the rate at which fine sediments are sequestrated in tidal wetlands or flushed out to sea. With the actual conditions of Luanda Bay in the shallows, the estimated residence time is over 30 days. The model results show that it is possible to significantly decrease the residence time of water to 9 days, in the shallows of Luanda Bay by opening a cut from the shallows to the sea. This simulates the natural, historical connections of the bay with the ocean near this area (Figure 2 top line) and allow colonization by propagules from the nearby coastal lagoon. This should improve the pollution problem of the shallows but has some negative socio-economic implications in that land will need to be reclaimed, and the cut will need maintenance. Excavating the channel over the land is straightforward engineering. This new channel would extend for some distance over the shallows to draw water in from the middle of the shallows at falling tide; this water is then exported to coastal waters by the tidal currents. The usual dredging ships cannot operate in the shallows because of the small depth. A barge would be needed with a sediment pump powered by a generator. The pump would suck the fine sediment out and pump it through pipes to a storage site. The storage site needs to be enclosed to prevent the sediment from moving out and returning to the dredged channel.

The cut should reduce the residence time of water in the shallows to about 9 days. In the wet season, there is presumably an inflow of water and its pollutants from the land, which would increase the concentration of pollutants in the shallows. Nevertheless, the cut should flush out the pollutants with the same residence time. The polluted water does not disappear in the presence of a dredged channel. Instead, it is transferred to coastal waters, where it is swiftly dispersed by the strong currents. Other measures will also be needed in the future in order to maintain the health of the catchment or drainage basin of Luanda Bay, directly determining a healthy coastal area.

*Figure 13. The native biodiversity of Luanda Bay that will be enhanced or restored with our nature-based solution, painted by the artist Sarita Deonilde Encarnação Camacho.*

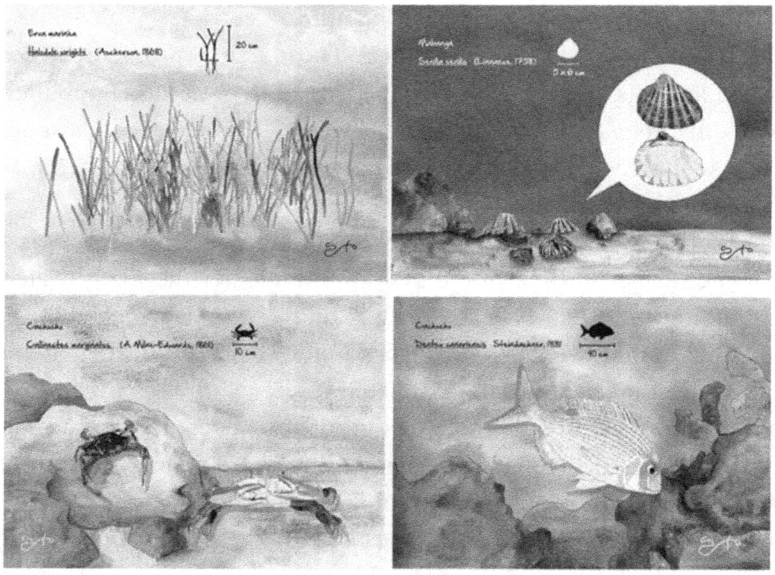

The proposed solution originated from the modelling exercise theoretically will increase the environmental health. The improved water quality and the increased connectivity between the bay, coastal waters and Mussulo lagoon will result in colonization by more key species (Figure 13), including propagules from seagrass and mangroves of Mussulo lagoon as well as invertebrate and vertebrate species. This should lead to the recovery of traditional artisanal fisheries that are now declining (e.g., clams like Mabanga -*Senilia senilis*), because residence time (RT 9 days) still allows post-larvae to settle since most of those planktonic phases have reduced pelagic larval duration, crucial in invertebrates or vertebrates' retention (Teodósio et al., 2016). This should return the system to a more natural state and increase the quality of marine food resources, reduce Harmful Algae Blooms, and improve the quality of life for people in the area, including opening a new waterfront for aquatic recreational activities.

Nevertheless, for long term (several decades) planning, remediation measures on land are also needed to decrease the pollutant load reaching the shallows. There is the experience in Western Australia of dredging a channel from the sea to an eutrophicated lagoon to increase the flushing rate of the lagoon (Wolanski, 2014). In the short-term this improved the ecosystem. However, the pollutant load was not addressed; it even increased over time; and in the long term (a few decades), the eutrophication problem came back. By analogy, the pollutant load to the shallows of Luanda Bay must be addressed. One way could be to build traps in the drains and streams flowing into Luanda Bay, to be cleaned annually. Some of these traps should be small artificial wetlands whose plants should be harvested after the wet season to remove the nutrients; this is a classical method in phytotechnology for urban watersheds (Zalewski et al, 2004). This would also be somewhat expensive and difficult to implement, as the experience with harbours worldwide has shown (Wolanski, 2006). Nevertheless, the nature-based solution we propose would measurably improve the system but not fully restore its historical environmental health.

## ACKNOWLEDGMENT

This study received Portuguese national funds from FCT - Foundation for Science and Technology through projects UIDB/04326/2020, UIDP/04326/2020 and LA/P/0101/2020; and the LuandaWaterFront Project – "Luanda Bay Ecological Assessment: A waterfront based approach to reduce environmental risks and increase quality of life" (333191101) supported by Aga Khan Network for Development (AKDN) and the FCT. Vânia Baptista was sustained by FCT Scientific Employment Stimulus (2021.00956.CEECIND; DOI:10.54499/2021.00956.CEECIND/CP1678/CT0001). Silvana Faria was funded by FCT with a Ph.D. fellowship (PRT/BD/154721/2022), and by AKDN and FCT through the LuandaWaterFront Project. Sustainable Horizons Project – SHEs - Universities designing the horizons of sustainability (Europe 101071300) and Preparation the new ocean Economy 2030- The blue Route of Discoveries Name: BLUE ROUTE Project ID: PT-INNOVATION-0072 (EEA GRANT) facilitated the connection of PhD candidates with stakeholders in associated partners. The bathymetry data from Luanda Bay was kindly supplied from the map (carta 16303) by the Portuguese Navy – Hydrographic Institute (Marinha Portuguesa) in agreement with the Angolan Luanda Port authorities. We are thankful to the artist Sarita Deonilde Encarnação Camacho, born in Luanda in 1975, for her beautiful aquarelles about the potential native biodiversity of Luanda Bay.

## REFERENCES

Abecasis, M. F. (1957). Les flèches de sable de la côte d'Angola [Sand spit formation on the coast of Angola]. *Journées de l'hydraulique, 4*(1), 100-108.

Andrade, E. (2022). *Zooplâncton como indicador de qualidade ambiental na Baía de Luanda* [Zooplankton as an indicator of environmental quality in Luanda Bay] [Unpublished BSc degree dissertation]. Universidade Agostinho Neto, Luanda, LD, Angola.

Baptista, V., Encarnação, J., Serrão, E. A., Wirtz, P., Pestana, L. B., Faria, S., & Teodósio, M. A. (2021). New Records of Fish Species from the Coast of Luanda, Angola. *Thalassas*, 37(2), 803–811. 10.1007/s41208-021-00297-1

Baptista, V., Leitão, F., Morais, P., Teodósio, M. A., & Wolanski, E. (2020). Modelling the ingress of a temperate fish larva into a nursery coastal lagoon. *Estuarine, Coastal and Shelf Science*, 235, 106601. 10.1016/j.ecss.2020.106601

Blanco, J., Livramento, F., & Rangel, I. M. (2010). Amnesic shellfish poisoning (ASP) toxins in plankton and molluscs from Luanda Bay. Angola. *Toxicon*, 55(2–3), 541–546. 10.1016/j.toxicon.2009.10.00819825382

Canda, R. (2022). *Contribuição para o conhecimento da ictiofauna da Baía de Luanda* [A contribution to the knowledge of the ichthyofauna of the Luanda Bay] [Unpublished BSc degree dissertation]. Universidade Agostinho Neto, Luanda, LD, Angola.

Chicharo, L., Ben Hamadou, R., Amaral, A., Range, P., Mateus, C., Piló, D., Marques, R., Morais, P., & Alexandra Chícharo, M. (2009). Application and demonstration of the Ecohydrology approach for the sustainable functioning of the Guadiana estuary (South Portugal) UNESCO-IHP Demonstration Projects in Ecohydrology. *Ecohydrology & Hydrobiology*, 9(1), 55–71. 10.2478/v10104-009-0039-3

Costa, S. (2022). *Incidência do fitoplâncton tóxico ou nocivo na Baía de Luanda com impacto na saúde pública e na qualidade do ambiente* [Incidence of toxic or harmful phytoplankton in Luanda Bay with an impact on public health and environmental quality] [Unpublished Master's dissertation]. Universidade Agostinho Neto, Luanda, LD, Angola.

Croese, S. (2018). Global urban policymaking in Africa: A view from Angola through the redevelopment of the Bay of Luanda. *International Journal of Urban and Regional Research*, 42(2), 198–209. 10.1111/1468-2427.12591

da Fonte, M. M. A. (2007). *Urbanismo e aquitectura em Angola - de Norton de Matos à Revolução* [Urbanism and architecture in Angola - from Norton de Matos to the Revolution]. [Doctoral thesis, Universidade de Lisboa]. ULisboa Repository.

Elliott, M., Mander, L., Mazik, K., Simenstad, C., Valesini, F., Whitfield, A., & Wolanski, E. (2016). Ecoengineering with Ecohydrology: successes and failures in estuarine restoration. *Estuarine, Coastal and Shelf Science 176*, 12-35.

Faria, S., Macuéria, M., Mosley, B. A., Teodósio, M. A., & Baptista, V. (2021). Characterization of small-scale fishing activity in Luanda Bay (Angola). *Journal of Spatial and Organizational Dynamics*, 9(3), 225–238.

Gouveia, L. (1970). Luanda. Estudo de Geografia Urbana: Ílidio do Amaral [Luanda. Study of urban geography: Ílidio do Amaral]. *Finisterra*, 5(10). Advance online publication. 10.18055/Finis2446

Leitão, A., Santos, A. M., & Boaventura, R. A. R. (2016). Complexation of lead by organic matter in Luanda Bay, Angola. *Environmental Monitoring and Assessment*, 188(10), 563. 10.1007/s10661-016-5557-z27624745

Lopes, L. F., & Eugénio, K. (2012, Aug 29). Inaugurada nova marginal da baía de Luanda [New waterfront inaugurated in Luanda Bay]. *Expresso.* https://expresso.pt/economia/inaugurada-nova-marginal-da-baia-de-luanda=f749660

Masifundise Development Trust. (2013). *Assessment report on small-scale fisheries in Africa*. J. Cox.

Melanidis, M. S., & Hagerman, S. (2022). Competing narratives of nature-based solutions: Leveraging the power of nature or dangerous distraction? *Environmental Science & Policy*, 132, 273–281. 10.1016/j.envsci.2022.02.028

Ministry of Fisheries. (2004). *Decreto-Lei n.º 6-A/04, de 8 de outubro* [Decree-Law No 6-A/04, of 8 October] (I series, Diário da República No. 81). Angolan National Government Press.

Ministry of Fisheries. (2005). *Decreto-Lei n.º 41/05, de 13 de Junho* [Decree-Law No 41/05, of 13 June] (I series Diário da República No. 70). Angolan National Government Press.

Ministry of Fisheries. (2015). *Decreto Presidencial n.º 28/15, de 13 de Janeiro* [Presidential Decree no. 28/15 of 13 January] (I series Diário da República No. 8). Angolan National Government Press.

Nicolau, S. (2016). *Caracterização da macrofauna bentónica da Baía de Luanda – Angola* [Characterisation of the benthic macrofauna of the Luanda Bay – Angola] [Master's dissertation, Universidade do Algarve]. Sapientia

Nogueira, P., Faria, S., Mosley, B. A., & Domingues, R. B. (2021). Environmental risks and quality of life of a coastal community (Luanda, Angola). *Journal of Spatial and Organizational Dynamics*, 9(3), 205–224.

Okubo, A. (1971). Oceanic diffusion diagrams. *Deep-Sea Research and Oceanographic Abstracts*, 18(8), 789–802. 10.1016/0011-7471(71)90046-5

Pestana, L. B., Dias, G. M., & Marques, A. C. (2020). Spatial and temporal diversity of non-native biofouling species associated with marinas in two Angolan bays. *African Journal of Marine Science*, 42(4), 413–422. 10.2989/1814232X.2020.1831954

Rangel, I., & Silva, S. (2006). First records of *Gymnodinium catenatum, Gambierdiscus toxicus* and *Pyrodinium bahamense* on northern Luanda coast, Angola. *Harmful Algae News*, 32, 10–11.

Rangel, I., & Silva, S. (2007). *Pseudo-nitzschia* spp and *Prorocentrum micans* blooms in Luanda Bay, Angola. *Harmful Algae News*, 33, 8–9.

Rangel, I., Silva, S., & Muai, J. (2004). Mortality caused by dinoflagellates bloom in Luanda Bay. *Harmful Algae News*, 26, 8–9.

Rosa, A. S. M. (2017). *Modelação Hidrodinâmica da Baía de Luanda.* [Hydrodynamic modelling of the Luanda Bay] [Master's thesis, Instituto Superior Técnico-Universidade de Lisboa]. Scholar.

Santos, A. M. P. A. (2012). *Estudo da qualidade da água da Baía de Luanda, distribuição de metais pesados na água, sólidos suspensos e sedimentos* [Study of the water quality of Luanda Bay, distribution of heavy metals in water, suspended solids and sediments] [Unpublished doctoral thesis]. Universidade Agostinho Neto, Luanda, LD, Angola.

Spagnol, S., Wolanski, E., Deleersnijder, E., Brinkman, R., McAllister, F., Cushman-Roisin, R., & Hanert, E. (2002). An error frequently made in the evaluation of advective transport in two-dimensional Lagrangian models of advection-diffusion in coral reef waters. *Marine Ecology Progress Series*, 235, 299–302. 10.3354/meps235299

Tavares, A. I., Assis, J., Patrício, A. R., Ferreira, R., Cheikh, M. A. S., Bandeira, S., Regalla, A., Santos, I., Potouroglou, M., Nicolau, S., Teodósio, M. A., Almada, C., Santos, R., Pearson, G. A., & Serrao, E. A. (2022). Seagrass Connectivity on the West Coast of Africa Supports the Hypothesis of Grazer-Mediated Seed Dispersal. *Frontiers in Marine Science*, 9, 809721. 10.3389/fmars.2022.809721

Teodósio, M. A., Paris, C. B., Wolanski, E., & Morais, P. (2016). Biophysical processes leading to the ingress of temperate fish larvae into estuarine nursery areas: A review. *Estuarine, Coastal and Shelf Science*, 183, 187–202. 10.1016/j.ecss.2016.10.022

Teodósio, M. A., Velho, F., & Santos, C. (Eds.). (2019). Colaboração luso angolana para formação de excelência em ciências do mar em África [Luso-Angolan collaboration for excellence in marine science training in Africa]. In *Proceedings of the XXIX Meeting of the Association of Portuguese Language Universities Conference on Art and Culture in the Identity of Peoples*. AULP.

Vale, P., Rangel, I., Silva, B., Coelho, P., & Vilar, A. (2009). Atypical profiles of paralytic shellfish poisoning toxins in shellfish from Luanda and Mussulo bays, Angola. *Toxicon*, 53(1), 176–183. 10.1016/j.toxicon.2008.10.02919041661

Viegas, S. L. (2016). Urbanisation and Peri-Urbanisation in Luanda: A Geopolitical and Socio-Spatial Perspective from the Late Colonial Period to the Present. *Journal of Southern African Studies*, 42(4), 595–618. 10.1080/03057070.2016.1192893

Wolanski, E. (1993). Water circulation in the Gulf of Carpentaria. *Journal of Marine Systems*, 4(5), 401–420. 10.1016/0924-7963(93)90024-G

Wolanski, E. (2006). Is harbour development ecologically sustainable. In Wolanski, E. (Ed.), *The environment in Asia Pacific harbours* (pp. 489–493). Springer. 10.1007/1-4020-3655-8_28

Wolanski, E. (2014). *Estuaries of Australia in 2050 and Beyond* (Wolanski, E., Ed.). Springer. 10.1007/978-94-007-7019-5

Wolanski, E., Chicharo, L., & Chicharo, M. A. (2008). Estuarine ecohydrology. In Jorgensen, S. E., & Fath, B. D. (Eds.), *Encyclopedia of Ecology* (pp. 1413–1422). Elsevier. 10.1016/B978-008045405-4.00044-6

Wolanski, E., & Elliott, M. (2015). *Estuarine Ecohydrology - An Introduction* (Wolanski, E., & Elliott, M., Eds.; 2nd ed.). Elsevier.

Wolanski, E. & Elliott, M. (2024). The Physics-Biology Links in Suspended and Settled Estuarine Mud Dynamics: A Review. *Sediment Transport Research – Further Recent Advances*: 10.5772/intechopen.1005423

Zalewski, M., Wagner-Lotkowska, I., & Robarts, R. D. (2004). *Integrated watershed management - ecohydrology and phytotechnology – manual* (1st ed.). UNESCO - IHP, UNESCO - ROSTE, UNEP IETC.

## KEY TERMS AND DEFINITIONS

**Biological Invasions:** Are recognised as a major economic and environmental threat worldwide. They occur when species colonise new geographical regions that are isolated from their existing populations. While many invasive species are exotic (non-native), there are cases where native species can also become invasive.

**Ecohydrology:** The science that relates hydrological processes to the biological dynamics of ecosystems at different spatial and temporal scales in order to develop effective solutions for restoring the health of aquatic ecosystems. The concept of ecohydrology was developed within the framework of UNESCO's International Hydrological Programme (IHP).

**Environmental Engineering Technology:** A specialised field within Engineering Technology that focuses on the application of scientific and engineering principles to address environmental challenges and concerns. Overall, it is a vital field that contributes to the sustainable development of societies by addressing environmental challenges and promoting responsible practices. Environmental Engineering Technology professionals combine technical knowledge with a commitment to environmental stewardship to create a healthier and more sustainable future.

**Eutrophication:** The process where an excess of nutrients in a water body leads to harmful algal blooms, reduced dissolved oxygen levels, and can result in the death of invertebrates and aquatic vertebrates, and in some cases, ecosystem collapse. This nutrient enrichment process, particularly compounds of nitrogen and/or phosphorus, accelerates algal growth, primary production and biomass, causing significant changes in nutrient balance, composition of aquatic organisms and degradation water quality.

**Exotic Species as Invasives:** Exotic species, introduced intentionally or unintentionally by humans, often become invasive. Global trade, tourism, and climate change contribute to their spread. These invasives disrupt ecosystems, alter biodiversity, and affect ecosystem services. Efforts to prevent and control invasions rely on understanding their physiology and ecology.

**Hydrodynamic Models:** Mathematical and computational tools used to simulate the movement and distribution of water and its constituents within diverse aquatic environments. They help understand and predict the behaviour of water bodies under different conditions, and are essential to study hydrodynamic problems, water resource management and environmental protection.

**Mangroves:** A coastal ecosystem from tropical and subtropical regions, characterized by salt-tolerant trees that grow in the transition between terrestrial and marine environments. They act as nurseries and buffers against climate change, playing ecological, economic and social roles.

**Native Species as Invasives:** Although less common, native species can also become invasive. A native plant that expands aggressively due to changes in environmental conditions (e.g., increased nutrients or altered disturbance regimes) can become invasive. The dynamics of native invasions differ from those of exotic invasions, but both impact ecosystems.

**Nature-Based Solution:** A holistic approach that integrates hydrological processes, trophic web dynamics and the historical evolution of natural systems to restore the ecological processes of a healthy ecosystem, including its fisheries resources and biodiversity, i.e. an ecohydrology solution.

**Residence Time:** The time it takes for 37% of the particles to remain in water body studied (the model domain) according to the oceanographic convention.

**Seagrasses:** A group of marine flowering plants that are found typically in shallow waters. These plants play an important role in coastal ecosystems supporting food security, mitigating climate change, enriching biodiversity, purifying water, protecting coastlines and controlling diseases.

**Subsistence Fishing:** Refers to fishing practices that is carried out primarily for the livelihood (to feed the family and relatives of the person who fishes) and income of the local fishing communities. This type of fishing generally implies the use of low-tech 'artisanal' fishing techniques and plays an essential role in food security, cultural heritage, and the livelihoods of many coastal and inland communities around the world.

**Watershed:** A hydrologic unit that encompasses all drainages flowing to a common water source. It includes streams, precipitation, and even groundwater and is used as a unit of measure for managing and delineating natural resources on a landscape scale.

# Chapter 9
# Analysing the Role of Artificial Reefs in Blue Tourism:
## Vilamoura Artificial Reef (South Portugal)

**Joana Costa**
*University of Algarve, Portugal*

**Jorge Ramos**
https://orcid.org/0000-0002-5115-7919
*CinTurs, University of Algarve, Portugal*

**Francisco Leitão**
https://orcid.org/0000-0003-4983-9782
*Center of Marine Science, University of Algarve, Portugal*

## ABSTRACT

*Artificial reefs (AR) are essentially implemented to increase marine biodiversity, which can indirectly benefit the local ecosystem and human communities. Their presence can attract a wide range of marine organisms and promote recreational activities, attracting anglers and scuba divers, improving local tourism. The aim of this work is to analyse how the Vilamoura AR (Southern Portugal) is useful to the sustainable coastal (blue) tourism in the area. Nowadays, with the help of tools such as vessels tracking intelligence and the use of direct georeferentiation, it is possible to infer about the use of artificial reefs by analysing patterns of recreational vessels around the AR area. The application of those tools reenforced the importance of combining methodologies such as not only monitoring artificial reefs but also to*

DOI: 10.4018/979-8-3693-2436-3.ch009

*have interviews with the stakeholders to better understand the recreational and blue tourism potential of the Vilamoura AR.*

## INTRODUCTION

Tourism plays a significant role in the economies in coastal and inland water regions (Picken, 2023). It leads to important economic and social benefits, including creating new jobs, income generation, and cultural exchange (Gormsen, 1997). However, the expansion of tourism activities can have negative environmental consequences by putting pressure on already fragile coastal ecosystems with eventually generating more waste and pollution (Mejjad et al., 2022).

From blue growth, the concept of economic growth using marine and coastal resources with a focus on sustainability, emerged the term blue tourism (Leposa, 2020). Involving a range of coastal and maritime recreational activities and emphasizing sustainability, blue tourism promotes the well-being of coastal ecosystems and local communities (Tonazzini et al., 2019; Wilks, 2021; Hanafiah, 2022).

The Algarve coast, in southern Portugal, is one of the most important touristic destinations in the country (Ortega et al., 2013). Due to its reliance on sun-and-beach tourism, like many coastal destinations, it faces challenges with seasonality (Valle et al., 2011; EC, 2023). Vilamoura, a resort in the Algarve, started building a marina in the 1960s with the primary goal of boosting tourist figures, while also aiming to leverage region's tourism appeal (Loo et al., 2009). Since then, it has grown to be a world-class marina with more attention to its impacts and sustainability. Beyond inland infrastructures there are also infrastructures offshore, namely artificial reefs (AR) that were deployed with the aim to support not only marine life, but also economic activities, including tourism (Ramos et al., 2007).

While the idea of using structures to attract fish is ancient, dating back 3,000 years, the modern concept of artificial reefs designed to improve commercial fishing originated in Japan. Europe adopted the idea much later, in the second half of the 20th century (Fabi et al., 2011; Ramm et al., 2021). It was not until that time that artificial reefs spread worldwide for the purpose of fishing, SCUBA diving and coastal recreation (Lima et al., 2019). Divers are drawn to locations displaying natural marine life and those featuring artificial reefs like shipwrecks, artistic sculptures and other sunken structures (Belhassen et al., 2017), while those who fish for leisure can find another option (McGlennon & Branden, 1994; Keller et al., 2016).

Assuming that artificial reefs are important 3D structures that enhance biocolonization of varied species by benefiting new food webs (Nauta et al., 2023) and are also useful by boosting coastal economy and human activities (Shani et al., 2012; Rendle & Rodwell, 2014). In that scope emerges the question of whether the artificial reefs off Vilamoura are being useful for human activities, specifically related to leisure/blue tourism.

This chapter discusses the concept of blue tourism for coastal regions, particularly in the Algarve where tourism is a major economic driver. The focus is on the development and evolution of the Vilamoura resort and its well-known marina, alongside with the contribution of the artificial reef off Vilamoura to support not only marine life, but also recreational human activities in a sustainable way. For this aim, a dataset was collected using Automatic Identification System (AIS) technology to record recreational vessels' activity over the Vilamoura artificial reef and in the inner area between the AR and the coast in the year 2022.

## BLUE TOURISM

Blue tourism refers to experiences related to coastal and maritime tourism while also focusing on sustainability. This economic sector is especially important for countries with attractive and accessible coasts (Tonazzini et al., 2019). Blue tourism also highlights the potential of coastal destinations for tourism, while ensuring the long-term health of coastal ecosystems and the economic and social well-being of local communities (Hanafiah, 2022).

Coastal and maritime tourism can be shore-based such as the recreational use of beaches (such as swimming, strolling and sunbathing) and sightseeing (Moreno & Amelung, 2009), but it can also include SCUBA diving, marine life observation and recreational fishing as offshore-based leisure. Blue tourism three main categories trending now are cruise ship travel, beach resort travel and ecotourism (Supriyanto, 2022).

Development in mass tourism destinations is characterized by an expansion of high-density hotels and resorts. Tourism destinations typically feature concentrated groups of retail stores, entertainment venues, and pre-organized tours for large numbers of tourists. Catering to a variety of visitor demands can lead to pressure on natural resources, making it difficult to maintain them in the long-term (Tonazzini et al., 2019). This pressure can go from an increase in air travel and noise pollution to biodiversity loss (Ghosh, 2012). By supporting sustainable maritime management practices, blue tourism can be more objectively planned out (Supriyanto, 2022).

### Tourism in the Algarve

The Algarve, has coastal waters with milder temperatures and calmer conditions compared to the windier and cooler western waters, likely due to the influence of the Mediterranean Sea (Ramos & Santos, 2015; Oliveira et al., 2019). This region of

Portugal is characterized by mild winters and pleasant summers, attracting tourists who want to enjoy beach activities such as sunbathing and swimming or practice water sports like kayaking and sailing.

Also known for its golf courses, gastronomy and historic and cultural heritage, the Algarve region in 2022 concentrated 33.4% of the total bed capacity available in the country. As one of the main destinations in the country, it was still mostly chosen for "leisure, recreation or vacation" trips, corresponding to 79.9% of domestic trips to the region (INE, 2023). In 2021, the main passengers arriving at Faro Airport were from the United Kingdom, Germany, France, and the Netherlands. Citizens of the United Kingdom alone added up to more than 1.1 million (VA, n.d.).

Algarve's climate puts it in advantage as a year-round destination for travellers, alongside other sunny nations in Southern Europe like Spain, Italy, Malta, and Greece. However, a challenge associated with coastal tourism destinations is seasonality. In summertime (peak season) there is a surge in tourist arrivals, concentrated in just a few months, particularly July and August (EC, 2023).

## The Vilamoura Case

In 1965, the vision of creating a tourist city dedicated to enhancing coastal tourism was focused on Vilamoura in the Algarve, where the private company LUSOTUR led by Cupertino de Miranda, purchased Quinta da Quarteira for this purpose. The "Urbanization Master Plan for Vilamoura," crafted by an international collaboration involving teams from the United States of America (Carver Baker & Associates, Los Angeles), France (SETAP, Paris), and Portugal (GEFEL, Lisbon), was approved in 1966 (Lobo, 2010). This project proposed a network of residential and tourist units managed as a single unit, setting a precedent for modern urban planning in Portugal (León, 2012). In Figure 1, through the chronological sequence of the photographs, it is possible to observe the evolution of the area from when there were practically no buildings, through the appearance of the Vilamoura marina and the growing infrastructure of the areas surrounding the resort.

*Figure 1. Vilamoura Marina: (a) in the 1960s, (b) in 1979, (c) in 1989, and (d) in 1999. Source: Algarve Real Estate (ARE, n.d.).*

Vilamoura nowadays is a modern seaside resort destination, known for its hotels, golf courses and marina with the capacity to accommodate yachts up to 60 meters, with 825 moorings points (RTA, 2021). The classification of marinas is done through different certification and standardization systems, such as the internationally recognized Gold Anchor Award by The Yacht Harbour Association (TYHA) (Sari et al., 2016). Marinas around the world hold awards from one to five gold anchors with the Vilamoura Marina becoming the first marina in Portugal to receive a 5 Gold Anchor Platinum, the highest certification. This distinction means exceptional on water and on shore facilities and infrastructure, customer service and sustainable practices (TYHA, n.d.). Notably, the Vilamoura Marina previously held the title of "Best International Marina" for three consecutive years from 2015 to 2017, further solidifying its reputation as a world-class destination (VW, n.d.).

The need to have projects focused on combating the seasonality of the Algarve emerged with the development and success of the marina over the years. The Vilamoura Sailing Hi Performance Centre project implemented in 2017 was born to promote the practice and improvement of sailing allowing high performance sailing athletes to find the ideal place to intern and train (RTA, 2021). Although the Sailing Centre operates year-round, the peak season is between October and May since other countries in Europe do not have the favourable conditions for team training like Vilamoura has (op. cit.). Later certified, in 2018, the Vilamoura Nautical Station supports a network of 64 businesses, promoting quality sailing, diving, and other water activities alongside complementary services like accommodation, restaurants, entertainment, and wellness options (op. cit.).

The negative impacts of coastal tourism on the environment must also be considered. Yachting activities and powered boat activities can have direct environmental impact related to use of anchors to ground recreational boats, for example, it can be harmful to the marine ecosystem and can lead to fewer fish, more algae and higher water turbidity (Mill, 2008). To maintain the levels of excellence for which the marina's reputation is known, it is important to stay on top of trends and obtain more certifications, to ensure compliance with sustainable practices, reinforcing the sustainability aspect of blue tourism.

Besides awarding the Gold Anchor certificate, The Yacht Harbour Association also has an international, voluntary accreditation system for marinas called The International Clean Marina Program, with the goal to reduce non-point source pollution (which occurs when water runs over land, picks up pollutants and then deposits them in surface waters) and to promote clean air and water (TYHA, n.d.). By signing the pledge to improve the environmental management system, the marina is committed to become a certified Clean Marina. In case the facilities are consistently implementing these practises, they will be designated a Clean Marina. This is the case of the Vilamoura Marina.

## AR Off Vilamoura

Due to heavy fishing pressure and a lack of natural reefs (Leitão et al., 2007) off the Algarve coast, the Vilamoura artificial reef was deployed in 1998 as part of a larger Algarve artificial reef project (Santos et al., 2006; Leitão et al., 2009; Moura et al. 2011; Santos et al., 2011). This reef, identically to the others previously deployed, was designed to mimic natural ecosystems and promote fish populations, complements the area's existing resources by offering sheltered areas for juvenile fish (protection reefs) and attracting a wider variety of species for recreational fishing and diving (exploitation reefs) (Santos & Monteiro, 1997, 1998, 2007; Santos et al., 2005).

Offshore Vilamoura, this artificial reef has the potential to boost coastal tourism originating from the Vilamoura Marina, upgrading its range and portfolio of activities. Since the artificial reefs that can be found offshore are open-access (Monteiro & Santos, 2000). Although underwater, and consequently not easily discernible, these modular reef structures offer a diversification of the alternatives available to divers and leisure anglers who want to try them out and eventually explore them further or more regularly.

## THE ROLE OF ARTIFICIAL REEFS

Two of the many reasons why ARs are deployed is to improve tourism, whether for recreational diving or fishing and also with the aim of diverting pressure from natural reefs (Firth et al., 2023). Self-Contained Underwater Breathing Apparatus (SCUBA) generically means diving with independent air support, requiring equipment such as mask, fins, weights, a wet suit as well as a portable air supply (e.g., Dimmock, 2007). This type of tourism encompasses promotional, sportive, and educational diving experiences aimed at observing, and photographing amidst the rich biodiversity and archaeological treasures lying underwater (Özgül & Lök, 2022). Depending on the level of experience of the divers, so is the type of depth to which they are exposed. Beginners dive at shallow depths in search of rewarding and adventurous experiences, while advanced divers can dive at greater depths and be looking for a separate way to allocate their capabilities. (Meisel-Lusby & Cottrell, 2008; Belhassen et al., 2017).

Since artificial reefs mimic natural reefs' characteristics and attract marine life, they can be a good substitute for natural reefs for SCUBA diving, by lessening the natural reefs' pressure while meeting tourism demand creating new spots (Treeck & Schuhmacher, 1999; Hsui & Wang, 2013). The biodiversity artificial reefs offer and even their depth range can be used for beginners (that might prefer shallower depths) and advanced divers (favouring deeper and more challenging waters) (Ramos, 2022).

The fishing of aquatic animals (e.g., fish, molluscs) as a leisure activity and not as a primary source of food or income, is named recreational fishing (FAO, 2012). The fish caught are usually released or kept for personal consumption, not sold. The most common recreational fishing technique is line fishing, where the fish is captured by a hook device (such as a baited hook mounted on line cast by a rod and reel), preventing it from escaping, thus allowing the fish to be removed from the water (Arlinghaus & Cooke, 2009). Other capture methods include using a spear, bow, rifle, trap or gillnet. Spearfishing for example, is done underwater using a speargun while freediving (Sbragaglia et al., 2019), unlike angling where the fisher is over the water in a boat.

By widening the range of spots and providing a location easy to access, artificial reefs can increase the diversity of species caught and even the catch rate of recreational fishing (Sutton & Bushnell 2007; Ramos and Santos, 2022).

### Artificial Reefs Off the South Coast of Portugal

The presence of Ria Formosa, a highly productive lagoon and estuarine system, together with the scarcity of natural reefs and high fishing intensity offshore resulted in a need to provide a compensation for the fishing effect in this area (Monteiro &

Santos, 2000). From this need resulted a pilot project of two artificial reef systems that were deployed in 1989-1990 with the intention of establishing not only by acting as nursery grounds, but also promoting an integrated strategy of management for the local fisheries resources, that could be used for gear such as purse seines and multigear purpose vessels (Santos & Monteiro, 1997). Two types of concrete reefs modules were deployed in offshore Faro and Olhão:

*Protection reefs*: Smaller modules deployed closer to the coast in shallow waters. These modules were called protective reefs because they create shelters due to their spread-out organization and inherent shape of each unit.*Exploitation reefs*: Larger units were established in deeper water. They were labelled exploitation reefs since they were designed to attract fish and overall did not provide shelter.

These initial AR systems were proven through the years to have increased fishing catches and diversity in fish assemblies (Santos & Monteiro, 1997; 1998). Following this positive evidence, a large-scale version of the artificial reef system was developed and its deployment occurred from 1998 to 2003 along the Algarve coast, with six new artificial reef systems being added: the extension of the Faro-Ancão artificial reef, Vilamoura, Oura, Alvor, Cacela and Tavira occupying in a total area of 43,5 $km^2$ (Santos et al, 2006; 2011; Leitão et al., 2009; Moura et al. 2011).

## METHODOLOGY

### Study Site

The Vilamoura artificial reef was deployed in 1998 after the promising results of the Faro and Olhão pilot artificial reef systems (Santos & Monteiro, 2007). Thus it was assembled following the same strategy, i.e., smaller cubic modules closer to the coast and larger octagonal modules further offshore, with the first being 35 modules of 2.7 $m^3$ and the second being 4 modules of 174 $m^3$ each (Figure 2). The two types of porous concrete structures were placed in dispersed groups and extending over an area of approximately 4 km (Ramos & Santos, 2015).

*Figure 2. The artificial reef module types: Used as protection reef, smaller cubic modules of 2.7 $m^3$ (left) and as exploitation reef, larger octagonal modules of 174 $m^3$ (right). Adapted from: Santos et al. (2011).*

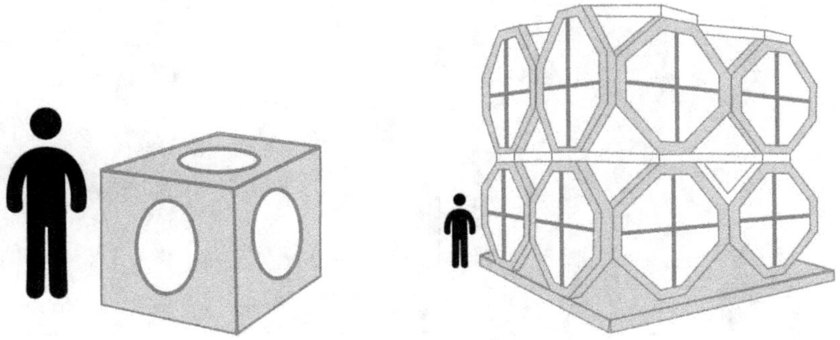

As in this investigation it is intended to analyze the importance of artificial reefs located off Vilamoura in promoting aquatic activities within the framework of blue tourism, it was decided to choose this reef complex as a case study. Two locations were considered for the study site: between the artificial reef and the coast (inner area) and over the artificial reef (Figure 3).

*Figure 3. Location of the Vilamoura artificial reef in Southern Portugal with the inner area highlighted in orange, next to the right of the Oura artificial reef along the 20m isobath. Source: Eurostat (n.d.).*

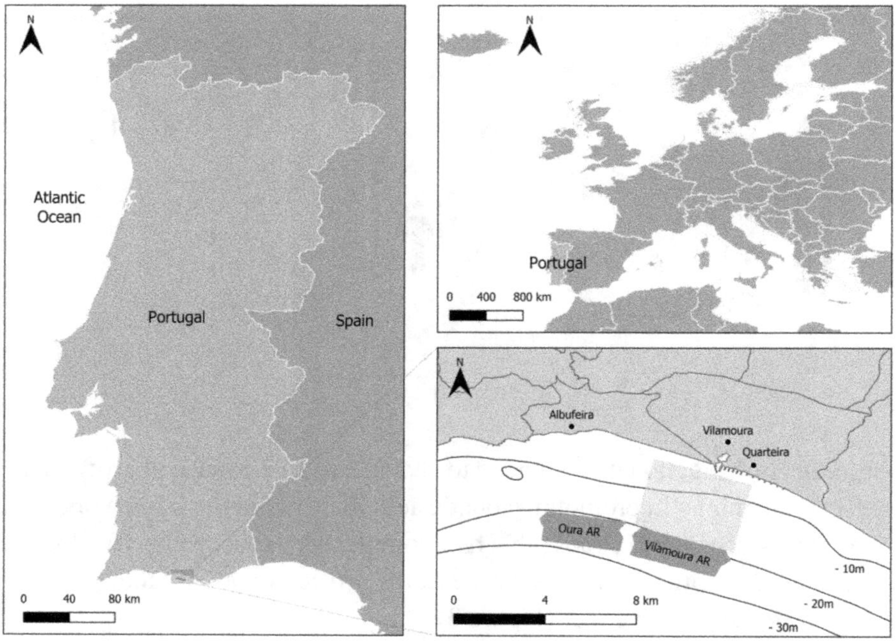


## Data Collection

MarineTraffic (MT, n.d.) provides real-time information on the trajectory of vessels and their current location using data from the Automatic Identification System (AIS). Being a ship-to-ship and ship-to-shore reporting system (Eriksen et al., 2006). The AIS uses global positioning systems (GPS), navigation sensors, and digital very high frequency (VHS) radio transceivers to exchange vessel information between vessels and stations on shore (Lee et al., 2019).

AIS uses the Maritime Mobile Service Identities (MMSI) number as identification to the messages sent and received by the transponders (Eriksen et al., 2006) and provides static information such as the type of vessel, length and beam and dynamic data such as the vessel's accurate position, time in universal time coordinated (UTC), the speed over ground (SOG) and navigational status. AIS transceivers can also provide voyage related data specifically details about the type of cargo, the port of destination and estimated time of arrival (ETA) (Kos et al., 2014).

Nevertheless, not all vessels with an AIS transponder are sending the same type of information since there are two types of AIS devices: AIS Class A is designed for commercial vessels above 300 gross tonnage, while AIS Class B is a more affordable but more limited version for smaller vessels (Figure 4). The positioning reporting rate is also different for both classes, whereas for Class A it is every 2-10 seconds, for Class B it can be from 5 to 30 seconds (Kos et al., 2014).

*Figure 4. Representation of the Information Exchange Between Vessels and AIS Stations*

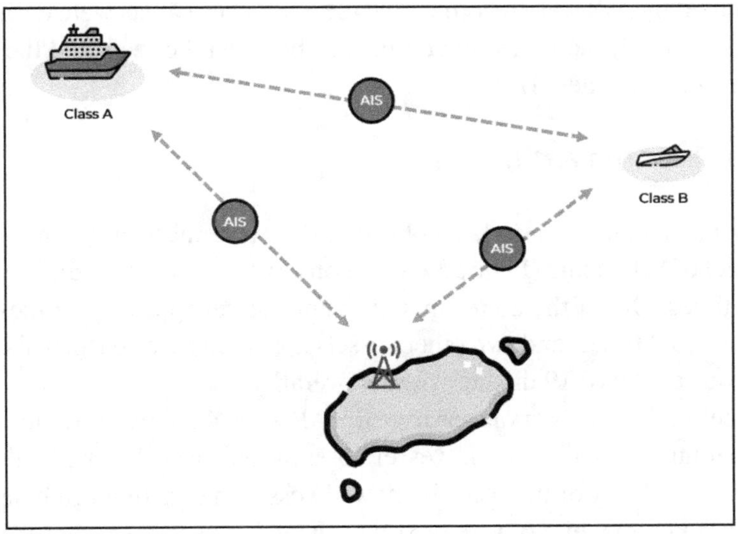


For this study it was decided to use information on the recreational activity over the artificial reef of Vilamoura and the inner area between the AR and the coast in winter (January, February and March) and summer (July, August and September) for the year 2022. The data relevant to the study was collected hourly and included the MMSI number of the vessels, their positioning (latitude and longitude coordinates), speed (in knots), date (year, month and day) and time (hours and minutes of the day).

From the unique number that the vessels have (i.e., the MMSI), it was possible to obtain information about the type of vessel, to retain only the vessels that carry out recreational activities that were in the data set. This included personal sailing and yacht vessels and charter vessels. From the MMSI the flag of the vessel's country was also obtained. With the latitude and longitude coordinates information, the vessels contained in the two areas, over the AR and in the inner area between the AR and the coast, were determined.

To infer on the use of the artificial reef, for the area over the reef, the information on the speed in knots was divided into two categories: drifting if the speed registered was lower than 5 knots and navigating if it was higher than 5 knots. Since the AR is underwater, recreational diving and fishing would require the boat involved to be at least drifting or going at a lower velocity, respectively.

## RESULTS

To monitor the use of the Vilamoura artificial reef, the positions of recreational vessels emitted by AIS were recorded in two areas. These areas were defined as: 1) over the reefs and 2) the inner area defined as between the reefs and the coastline (Figure 3 right below detail).

### Records Over the Artificial Reef

During the sampling period, 26 observations were made in the area over the artificial reef of Vilamoura (Figure 5). These observations involved different vessels at various times. Out of the 26 records, the same vessel appeared 4 times, another vessel appeared 3 times, and two other vessels appeared twice each. This indicates that there was a total of 19 distinct vessels overall.

From these 26 total observations over the AR area, 4 records were in the winter and 22 in summer. There were no vessel observations over the artificial reef area in January 2022. Most of the records showed vessels navigating (going at a speed higher than 5 knots) while 6 were drifting (under 5 knots) over the artificial reef. The types of vessels recorded over the AR were all personal sailing vessels but one that was a rented sailing vessel.

*Figure 5. Number of observations of recreational vessels by type of motion (drifting or navigating) per month in winter (January, February and March) and summer (July, August and September) 2022 over the artificial reef of Vilamoura area*

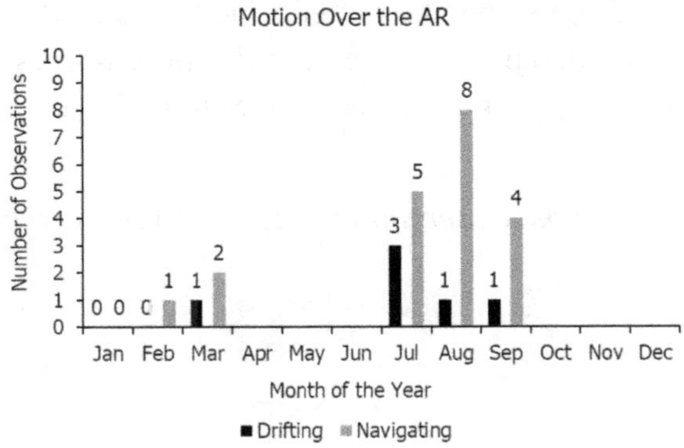


## Records Between the Artificial Reef and the Coastline

Regarding the inner area between the AR and the coast, 461 observations were made (Figure 6). From these records, 94 were in winter and 367 were in summer. Most of these vessels were originating from the Vilamoura Marina and were personal recreation vessels such as sailing vessels and yachts. These records came from 75 different boats.

*Figure 6. Number of observations of recreational vessels per month in winter (January, February, and March) and summer (July, August, and September) 2022 in the inner area between the artificial reef and the coast*

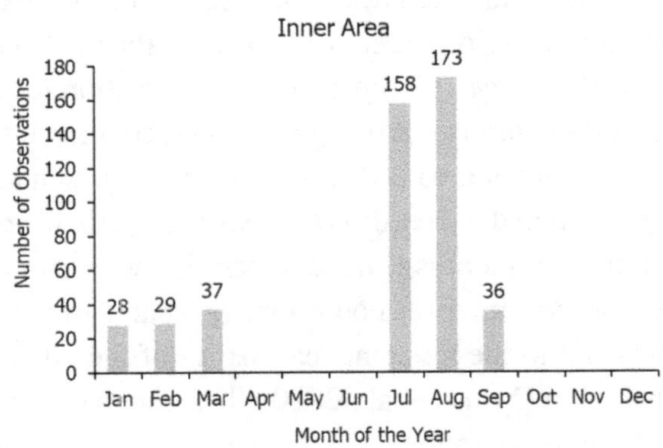

## Nationality of Origin of Pleasure Craft

Disregarding area, during the sampling period over the AR and in between the AR and the coast 94 different vessels were observed. From these 94 vessels, 33 were Portuguese flagged, 20 were from the United Kingdom and 10 from Spain (Figure 7). Other nationalities with only one vessel recorded in both areas were mostly from Europe except for Canada, Cayman Islands and South Africa.

*Figure 7. Vessels' Flag in the Observation Over the AR Area and in the Inner Area*

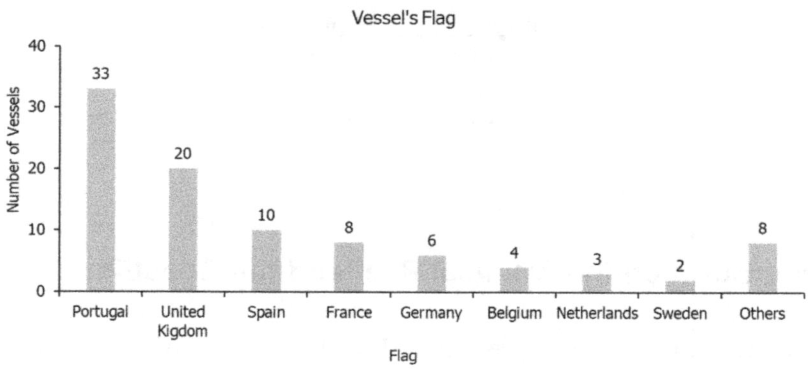


## DISCUSSION

The records show a higher number of vessels in the inner area between the artificial reef and the coast when compared to the area over the AR, regardless of season. Drifting is essential when accessing the use of the AR, since either for diving, angling and other types of recreational fishing, the boat needs to be drifting or stopped near the artificial reef. Comparing the observation results, it shows that the artificial reef of Vilamoura is not being used by recreational boats, at the time of observations while the inner area is being used for navigation from and to shore.

Another aspect evident in the records is the seasonality, there are up to five times more records in the summer months while compared to winter months. This can be due to the roughness of the winter sea conditions (Fernandes et al., 2018), to vessel management periods and to the inherent seasonality of the Algarve which brings more people to this region (Nunes et al., 2020). The country's flag most abundant in vessels was the Portuguese, followed by the United Kingdom and Spain.

The Vilamoura resort and marina has evolved throughout the years to become a major tourist destination, offering various activities for distinct types of tourists (VW, n.d.). The deployment of the artificial reefs off Vilamoura to enhance marine life and resources presented an enhanced opportunity for the marina firms that explore the niche of maritime activities such as diving and recreational fishing to add to their portfolio some potential spots for coastal and land-based activities.

One of the constraints of practicing the activities engaged with the AR is that they require specific skills, unlike other water-based activities such as boarding a passenger boat for wildlife watching. SCUBA diving requires certification to participate (Johansen, 2013), and recreational fishing involves selecting appropriate gear and having good handling skills (Palme et al., 2016). This requirement for specialized skills could be an impediment for tourists passing by Vilamoura, even if they are interested in interacting with the artificial reefs. Another factor is the lack of public awareness regarding their existence. Potential users perceive the siting of the artificial reefs to be either generally unknown (Ramos et al., 2007).

As for divers, one reason for less use is the lack of attractiveness about using the modular AR off Vilamoura as a diving site as it can appear monotonous (Ramos et al., 2022). In comparison, the four sunken ships donated by the Portuguese Navy, called Ocean Revival, off Portimão (Guerra et al., 2015; OR, n.d.) may present a more challenging and enigmatic dive (Bideci & Cater, 2019).

The artificial reefs structures are deployed at over 20 m depth underwater and even thought they can be used by the public; they are not easily accessible. Since they are roughly 5 km off coast, the enthusiasts would have to get near the AR by boat either for angling, diving or spearfishing. This type of boat is not usually large. The dataset used in this chapter came from AIS data, that while valuable and non-intrusive, only captures information from vessels equipped with transponders. This means the records might not fully represent the reality, especially for smaller boats that often do not have AIS (Harati-Mokhtari et al., 2007; Pelich et al., 2019).

The Vilamoura artificial reef was deployed to promote marine life and fisheries, with success (Ramos et al., 2015). However, information about recreational fishing activity in the RA is limited, as recreational fishermen and spear fishers often keep their fishing locations confidential for a variety of reasons. While the data from this study does not show recreational use, the AR's potential to enhance the marina's portfolio of activities remains, particularly if advertised by charter, angling or diving companies. These activities could contribute to the enhancement of the Vilamoura marina's blue tourism commitments.

To gain a more accurate description of the connection between artificial reefs and the recreational activities near Vilamoura, questionnaires and interviews with stakeholders could be conducted (Milon et al., 2000; Ramos et al., 2007). This would provide valuable insights from their perspectives, complementing the existing AIS data of the vessels. Another way of getting better understanding of this concept is to have a broader sample of data, in this case to compare the data from different years.

## CONCLUSION

This chapter investigated whether the artificial reefs off Vilamoura contribute to leisure activities associated with blue tourism. The Vilamoura AR complements the area's existing resources by attracting a wider variety of fish species, making the area more attractive for recreational activities like SCUBA diving and recreational fishing. The data showed minimal recreational activity directly over the artificial reef area. This finding does not directly indicate that the AR is not currently being used, but solely that at least there were not detected vessels equipped with an AIS transponder, despite its potential to enhance blue tourism experiences in Vilamoura.

The limitations that emerged from this chapter indicated that a larger sample size could have revealed more about the reality and additional methods such as surveys and interviews with divers and recreational fishers could eventually elucidate on the use of the artificial reef of Vilamoura and its contribution to blue tourism in the area.

## FUNDING

This work is funded by Portuguese national funds through FCT—Foundation for Science and Technology within the scope of the projects UIDB/04020/2020 (https://doi.org/10.54499/UIDB/04020/2020), UIDP/04020/2020 (https://doi.org/10.54499/UIDP/04020/2020), UIDB/04326/2020, UIDP/04326/2020, LA/P/0101/2020 and 2022.01783.PTDC and the Individual Calls to Scientific Employment Stimulus 2021.02180.CEECIND (https://doi.org/10.54499/2021.02180.CEECIND/CP1672/CT0003) (JR) and 2022.04803.CEECIND (FL).

## ACKNOWLEDGMENT

The authors would like to thank three anonymous reviewers whose comments substantially helped to improve the manuscript.

## REFERENCES

ARE. (n.d.). Algarve Real Estate. Retrieved from https://algarverealestate.com/

Arlinghaus, R., & Cooke, S. J. (2009). Recreational fisheries: socioeconomic importance, conservation issues and management challenges. *Recreational hunting, conservation and rural livelihoods: science and practice*, 39-58

Belhassen, Y., Rousseau, M., Tynyakov, J., & Shashar, N. (2017). Evaluating the attractiveness and effectiveness of artificial coral reefs as a recreational ecosystem service. *Journal of Environmental Management*, 203, 448–456. 10.1016/j.jenvman.2017.08.02028837911

Bideci, C., & Cater, C. (2019). In search of underwater atmosphere: A new diving world on artificial reefs. In *Atmospheric turn in culture and tourism: Place, design and process impacts on customer behaviour, marketing and branding* (pp. 245–257). Emerald Publishing Limited. 10.1108/S1871-317320190000016023

Dimmock, K. (2007). SCUBA diving, snorkeling, and free diving. In *Water-based tourism, sport, leisure, and recreation experiences* (pp. 128–147). Routledge. 10.1016/B978-0-7506-6181-2.50013-0

EC. (2023). *European Commission. The EU Blue Economy Report. 2023*. Publications Office of the European Union.

Eriksen, T., Høye, G., Narheim, B., & Meland, B. J. (2006). Maritime traffic monitoring using a space-based AIS receiver. *Acta Astronautica*, 58(10), 537–549. 10.1016/j.actaastro.2005.12.016

Eurostat. (n.d.). GISCO - Geographic Information System for Cooperation. Retrieved from https://ec.europa.eu/eurostat/web/gisco/geodata/reference-data

Fabi, G., Spagnolo, A., Bellan-Santini, D., Charbonnel, E., Çiçek, B. A., García, J. J. G., Jensen, A. C., Kallianiotis, A., & Santos, M. (2011). Overview on artificial reefs in Europe. *Brazilian Journal of Oceanography*, 59(spe1), 155–166. 10.1590/S1679-87592011000500017

FAO. (2012). *Food and Agriculture Organization. Technical Guidelines for Responsible Fisheries, No. 13*. FAO.

Fernandes, M., Fernandes, C., Barroqueiro, T., Agostinho, P., Martins, N., & Alonso-Martirena, A. (2018). Extreme wave height events in Algarve (Portugal): Comparison between HF radar systems and wave buoys. *Proceedings of the 5th Jornadas Engenharia Hidrográfica*, 19-21.

Firth, L. B., Farnworth, M., Fraser, K. P., & McQuatters-Gollop, A. (2023). Make a difference: Choose artificial reefs over natural reefs to compensate for the environmental impacts of dive tourism. *The Science of the Total Environment*, 901, 165488. 10.1016/j.scitotenv.2023.16548837524181

Fosse, J., Tonazzini, D., Morales, E., Gonzalez, A., Klarwein, S., Moukaddem, K., & Louveau, O. (2019). Blue Tourism: The Transition Towards Sustainable Coastal and Maritime Tourism in World Marine Regions.

Ghosh, T. (2012). Sustainable coastal tourism: Problems and management options. *Journal of Geography and Geology*, 4(1), 163–169. 10.5539/jgg.v4n1p163

Gormsen, E. (1997). The impact of tourism on coastal areas. *GeoJournal*, 42(1), 39–54. 10.1023/A:1006840622450

Guerra, F., Grilo, C., Pedroso, N. M., & Cabral, H. (2015). Environmental Impact Assessment in the marine environment: A comparison of legal frameworks. *Environmental Impact Assessment Review*, 55, 182–194. 10.1016/j.eiar.2015.08.003

Hanafiah, M. (2022). Framing The Future Agenda Of Blue Tourism In Sustainable Coastal Tourism Destinations. *Tourism and Hospitality Management*, 28(2), 465–470. 10.20867/thm.28.2.6

Harati-Mokhtari, A., Wall, A., Brooks, P., & Wang, J. (2007). Automatic Identification System (AIS): data reliability and human error implications. The Journal of Navigation, 60(3), 373-389.

Hsui, C. Y., & Wang, C. C. (2013). Synergy between fractal dimension and lacunarity index in design of artificial habitat for alternative SCUBA diving site. *Ecological Engineering*, 53, 6–14. 10.1016/j.ecoleng.2013.01.014

INE. (2023). Instituto Nacional de Estatística. Estatísticas do Turismo: 2022. Lisboa.

Jakšić, S., Stamenković, I., & Đorđević, J. (2013). Impacts of artificial reefs and diving tourism. *Turizam*, 17(4), 155–165. 10.5937/Turizam1304155J

Johansen, K. (2013). SCUBA diving education and training. In *SCUBA diving tourism* (pp. 81–96). Routledge.

Keller, K., Steffe, A. S., Lowry, M., Murphy, J. J., & Suthers, I. M. (2016). Monitoring boat-based recreational fishing effort at a nearshore artificial reef with a shore-based camera. *Fisheries Research*, 181, 84–92. 10.1016/j.fishres.2016.03.025

Kos, S., Vojković, L., & Brčić, D. (2014, May). Development of AIS and its influence on marine traffic control. In *8 Proceedings of the 8th Annual Baška GNSS Conference* (pp. 7-9).

Lee, E., Mokashi, A. J., Moon, S. Y., & Kim, G. (2019). The maturity of automatic identification systems (AIS) and its implications for innovation. *Journal of Marine Science and Engineering*, 7(9), 287. 10.3390/jmse7090287

Leitão, F., Santos, M. N., Erzini, K., & Monteiro, C. C. (2009). Diplodus spp. assemblages on artificial reefs: Importance for near shore fisheries. *Fisheries Management and Ecology*, 16(2), 88–99. 10.1111/j.1365-2400.2008.00646.x

Leitão, F., Santos, M. N., & Monteiro, C. C. (2007). Contribution of artificial reefs to the diet of the white sea bream (Diplodus sargus). *ICES Journal of Marine Science*, 64(3), 473–478. 10.1093/icesjms/fsm027

León, I. (2012). The tourist project of the SETAP Group for the International Bid of Ideas for Maspalomas Costa Canaria. In *6th Conference of the International Forum on Urbanism (IFoU): TOURBANISM, Barcelona, 25-27 gener* (pp. 1-10). International Forum on Urbanism.

Leposa, N. (2020). Problematic blue growth: A thematic synthesis of social sustainability problems related to growth in the marine and coastal tourism. *Sustainability Science*, 15(4), 1233–1244. 10.1007/s11625-020-00796-9

Lima, J. S., Zalmon, I. R., & Love, M. (2019). Overview and trends of ecological and socioeconomic research on artificial reefs. *Marine Environmental Research*, 145, 81–96. 10.1016/j.marenvres.2019.01.01030837123

Lobo, S. (2010). Sun, Sand, Sea and Bikini. Arquitectura e turismo: Portugal anos 60. [in Portuguese]. *Revista Critica de Ciencias Sociais*, (91), 91–106. 10.4000/rccs.4170

Loo, M. J., Demirel, M. C., Garg, A., & Jayasinghe, R. P. (2009). The Impact of Marinas on the Coastal Zone. INUAF STUDIA-Instituto Universitário Dom Afonso III. Suplemento, 16, 25-27.

McGlennon, D., & Branden, K. L. (1994). Comparison of catch and recreational anglers fishing on artificial reefs and natural seabed in Gulf St. Vincent, South Australia. *Bulletin of Marine Science*, 55(2-3), 510–523.

Meisel-Lusby, C., & Cottrell, S. (2008). Understanding motivations and expectations of SCUBA divers. *Tourism in Marine Environments*, 5(1), 1–14. 10.3727/154427308785855279

Mejjad, N., Rossi, A., & Pavel, A. B. (2022). The coastal tourism industry in the Mediterranean: A critical review of the socio-economic and environmental pressures and impacts. *Tourism Management Perspectives*, 44, 101007. 10.1016/j.tmp.2022.101007

Mill, R. C. (2008). *Resorts: management and operation*. John Wiley and Sons.

Milon, J. W., Holland, S. M., & Whitmarsh, D. J. (2000). *Social and economic evaluation methods*. CRC Press LLC.

Monteiro, C. C., & Santos, M. N. (2000). Portuguese artificial reefs. In *Artificial Reefs in European Seas* (pp. 249–261). Springer Netherlands. 10.1007/978-94-011-4215-1_15

Moreno, A., & Amelung, B. (2009). Climate change and coastal and marine tourism: Review and analysis. *Journal of Coastal Research*, 1140–1144.

Moura, A., Fonseca, L. C. D., Boaventura, D., Santos, M. N., & Monteiro, C. C. (2011). Estimation of secondary production of the Faro/Ancão artificial reefs. *Brazilian Journal of Oceanography*, 59(spe1), 91–94. 10.1590/S1679-87592011000500011

MT. (n.d.). MarineTraffic - Global Ship Tracking Intelligence. https://www.marinetraffic.com/

Nauta, J., Christianen, M. J., Temmink, R. J., Fivash, G. S., Marin-Diaz, B., Reijers, V. C., Didderen, K., Penning, E., Borst, A. C., Heusinkveld, J. H., Zwarts, M., Cruijsen, P. M., Hijner, N., Lengkeek, W., Lamers, L. P., van der Heide, T., Bouma, T. J., van der Wal, D., Olff, H., & Govers, L. L. (2023). Biodegradable artificial reefs enhance food web complexity and biodiversity in an intertidal soft-sediment ecosystem. *Journal of Applied Ecology*, 60(3), 541–552. 10.1111/1365-2664.14348

Nunes, L. J., Raposo, M. A., & Gomes, C. J. P. (2020). The impact of tourism activity on coastal biodiversity: A case study at praia da cova redonda (Algarve—Portugal). *Environments (Basel, Switzerland)*, 7(10), 88. 10.3390/environments7100088

Oliveira, M. J., Farinha, F., da Silva, E. M., Lança, R., Pinheiro, M. D., & Miguel, C. (2019). Observatory of sustainability of the Algarve Region for tourism: Proposal for environmental and sociocultural indicators. *International Journal of Humanities and Social Science*, 13, 1237–1244.

OR. (n.d.) Ocean Revival. Retrieved from http://www. oceanrevival .pt/pt/

Ortega, C., Nogueira, C., & Pinto, H. (2013). Sea and littoral localities' economy: Exploring potentialities for a maritime cluster-An integrated analysis of Huelva, Spain and Algarve, Portugal. *Journal for Maritime Research*, 10(2), 35–42.

Özgül, A., & Lök, A. (2022). Artificial Reef Applications for Diving Tourism: Artificial Wreck Reefs in Turkey. In *Impact of Artificial Reefs on the Environment and Communities* (pp. 151-168). IGI Global.

Palme, A. C., Nguyen, V. M., Gutowsky, L. F., & Cooke, S. J. (2016). Do fishing education programs effectively transfer 'catch-and-release'best practices to youth anglers yielding measurable improvements in fish condition and survival? *Knowledge and Management of Aquatic Ecosystems*, (417), 42. 10.1051/kmae/2016029

Pelich, R., Chini, M., Hostache, R., Matgen, P., Lopez-Martinez, C., Nuevo, M., Ries, P., & Eiden, G. (2019). Large-scale automatic vessel monitoring based on dual-polarization sentinel-1 and AIS data. *Remote Sensing (Basel)*, 11(9), 1078. 10.3390/rs11091078

Picken, F. (2023). Tourism and the blue economy. *Tourism Geographies*, 1–9. 10.1080/14616688.2023.2291821

Ramm, L. A., Florisson, J. H., Watts, S. L., Becker, A., & Tweedley, J. R. (2021). Artificial reefs in the Anthropocene: A review of geographical and historical trends in their design, purpose, and monitoring. *Bulletin of Marine Science*, 97(4), 699–728. 10.5343/bms.2020.0046

Ramos, J. (2022). Portuguese Artificial Reef Modules: Are They Suitable for Dive Tourism? In *Impact of Artificial Reefs on the Environment and Communities* (pp. 127-150). IGI Global.

Ramos, J., & dos Santos, M. (2022). Efficiency of Artificial Reefs for Local Communities Over Time. In *Impact of Artificial Reefs on the Environment and Communities* (pp. 1–20). IGI Global. 10.4018/978-1-6684-2344-8.ch001

Ramos, J., & Santos, M. N. (2015). Facilitating fishing decisions in an artificial reef area off southern Portugal: A case study using generalized additive models. *Journal of Applied Ichthyology*, 31, 24–34. 10.1111/jai.12948

Ramos, J., Santos, M. N., Whitmarsh, D., & Monteiro, C. C. (2007). Stakeholder perceptions regarding the environmental and socio-economic impacts of the Algarve artificial reefs. In *Biodiversity in Enclosed Seas and Artificial Marine Habitats: Proceedings of the 39th European Marine Biology Symposium, held in Genoa, Italy, 21–24 July 2004* (pp. 181-191). Springer Netherlands. 10.1007/978-1-4020-6156-1_16

Rendle, E. J., & Rodwell, L. D. (2014). Artificial surf reefs: A preliminary assessment of the potential to enhance a coastal economy. *Marine Policy*, 45, 349–358. 10.1016/j.marpol.2013.09.004

RTA. (2021). Região de Turismo do Algarve. Plano de Marketing Estratégico do Turismo do Algarve: 2020-2023.

Santos, M. N., Lino, P. G., Pousão-Ferreira, P., & Monteiro, C. C. (2006). Preliminary results of hatchery-reared seabreams released at artificial reefs off the Algarve coast (southern Portugal): A pilot study. *Bulletin of Marine Science*, 78(1), 177–184.

Santos, M. N., & Monteiro, C. C. (1997). The Olhao artificial reef system (south Portugal): Fish assemblages and fishing yield. *Fisheries Research*, 30(1-2), 33–41. 10.1016/S0165-7836(96)00567-X

Santos, M. N., & Monteiro, C. C. (1998). Comparison of the catch and fishing yield from an artificial reef system and neighbouring areas off Faro (Algarve, south Portugal). *Fisheries Research*, 39(1), 55–65. 10.1016/S0165-7836(98)00169-6

Santos, M. N., & Monteiro, C. C. (2007). A fourteen-year overview of the fish assemblages and yield of the two oldest Algarve artificial reefs (southern Portugal). In *Biodiversity in Enclosed Seas and Artificial Marine Habitats:Proceedings of the 39th European Marine Biology Symposium, held in Genoa,Italy,21–24 July 2004* (pp. 225-231). Springer Netherlands.

Santos, M. N., Monteiro, C. C., & Lasserre, G. (2005). Observations and trends on the intra-annual variation of the fish assemblages on two artificial reefs in Algarve coastal waters (southern Portugal). *Scientia Marina*, 69(3), 415–426. 10.3989/scimar.2005.69n3415

Santos, M. N., Monteiro, C. C., & Leitão, F. (2011). The role of artificial reefs in the sustainability of artisanal fisheries. Artificial reefs in Fisheries Management. CRC Press, Taylor and Francis Group.

Sari, F. O., Bulut, C., & Pirnar, I. (2016). Adaptation of hospitality service quality scales for marina services. *International Journal of Hospitality Management*, 54, 95–103. 10.1016/j.ijhm.2016.02.004

Sbragaglia, V., Morroni, L., Bramanti, L., Weitzmann, B., Arlinghaus, R., & Azzurro, E. (2018). Spearfishing modulates flight initiation distance of fishes: The effects of protection, individual size, and bearing a speargun. *ICES Journal of Marine Science*, 75(5), 1779–1789. 10.1093/icesjms/fsy059

Shani, A., Polak, O., & Shashar, N. (2012). Artificial reefs and mass marine ecotourism. *Tourism Geographies*, 14(3), 361–382. 10.1080/14616688.2011.610350

Supriyanto, E. E. (2022). Blue tourism: Treating marine ecosystems and increasing the potential of maritime tourism in Indonesia. *Jurnal Kepariwisataan Indonesia: Jurnal Penelitian dan Pengembangan Kepariwisataan Indonesia, 16*(2), 138-148.

Sutton, S. G., & Bushnell, S. L. (2007). Socio-economic aspects of artificial reefs: Considerations for the Great Barrier Reef Marine Park. *Ocean and Coastal Management*, 50(10), 829–846. 10.1016/j.ocecoaman.2007.01.003

Tonazzini, D., Fosse, J., Morales, E., González, A., Klarwein, S., Moukaddem, K., & Louveau, O. (2019). *Blue Tourism. Towards a sustainable coastal and maritime tourism in world marine regions. Edited by eco-union.*

Treeck, P., & Schuhmacher, H. (1999). Mass diving tourism–a new dimension calls for new management approaches. *Marine Pollution Bulletin*, 37(8-12), 499–504. 10.1016/S0025-326X(99)00077-6

TYHA. (n.d.). The Yacht Harbour Association. Retrieved from https://www.tyha.co.uk/

VA. (2021). *VisitAlgarve*. Boletim Conjuntura Turística.

Valle, P. O. D., Guerreiro, M., Mendes, J., & Silva, J. A. (2011). The cultural offer as a tourist product in coastal destinations: The Case of Algarve, Portugal. *Tourism and Hospitality Research*, 11(4), 233–247. 10.1177/1467358411420623

VW. (n.d.). Vilamoura World. Retrieved from https://www.vilamouraworld.com/

Wilks, J. (2021). Safety in coastal and marine tourism. In *Tourist health, safety and wellbeing in the new normal* (pp. 411–442). Springer Singapore. 10.1007/978-981-16-5415-2_17

## KEY TERMS AND DEFINITIONS

**Algarve Coast:** Southernmost part of Portugal, known for its coastline of beaches, caves and cliffs overlooking the Atlantic Ocean. The coast to the east (leeward) is mainly made up of large beaches with fine, light sand without any prominent cliffs, while to the west (windward) there are several rocky areas whose cliffs intertwine with immense small sandy beaches.

**Artificial Reefs:** Submerged man-made structures designed to mimic some of the characteristics of natural reefs, usually to promote biodiversity or recreational activities.

**Automatic Identification System:** Tracking system used on vessels for exchanges of information, useful for avoiding collisions, reporting positions and monitoring.

**Blue Growth:** Economic improvement using marine and coastal resources in a sustainable way.

**Blue Tourism:** Tourism activities associated with marine and coastal areas related to sustainability.

**MMSI:** Maritime Mobile Service Identity (MMSI) numbers are a number unique to a given vessel and consist of a series of nine digits used to uniquely identify a radio with a technology associated with VHF radios and called Digital Selective Calling (DSC), and that exists on board that vessel.

**Observation:** When the researcher uses their senses to gather information about the surroundings. Observation involves the systematic and intentional monitoring of a subject, object or event. During observation, records are made/obtained to document the observations made during the observation process.

**Recreational Fishing:** The activity of fishing for leisure and not as a primary source of food, there are distinct types such as angling (the most common, using a rod and fishing line) or spearfishing (using a speargun).

**SCUBA Diving:** The activity of moving or swimming underwater using a self-contained underwater apparatus (SCUBA) which is an air tank strapped to the body.

**Sustainability:** The ability to maintain natural resources long-term without damaging the environment.

**UTC:** This abbreviation refers to Coordinated Universal Time or Universal Time Coordinated. It consists of a coordinated time scale maintained by the *Bureau International des Poids et Mesures* (BIPM).

**Vilamoura Artificial Reef:** Consists of an artificial reef area composed of concrete modules with some porosity, in which two types of modules were used grouped in regular sets. This modular grouping consists of three parallel lines arranged at a shallower depth with sets of 105 units of smaller modules and a line at a greater depth for sets of 4 larger modules. The general arrangement of the modules extends longitudinally for approximately 4 km in bathymetric areas between 18 and 36 meters deep.

# Chapter 10
# Sport Events' Strategic Multi-Leverage:
## Linking Beach Soccer to Big Waves to Stimulate Coastal Tourism

**Margarida Mascarenhas**
*Faculty of Human Kinetics, Universidade de Lisboa, Portugal & CinTurs, Universidade do Algarve, Portugal*

**Édi Milhazes**
*Faculty of Human Kinetics, Universidade de Lisboa, Portugal*

**Rute Martins**
*CinTurs, Universidade do Algarve, Portugal*

## ABSTRACT

*Nazaré is internationally recognized as a big wave surfing destination. Combined with the media dynamics surrounding that phenomenon, the local government has pursued the strategy of promoting Nazaré as a prime location for beach sports. Subsequently, this research aimed to understand and stimulate the strategic leveraging process of a portfolio of beach soccer events hosted in 2021, in Nazaré, Portugal. The research strategy adopted the case study through an action-research approach. Although strategic goals and actions were implemented pointing to economic, social, environmental, sport, and political dimensions, the results showed that the long-term strategic economic objectives were the most pursued, allowing the image of Nazaré to be anchored to the worldwide phenomenon of big waves. To maximize the host community's benefits, greater attention must be paid to portfolio design and the integration of all local stakeholders representing the destination's areas which are intended to be strategically leveraged in different dimensions.*

DOI: 10.4018/979-8-3693-2436-3.ch010

## 1. INTRODUCTION

Sport has social and economic value for the community and should therefore be understood as a long-term public investment (Soares et al., 2016). In this sense, sport events should act as an instrument, allowing: (a) widespread access to sport, also functioning as an excellent tool for promoting the sector itself; and (b) the creation of opportunities for community development at various levels, such as economic, social, environmental, sporting and political (Chalip, 2004; Chalip et al., 2017; O'Brien & Chalip, 2008; Pereira et al., 2015, 2020; Sant et al., 2019; Ziakas, 2023).

The perspective of strategic leverage presupposes the proactiveness of organizers to stimulate a diversity of benefits for the community that can be maximized by events (O'Brien & Chalip, 2008). In this way, sport events have been strongly associated with strategies to combat seasonality, given their potential resolving effect in the face of frequent fluctuations in demand for destinations (Ziakas, 2023). Many destinations decide to host sport events, including non-mega sport events, precisely because of their potential benefits for developing the host communities (Duglio & Beltramo, 2017; Malchrowicz-Mośko & Poczta, 2018).

Some authors (e.g., Getz & Page, 2016; Ziakas, 2023) argue that, when there is a concerted logic between events (i.e., portfolio), the complementarity effect created around them can capitalize on the ability to create new markets. Furthermore, when they are integrated into the human capital and existing infrastructures of the host community, events produce sustainable forms of tourism promotion (Schulenkorf et al., 2019; Ziakas & Getz, 2020). The portfolio concept incorporates the complexity of achieving synergies between stakeholders, which, together, maximize the benefits associated with events, whether for their own benefit or that of third parties (Salgado-Barandela et al., 2022; Ziakas, 2023).

Nazaré, a small Portuguese village, is a coastal destination whose local government has focused on promoting beach sport events leveraged by Nazaré's global tourism strategy as a big wave surfing destination. Therefore, the local tourism strategy is based on planning and carrying out a portfolio of beach sport events. Given the prominence of standalone events as a research context, which contrasts with the need to deepen knowledge about the strategic leverage of event portfolios (Schulenkorf et al., 2022), the contributions of this study are particularly relevant. Additionally, this study follows the recommendation for alternative approaches to the production of knowledge about strategic leverage, such as action-research (Chalip et al., 2017).

Following the above, this study performed an action-research aiming to understand and stimulate the strategic leveraging process of a portfolio of beach soccer events hosted in Nazaré in five dimensions: economic, social, environmental, sporting and political. To this end, the following specific objectives were defined: (a) identify the objectives and strategic leverage actions implemented in previous editions of the sport events portfolio; and (b) analyse the implementation of the strategic leverage objectives and actions of the sport events portfolio, as well as emerging strategic leverage actions.

# 2. LITERATURE REVIEW

## 2.1. Sport Events: Portfolio Design

The creation of a portfolio of events aims to share resources and common elements, providing the generation of complementary markets and the transfer of practical knowledge between events (Antchak et al., 2019; Salgado-Barandela et al., 2022; Ziakas, 2023). Consequently, although the construction of the event portfolio may take on a more formal or informal nature, both the strategic design for its composition and the creation of synergies between stakeholders are fundamental to achieving sustained development (Salgado-Barandela et al., 2022). In addition, the strategic design of the portfolio presupposes the development of synergies between events and their use to achieve multiple objectives (Ziakas, 2016).

From this perspective, Ziakas (2019) argues that the portfolio of events can serve as a political instrument, supporting the discourse for sustainable development. On the other hand, Pereira et al. (2020) reinforce that coordination between political and sporting entities and institutions allows for "capitalization of the capacity (know-how), resources, volunteer pools, or the markets that might be engendered and maintained by the array of events (...) understanding interrelationships, fostering synergies, and enhancing complementarities" (p. 71).

## 2.2. Sport Events: Strategic Leveraging

Strategic leveraging of sport events (Chalip, 2004) considers events / event portfolios as leverageable resources, capable of maximizing potential benefits for the host community. Research has highlighted interest in studying the strategic leverage of events (Schulenkorf et al., 2022), reinforcing the need to plan, synergistically and strategically, the effects that are desired to be produced in destinations (Chalip, 2004; Chalip et al., 2017; O'Brien & Chalip, 2008; Pereira et al., 2015, 2020; Sant et al., 2019; Ziakas, 2023). Strategic leverage mitigates the limited perspective of the concepts of impact and legacy, adding the pre-event phase to the study of sport events. While leverage requires event stakeholders to be proactive in maximizing benefits in the short, medium and long term, impact and legacy seem to be measuring instruments - even used in the strategic leverage process - to justify or repudiate the hosting of sport events (Chalip, 2018). This premise means that the focus of strategic leverage is not solely on the organization of the sport event, nor is it limited in time, but is actually a complex process of community learning, and increasingly, in concert with the local development of communities hosting sport events (Chalip, 2004; Ziakas, 2019).

Research has advanced with several strategic leverage objectives, namely: (a) economic (Chalip, 2004)); (b) social / environmental (O'Brien & Chalip, 2008); (c) sport participation (Chalip et al., 2017; Pereira et al., 2020); and (d) political (Pereira et al., 2020).

### 2.2.1. Economic Leverage

The economic leverage model for sport events developed by Chalip (2004) is based on two fundamental premises: (a) the way in which visitors become an opportunity for leverage; and (b) the intervention of the media as catalysts for positive and lasting impacts on destinations; i.e., maximizing business/income, through the increase in tourist flows provided by events; and strengthening the brand as a destination / community that hosts one or more events through media coverage (Pereira et al., 2015).

According to Chalip (2004), there are strategic actions that can maximize benefits for the local community in the short and long term; while visitors presuppose immediate and direct gains in the local economy, the media induce a longer-term return, mainly by propagating the image of the destination. Thus, to optimize business and revenue, Chalip (2004) argues that events must: (1) induce expenditure; (2) extend the stay; (3) retain event expenses in the local economy; and (4) create 'new' business opportunities. There are several examples of studies demonstrating how these strategic actions can be implemented in destinations: (a) the Spetses (Greece) mini marathon triggered the development of commercial partnerships with bars, cafes, restaurants and other local establishments, with the aim of providing special offers and prices to race participants (Schulenkorf et al., 2019); (b) aiming to extend visitors' stay, nautical events in Portimão (Portugal) provided information to external event organizers about the specific characteristics of the destination - culture, gastronomy, tourist attractions, etc. - so that they could be publicized in the media of the events themselves (Pereira et al., 2015); and (c) Mhanna et al. (2017) identified tourism promotion actions developed at the London Olympic Games in 2012, where several discount campaigns were created to economically boost the destination.

In the long term, according to Chalip (2004), economic benefits are achieved through the media, based on two tactics: (1) showcase via event advertising and reporting; and (2) use the event in advertising and promotions. Recent research (Duignan et al., 2023; Mascarenhas et al., 2021; Salgado-Barandela et al., 2022; Ziakas, 2023) has reinforced that synergies between entities are crucial to achieving event sustainability and maximizing benefits to the host community. Sport events also have the ability to increase the favourable perception of the travel destination, since a good experience induces the desire to return to the place in the sport participant / visitor (Mainolfi & Marino, 2020; Tavakkoli, 2016).

### 2.2.2. Social Leverage

At a social level, the strategic leverage of sport events applies the concepts of liminality and *communitas* (O'Brien & Chalip, 2008) in which they are considered as drivers of promoting social interaction and the spirit of celebration (Chalip, 2006). Taking advantage of this 'invisible wave of solidarity' is precisely what is at the basis of this leverage model, which must be aligned with the social values of the local community and the specific sport subculture of the event's type of sport (O'Brien & Chalip, 2008). For example, Richardson and Fletcher (2020) studied the Premier League Kicks project in which football was used as a tool for social inclusion and the fight against racism; the authors argue that the more sport events focus on community development, especially in a youth-centred approach, the better they will promote the acquisition of social capital (i.e., the desire for association / volunteering), increasing the efficiency of communities and allowing for greater effectiveness in the pursuit of collective interests.

According to Richelieu (2018), both sports, in general, and sport events, in particular, must be organized around social cohesion, so that a sense of community around sport and horizontal social mobility can be created, enabling the transition of an individual from one social group to another. In turn, Ziakas (2016, 2019) argues that it is crucial to consider a broad approach to the intended social effect, development processes and maintenance of the social order of the host community, especially because all of them are influenced (and influence) directly by events.

### 2.2.3. Environmental Leverage

Currently, sport events are not only faced with the need to ensure greater care regarding the inclusion of aspects of social and economic development of communities, but also with the protection and management of the environment through ecologically sustainable practices (Martins et al., 2022; Melovic et al., 2019).

However, environmental management should not be confused with environmental leverage. While the first presupposes control of the harmfulness produced in the environment, the second foresees, after the sport event, a sustained change in terms of knowledge and individual and collective attitudes and behaviours (O'Brien & Chalip, 2008). Therefore, the organization of sport events must encourage environmental responsibility (Malchrowicz-Mośko et al., 2019; Martins et al., 2022).

Additionally, as highlighted by Mascarenhas et al. (2021), within the sports tourism market, there is both a need for collaboration between event/destination stakeholders for the development of environmental strategies and initiatives, and for the promotion of environmental education of the respective consumers and producers. In this aspect, stakeholders play a fundamental role as they are often the main agents of pressure on sport organizations to adopt sustainable practices (Mascarenhas et al., 2021; McCullough et al., 2016).

This problem led O'Brien and Chalip (2008) to integrate the triple bottom line (TBL) approach into strategic leverage, arguing that sport events can stimulate economic development, and at the same time, changes in the social and environmental agendas of destinations. To this end, O'Brien and Chalip (2008) defined the following strategic objectives: (1) focus stakeholders' attention on specific social and/or environmental themes previously defined; and (2) set and/or change the community agenda in accordance with social and/or environmental issues.

### 2.2.4. Sport Leverage

Sports participation is another dimension that can be leveraged by sport events (Chalip et al., 2017; Weed, 2017). The findings of Potwarka et al. (2020) proved that exposure to a leverage initiative at a track cycling event - offering a voucher to practice the sport after the event - can stimulate participation in the sport; also, due to the attendance of competitions, event spectators tend to consume the respective sport discipline in the future; lastly, the voucher produced positive effects on spectators who were more reticent about practicing this type of sport, showing a greater likelihood of experiencing track cycling again compared to those who did not receive the voucher, as a result of the experience of spectating the sport event. From a different perspective, according to the study of a sport event by Taks et al. (2014), in the opinion of the respective organizers, the 'legacy' effect of the facility built for its holding and the emphasis given by the local media to the sport discipline of the event would automatically generate greater interest and sport participation in the community; however, the evidence did not corroborate the opinion of the event organizers, not least because specific actions were not implemented to effectively stimulate sport participation and development in the local community.

The model of strategic leverage of sport participation (Chalip et al., 2017) is not limited to reinforcing sports practice; rather, it is a plan that encompasses a whole set of systems, structures and attitudes designed to increase the involvement of organizations, both those directly related to sport events / sport sector and those in non-sporting sectors. From this perspective, the exploration of partnerships is crucial to support the ability to implement strategic leverage actions in sport events (Misener et al., 2020), since, normally, organizers limit their intervention to the holding of events, leaving the responsibility of leverage to a secondary level. According to Chalip et al. (2017), sport organizations need the support of event organizers to guarantee the leverage effect, and both event organizers and sport organizations need the collaboration of non-sport organizations.

### 2.2.5. Political Leverage

The conceptualization of political leverage objectives for sport events, according to empirical research by Pereira et al. (2020), identifies two tactics aiming to reinforce the political advantage of decision-makers: (1) strengthen the political design; and (2) enhance the visibility of political actors. The political dimension underlies these purposes, since the strategic planning of sport events can produce very positive effects on the local community, and even serve as a political justification for radical changes at various levels, such as economic/financial, social, urban, tourist, landscape, environmental and political (Sant et al., 2019; Ziakas, 2019).

The decision to organize a sport event is the first of many political decisions resulting from this entire process, which generally leads to large campaigns to promote the destination, especially in sport events with extensive media coverage. In this sequence, the approximation of external organizing entities of sporting events to local authorities is inevitable (Antchak, 2017).

Furthermore, sport events can serve to justify investment in infrastructure with the aim of attracting tourists, encouraging inward investment and changing the image of the destination (Rocha, 2017; Sroka, 2021). Therefore, the political approach in the decision-making process regarding events gives local organizing committees a decisive role, as they are responsible for reconciling the event's framework with the political agendas of local governments (Pereira et al., 2020). This symbiosis is essential to foster effective synergies between local industry, sport organizations and local residents, which can be leveraged for the set of events that make up a given event portfolio (Pereira et al., 2020; Ziakas, 2019). The study by Lu and Lin (2020) attested that the successful organization of the sport event boosted the image of the destination, and reinforced infrastructure and national identity, thus producing social benefits in terms of urban development and the formation / increase of national pride; these benefits resulted from the establishment of good relationships between public and private entities, both vertically and horizontally; The results of this study also verified the political legacy left by the success of the event in terms of global visibility and international reputation.

### 2.2.6. Cross-Leverage

The concept of cross-leverage encourages the development of actions with the potential to simultaneously stimulate more than one of the strategic leverage objectives, and if possible, transversal to a set of events strategically selected and coordinated between stakeholders, mainly from the community that hosts the event (Chalip et al., 2017; Ziakas, 2019).

To this end, it is crucial that the strategic planning process triggered by leverage is guided by the selection of tactics that guarantee cross-leverage inter-events (Antchak et al., 2019; Ziakas, 2023) and intra-events (Pereira et al., 2015), aiming at the reuse (efficiency) of resources and the full exploitation of potential opportunities. From this perspective, studies that argue that an event, in itself, does not influence destination revisit behaviour (Aicher et al., 2020; Newland & Yoo, 2021), also underline the premise of multidisciplinary leverage, highlighting that the destination should not depend on just one event to attract future tourists (Newland & Yoo, 2021).

Cross-leveraging has the potential to extend the benefits of events beyond the destination's borders (Ziakas, 2019). For this reason, the concertation of sport events at regional and national level is increasingly frequent (Antchak et al., 2019), as a tool for equitable promotion of the territory (Pereira et al., 2015, 2020; Sant et al., 2019).

## 3. METHODOLOGY

### 3.1. Characterization and Contextualization of Events

The portfolio studied in this research was composed of three sport events (namely, beach soccer events), two international competitions - Euro Beach Soccer League Stage (EEBSL) and Euro Winners Cup (EWC) -, and one national competition - National Championship-Elite Division (CN-DE) -, hosted in Nazaré, Portugal. All these events have been taking place in Nazaré since 2017, at the "Viveiro Jordan Santos" (VJS) stadium. The three events have different characteristics, both from the perspective of participation and public investment. Among the three events, the EWC is the one that involves a greater share of public funds, while simultaneously attracting a greater number of teams (Table 1).

*Table 1. Characterization of the Evolution of Beach Football Events Hosted in Nazaré*

| Event | Year | Teams (n) | Public investment (€: euros) | Participant countries (n) |
|---|---|---|---|---|
| EEBSL | 2017 | 14 | 100 000 | 12 |
| | 2018 | 17 | 100 000 | 15 |
| | 2019 | 18 | 130 000 | 15 |
| | 2020 | 5 | 130 000 | 5 |
| | 2021 | 19 | 120 000 | 15 |
| EWC | 2017 | 84 | 200 000 | 26 |
| | 2018 | 78 | 210 000 | 28 |
| | 2019 | 82 | 210 000 | 25 |
| | 2020 | 32 | 230 000 | 13 |
| | 2021 | 68 | 220 000 | 17 |
| CN-DE | 2017 | 16 | 0 | 1 |
| | 2018 | 16 | 0 | 1 |
| | 2019 | 12 | 0 | 1 |
| | 2020 | 16 | 0 | 1 |
| | 2021 | 20 | 0 | 1 |

*Note.* EEBLS - Euro Beach Soccer League Stage; EWC - Euro Winners Cup; CN-DE - National Championship-Elite Division

All sport events included in the studied portfolio took place in 2021, on the following dates: (a) EEBSL, from June 17th to 19th; (b) EWC, from July 12th to 18th; and (c) CN-DE, from August 5th to 8th. The events were organized by four entities / organizations, namely, The Nazaré City Council (CMN), Nazaré Qualifica (NQ) - a municipal company created to develop the destination's brand image associated with hosting beach sport events (NQ, 2020) -, Beach Soccer World Wide (BSWW), and Portuguese Football Federation (FPF). In 2017, CMN acquired the VJS stadium, guaranteeing a single infrastructure for the progress of beach sports in the destination, which implies that the decision to host sport events was fundamental in making the investment decision in the sport facility. These decisions outlined the integrated strategy that CMN has been implementing: promoting the destination as a prime location for beach sports in combination with the media dynamics surrounding the phenomenon of big waves, as stated by the president of CMN in an interview with the digital newspaper E24 (E24, 2021). The media manifestation of big waves was even more sensational due to the fact that Nazaré is one of the few destinations in the world where one of the stages of the Big Wave Challenge world surfing event has been annually hosted since 2016.

## 3.2. Research Design

The methodological design of this study adopted the case study as a research strategy (Yin, 2018), carrying out a qualitative analysis of the phenomenon (Creswell & Creswell, 2018), in combination with the action-research methodology (Chevalier & Buckles, 2019). Aiming to search for alternatives to the reality that is to be transformed and developed (Chevalier & Buckles, 2019), the research team took on a proactive role, involving the sport events' key actors in order to debate and select strategies for leverage that could potentially promote various benefits for the local community. Thus, the study evolved through the following phases:

1. Research of the four editions of the studied events, which took place in Nazaré before 2021, to identify possible strategic leverage actions undertaken in previous editions;
2. Development of a seminar by researchers on the topic of strategic multi-leverage of sport events, in which representatives of organizing entities and other stakeholders of the events participated, in order to raise awareness / train key actors on strategic leverage models;
3. Execution of two focus group sessions, integrating representatives of the organizing entities and other stakeholders of the events, to prioritize the strategic objectives of the leverage process, as well as outline and elect the corresponding actions to be implemented at the events;
4. On-site observation of events to assess leverage actions actually implemented (planned / emerging), as well as the interaction of the actors involved in their implementation;
5. Lastly, semi-structured interviews were carried out with representatives of the main entities organizing the events to analyse the success / failure of the implemented actions (planned / emerging).

## 3.3. Data Collection and Analysis

Data was collected from:

i) Documentary sources (Creswell & Creswell, 2018), including research into official organizational documents, media communications, and publications on official social networks (Table 2);

*Table 2. Documentary Sources Analysed by Event*

| Documentary sources | EEBLS (n) | EWC (n) | CN-DE (n) |
|---|---|---|---|
| **Official organizational documents** | 6 | 8 | 3 |
| **News from official websites** | 21 | 32 | 22 |
| **Publications on official Facebook pages** | 36 | 74 | 28 |
| **Total (n)** | 62 | 114 | 53 |

*Note.* EEBLS - Euro Beach Soccer League Stage; EWC - Euro Winners Cup; CN-DE - National Championship-Elite Division

ii) Two focus group sessions (Hennink, 2014), on the 5th and 12th of May 2021, whose constitution was carried out according to the areas of intervention / expertise of each of the entities involved in the leverage process (Table 3). Brainstorming exercises and the consensus technique (Chevalier & Buckles, 2019) were applied in these sessions;

*Table 3. Characterization of Focus Groups' Participants*

| | Institution | Participants (n) | Role / Department |
|---|---|---|---|
| **Focus Group 1** | CMN | 2 | Technical Coordinator / Physical Activity and Sport;<br>Technical Manager / VJS Stadium. |
| | NQ | 1 | Technical / International Events. |
| | BSWW | 3 | Vice President;<br>Technical / Logistics;<br>Technical / Accommodation and Transport. |
| | FPF | 2 | Technical Coordinator / Soccer Beach;<br>Sport Journalist / Soccer Beach. |
| | Regional Football Association | 1 | President |
| | Local sport association | 1 | President |

continued on following page

*Table 3. Continued*

| | | | |
|---|---|---|---|
| **Focus Group 2** | CMN | 2 | Technical Coordinator / Tourism; Technical / Tourism. |
| | Tourism of Portugal - Central Zone | 1 | Technical |
| | Local Commerce Association | 1 | President |
| | Local Accommodation Unit A | 1 | Manager |
| | Local Accommodation Unit B | 1 | Manager |

*Note.* CMN - Nazaré City Council; NQ - Nazaré Qualifica; BSWW - Beach Soccer World Wide; FPF - Portuguese Football Federation.

iii) Field observation (Creswell & Creswell, 2018) during all days of the events;
iv) Seven semi-structured interviews (#1 - #7) carried out after the events, with an average duration of 45 minutes / each (Table 4). Only one of the interviewees had not previously participated in focus group sessions.

*Table 4. List of Interviewees*

| Institution | Department | Interviewees (n) |
|---|---|---|
| **CMS** | Physical Activity and Sport; Communication. | 3 |
| **NQ** | International Events | 1 |
| **BSWW** | Management | 1 |
| **FPF** | Soccer Beach (Coordination and Media) | 2 |

*Note.* CMN - Nazaré City Council; NQ - Nazaré Qualifica; BSWW - Beach Soccer World Wide; FPF - Portuguese Football Federation.

Data analysis was carried out using content analysis (Bardin, 2018), deductively, in which data categorization was guided by the theoretical conceptualization adopted in this study (Chalip, 2004, 2006; Chalip et al., 2017; O'Brien & Chalip, 2008; Pereira et al., 2020). Validation through data triangulation (Creswell & Creswell, 2018; Yin, 2018) was also performed.

## 4. RESULTS

### 4.1. Strategic Objectives and Corresponding Leverage Actions Implemented in Previous Editions (2017-2020) of the Events Included in the Portfolio

The strategic leverage actions implemented in previous editions of beach soccer events hosted in Nazaré (i.e., from 2017 to 2020), have a greater impact on the economic strategic objective, 24 actions out of a total of 28 (Table 5). In particular, with regard to optimizing revenue for the local community, the following actions were developed: (i) reservation of all hotel units on behalf of the event organization; (ii) creation of a pre-qualifying event for the main event; (iii) promotion of official training to adapt to the playing areas; (iv) merchandising point of sale for BSWW and Nazaré; (v) provision of spaces for the commercialization of goods and services in the hosting event zone (HEZ); and (vi) sealing the playing areas attached to the HEZ to rent spaces for training and private games for teams / selections. Additionally, the EWC - i.e., the event that lasts the longest and encompasses the largest number of participants, especially foreigners - was the target of all actions taken to achieve the short-term strategic economic objective.

*Table 5. Strategic Objectives and Leverage Actions Implemented in Previous Editions of the Soccer Beach Events (2017-2020)*

| **Strategic objectives** | | **Actions (n)** | | |
|---|---|---|---|---|
| | | **EWC** | **EEBSL** | **CN-DE** |
| **Economic** | Optimize business and revenue | 6 | 4 | 2 |
| | Intensify the destination's image | 4 | 4 | 4 |
| **Sport Participation** | Promote sport participation | 1 | | |
| **Political** | Increase political visibility | 1 | 1 | 1 |

*Note.* EEBLS - Euro Beach Soccer League Stage; EWC - Euro Winners Cup; CN-DE - National Championship-Elite Division

To intensify the image of Nazaré, from a long-term economic strategy perspective, four actions were implemented across beach soccer events: (i) inclusion of key elements of the destination in the event logo; (ii) dissemination of panoramic photographs of Nazaré in various media to promote events; (iii) integration of videos produced in emblematic areas of Nazaré to promote the events' portfolio; and (iv) television broadcast / streaming.

The importance of destination advertising in event communications is highlighted by #2,

*An extremely important way of promoting Nazaré, and perhaps the one with the greatest attendance, concerns the broadcasts of the games (…) in addition to paying attention to the game itself, [the spectator] also views/consumes the place where the competition takes place. In this sense, there is always a concern in placing advertising signs in all fields where games are held, with information about Nazaré - the first line of advertising; and, in addition, when collecting images, care is always taken to capture as background the beauty of the emblematic promontory of Nazaré and the bathing activity of Beach of Nazaré. (#2)*

Regarding the dimension of sport participation, a single leveraging action was implemented in the events' editions of 2017-2020, which aimed to raise awareness and promote the sport itself in the local community. In particular, all collaboration protocols established for the EWC included the offer of a wild-card to a sport club in Nazaré, aiming to bring sport closer to the local community, in general, and to the practitioner, in particular. According to #1, this action increased interest in the sport in the local community,

*In fact, after the first editions of beach soccer events, there was a local club that opened a beach soccer section, creating teams at the senior level in both genders. (...) In terms of sponsors, it is clear that beach football provides exposure that few sports do, largely due to the guarantee of television broadcasts, which is why it is desirable and, perhaps, simpler to establish agreements with sponsors, especially for local clubs. (#1)*

The ability of the beach soccer events to serve as political leverage, through the platform it provides for local political decision-makers (#3; #5) was operationalized through their presence at key moments, such as prize draws, press conferences, ceremonies, awards and interviews with the media, as stated by #3: "The institutional communication of these events (...) has above all the purpose of reinforcing the political advantage of political decision-makers, with a sense of advertising and tourist promotion of the host location".

## 4.2. Strategic Objectives and Corresponding Leverage Actions Implemented in the Portfolio of Beach Soccer Events Hosted in 2021

In 2021, aiming to achieve four strategic objectives, seven strategic leverage actions previously planned in the focus group sessions were implemented in the portfolio of beach soccer events hosted in Nazaré (Table 6). All actions were implemented in EEBSL; three of which, in EWC; and two of which, in the CN-DE.

*Table 6. Strategic Objectives and Corresponding Leverage Actions Planned in the Focus Group Sessions and Implemented in the Portfolio of Beach Soccer Events Hosted in 2021*

| Strategic objectives | Planned and implemented actions | EEBSL | EWC | CN-DE |
|---|---|---|---|---|
| **Economic** | Creation of a destination tourist guide - website "Find Out Nazaré" | √ | √ | √ |
| | Promotion of the slogan "Nazaré, Capital of Beach Soccer Events", in all event materials and communication | √ | - | - |
| **Sport Participation** | Promotion of an interaction session (via zoom) between elite athletes and students from the local basic school | √ | - | - |
| **Social** | Use of the local mascot at HEZ | √ | - | - |
| **Environmental** | Promotion of 'educational games' about plastic recycling | √ | √ | √ |
| **Multidimensional Social/Sport Participation/ Economic** | Providing services with recreational/social activities | √ | - | - |
| **Multidimensional Social/Sport Participation** | Offer of game areas along the beach, before, during and after events | √ | √ | - |

*Note.* EEBLS - Euro Beach Soccer League Stage; EWC - Euro Winners Cup; CN-DE - National Championship-Elite Division

The two actions implemented for the strategic economic leverage of the destination were adapted to the existing resources of the CMN, which was the entity responsible for their operationalization. Therefore, in order to reach the largest number of visitors possible, CMN added the embryonic project of the "Find Out Nazaré" website, developed by its tourism department, to the beach soccer events. Specifically, the website was created so that its dissemination to the public coincided with the start of the first beach soccer event in Nazaré (i.e., EEBSL), serving, from then on, as a tourist guide to Nazaré, and ensuring, at the same time, transversality to the portfolio of beach soccer events hosted in 2021 (Figure 1).

*Figure 1. Promotion of the Project "Find Out Nazaré" at the Events*

As for the slogan "Nazaré, Capital of Beach Football Events", the results verified that its insertion in some decorative/interactive materials and its promotion in the media only occurred in the first event (i.e., EEBSL).

Regarding leveraging sport participation, the interaction session (via zoom), between athletes from the national beach soccer team and students from a local basic school, at EEBSL, was poorly attended. Field observation verified that the online format was a constraint on greater interaction between students and elite athletes. Additionally, the lack of timely dissemination and graphic support led to poor student participation.

From the perspective of the social dimension, aiming to leverage the spirit of celebration in the destination, and thus reinforce liminality, the mascot associated with the municipal holiday was used in a photo shoot, using a very large frame (Figure 2).

*Figure 2. Local Mascot at the Host Event Zone*

Regarding the action planned to achieve the strategic environmental objective, the event organizers created an interactive area for the public, especially younger people, where activities were carried out in a marker that was used to deposit plastic materials / packaging of consumed products (Figure 3).

*Figure 3. Educational Games About Plastic Recycling*

In line with the action, the entire HEZ surroundings were equipped with containers for storing recyclable plastic. Although the action has been implemented in all the portfolio's beach soccer events, as mentioned by #2, more actions of this type are necessary so that the message can have the right effect with the target audience,

*We took action to reduce and recycle plastic and to some extent the objective was achieved, However, there is still a lot of work to be done. (...) for example, through awareness raising actions among the employees of each of the entities. (#2)*

Concerning to multidimensional strategic leverage, two actions were implemented: (a) one with a triple strategic objective - social, economic and sport participation - aiming to disseminate recreational services provided to the local population and simultaneously, promote social interaction of events' participants; and (b) another with a double strategic objective - social and sport participation - pointing to the provision of small beach soccer fields, along the beach, before, during and after the events.

Specifically, to publicize local recreational services and increase the social interaction of participants in the event, CMN organized a walk in a natural space. The implementation of this action at EEBSL provided publicity for one of the destination's pedestrian routes and the physical activity services provided by CMN. The action was publicized through the local Senior University and a local social solidarity institution, promoting both the participation of older local residents and orphaned teenagers in the care of the social institution. This action should continue to be implemented due to the long-term benefits, as supported by #1,

*It is an action that will not bring an immediate return, but will have an indirect return that can bring very significant results, because the person stops seeing Nazaré only as a destination for beach sports (...) and starts to understand the most assets that, in this case, Nazaré has in other areas (cultural, gastronomic, landscape, nature), encouraging later return to the destination as a tourist. (#1)*

As for multidimensional strategic leverage aiming, simultaneously, at the social and sport participation dimensions, several small playing fields were made available along the beach of Nazaré. The results showed that the action provided an opportunity for social interaction and activation of the sport subculture, as parents, children and event participants became involved in a symbiotic environment around beach soccer. Although this action had the intended effect, it must be adjusted in order to integrate local sports entities,

*We set up small beach soccer fields, next to the stadium's adjacent fields, with the aim of allowing children to informally try out the sport. What is certain is that this could be an action developed by local clubs, which would be responsible for the part most targeted at the population. (#4)*

In addition to the actions planned in the focus group sessions, the results verified that four emerging actions were implemented at international events, i.e., EEBSL and EWC (Table 7).

*Table 7. Emergent Actions Implemented in the Portfolio of Beach Soccer Events Hosted in 2021*

| Strategic objectives | Emergent actions | EEBSL | EWC | CN-DE |
|---|---|---|---|---|
| **Economic** | Offer of a Nazaré promotional kit to event participants | - | √ | - |
| **Sport Participation** | Changing the rules of the beach soccer game | √ | √ | - |
| **Multidimensional Economic/Social** | Creation of a cultural event complementary to sport events' portfolio | √ | - | - |

continued on following page

*Table 7. Continued*

| Strategic objectives | Emergent actions | EEBSL | EWC | CN-DE |
|---|---|---|---|---|
| **Multidimensional Economic/ Environmental** | Sale of reusable bottles | - | √ | - |

*Note.* EEBLS - Euro Beach Soccer League Stage; EWC - Euro Winners Cup; CN-DE - National Championship-Elite Division

With the economic objective of reinforcing the destination's image, CMN provided a Nazaré promotional kit that was delivered to each of the teams present at the EWC, containing information about culture, shopping, gastronomy and other attractions of the destination, as well as courtesy offers.

To strategically leverage the development of sport dynamics, and consequently, increase sport participation and the intensity of the spectacle of beach soccer games, BSWW produced a set of changes to the rules of the game itself, as verified by observation on the field and corroborated by #6, who said "The changes came to level the teams, especially because the role of goalkeepers is no longer so interventionist as it was until now" (#6), and #7 (both former beach soccer athletes),

*Without a doubt, these changes have given theoretically weaker teams an additional dose of ambition. Beach soccer has become more attractive and, in television terms, there are much fewer breaks in the game. (…) with these changes the game flows and spectators win. (#7)*

During one of the EEBSL days, a cultural side-event was held, in which several musical groups performed at specific points in the historic centre of Nazaré. The intention of making the date of the cultural event coincide with the weekend in which the EEBSL took place was clear and expressed, so that participants could experience the event, enjoy the culture of the destination and consequently, promote the destination,

*This interrelationship between culture and sport is fundamental, also because tourism is increasingly done in the family and the family comes with the athlete who is going to participate in the competition, with the aim of watching the games, but then there is a gap in the entertainment. It cannot just be the beach and the landscape that provide entertainment. There must be a cultural part that does this, through concerts in the theatre, on the streets, etc. (#1)*

*This year we tried to associate a traditional cultural event from Nazaré - the 'Esquininhas' - with the EEBSL. (…) we can 'sell' these [local] events at the stadium and these events can 'sell' the activities that take place at the stadium. (#4)*

Another emerging action, developed by BSWW (i.e., one of the entities organizing the events), aimed to focus stakeholders' attention on environmental issues, through the sale of a reusable bottle containing the message 'Ocean Bottle - Small Bottle, Big Impact' (Figure 4). However, the event organization itself distributed around 22 thousand disposable plastic water bottles, and for this reason, the environmental objective underlying the action was considered incoherent (#3). The same action also intended to induce spending among visitors, as it was a merchandising action by BSWW itself. Yet, field observation verified that the sales volume of bottles was clearly lower in relation to other merchandising materials (e.g., t-shirts, jackets, caps, beach towels, balls, etc.).

*Figure 4. Beach Soccer Eco Bottle*

Finally, the analysis of event communication was carried out by quantifying the references identified on strategic leverage actions planned and implemented in each of the beach soccer events in the portfolio hosted in 2021 (Table 8).

*Table 8. Number of Actions Planned and Implemented by Strategic Objective in the Communication of Portfolio Events Hosted in 2021*

| **Strategic objectives** | | **References (n)** | | | |
|---|---|---|---|---|---|
| | | **EEBSL** | **EWC** | **CN-DE** | **Total** |
| **Economic** | Revenue/business induction | 5 | 2 | - | 7 |
| | Destination promotion | 17 | 18 | 8 | 43 |
| **Social** | Social interaction and/or spirit of celebration | 2 | 2 | - | 4 |
| | Social problems | - | 3 | 1 | 4 |

continued on following page

*Table 8. Continued*

| Strategic objectives | | References (n) | | | |
|---|---|---|---|---|---|
| | | EEBSL | EWC | CN-DE | Total |
| **Environmental** | | 1 | 2 | - | 3 |
| **Sport Participation** | | 9 | 11 | 8 | 28 |
| **Political** | | 3 | 2 | - | 5 |
| **Total** | | 37 | 40 | 17 | 94 |

*Note.* EEBLS - Euro Beach Soccer League Stage; EWC - Euro Winners Cup; CN-DE - National Championship-Elite Division

The communication of events predominantly focused on actions that strategically aimed to promote Nazaré as a tourist destination, especially in communication regarding international events. Furthermore, in the majority of cases classified as merely informative references, there were details that, in a certain way, aimed at promoting Nazaré, such as the use of the slogan 'Capital of Beach Sports', and hashtags (#) with the word Nazaré or the insertion of emojis related to beach sports (sun, sun hat, soccer ball, sea wave, etc.).

Secondly, the communication of the events also highlighted the actions that aimed to strategically leverage sport participation. Finally, the data contained in table 8 also highlights the greater number of references to actions implemented in international events compared to the national event.

## 5. DISCUSSION

The strategic objective pursued by CMN, which was the basis for the decision to host the portfolio of beach soccer events, was to affirm the destination as a notable place for practicing beach sports. The fact that Nazaré has been hosting this portfolio of events uninterruptedly for five years reveals its importance for the local community, corroborating the findings of other studies (e.g., Gibson et al., 2012; Ziakas & Getz, 2020) that instil sustainable forms in events to promote the tourist destination.

The study of the strategic leverage process of the portfolio of events hosted in Nazaré, in 2021, demonstrated the supremacy of the implementation of actions that aimed to achieve strategic economic objectives, in accordance with the greater attention given to the economic dimension compared to the remaining dimensions of leverage, reinforcing the conclusions of other authors (e.g., Chersulich Tomino et al., 2020; Schulenkorf et al., 2022). The actions implemented in the various events to obtain short-term economic benefits focused on two fundamental premises, namely, prolonging the participants' stay and inducing visitors to spend, consequently intending the development of the local economy (Antchak, 2017; Mascarenhas et al., 2023; Schulenkorf et al., 2019). The pre-event phase was used to extend the participants' stay, through the creation of official adaptation training. The implementation of actions aimed at persuading visitors' consumption produced direct gains for the entities that organized the events, such as, for example, selling merchandising, renting fields for training and unofficial games, and selling goods and services around the VJS stadium. Nevertheless, the possibilities for economic leverage could have been better explored, namely, through greater activation of potential partnerships with local businesses (Mhanna et al., 2017; Schulenkorf et al., 2019; Xing & Chalip, 2006).

In turn, the launch of the Find Out Nazaré website (i.e., guide to Nazaré tourist offer) was an inter-event cross leverage action, aiming at greater reach in the dissemination / use of the digital tool, and consequently, the stimulation of opportunities for local business. Increasing the same purpose, the results highlighted the action related to the offer of a Nazaré promotional kit, which emerged at the EWC, taking advantage of the presence in the destination of 17 sports teams from different countries. This action underlines the context as a determining factor for the execution of strategies, sometimes providing the development of emerging actions arising from the full use of the favourable conditions that such a context offers for their implementation (Pereira et al., 2015, 2020).

The strategic economic leverage actions developed in the portfolio of beach soccer events to intensify the destination's image included key attributes of Nazaré (Salgado-Barandela et al., 2022; Ziakas, 2019), namely, elements directly related to the promontory of Nazaré and the Lighthouse of Forte de São Miguel Arcanjo, whose beach is globally recognized as a place of excellence for big wave surfing and has hosted, since 2016, the world big wave surfing competition.

The intergenerational walk in the natural space of Nazaré promoted during the EEBSL aimed to reach several strategic objectives with its implementation, which is in line with intra-event cross-leveraging (Pereira et al., 2015). In particular, the action aimed to: increase the number of consumers of sport services provided by CMN; promote the destination; and also, enhance the social interaction of the local population of different generations and social realities. In addition, although increasing sport participation (Chalip et al., 2017), in particular with regard to the practice of recreational / leisure sport, was not one of intentions of this action in the strategic planning phase of leveraging beach soccer events, its implementation induced this potential outcome. However, it is important to highlight that the strategic scope of this action could still have gone further, since, as argued by Mascarenhas et al. (2021), the promotion of sport tourism activities carried out in natural spaces tends to influence ecologically sustainable behaviours. At the same event (i.e., EEBSL), the cultural side-event that was held also constituted an intra-event cross leverage action, as it simultaneously promoted the image of the destination and strengthened the feeling of celebration. Thus, the cultural side-event favoured the positive predisposition of visitors to get involved in activities and exposed the liminality generated through the event (Musikavanhu et al., 2021; Ziakas, 2016). Furthermore, the simultaneous use of the Nazaré mascot proved to be a very important moment of social interaction between participants, visitors and the local population, generating an atmosphere of celebration, i.e., of liminality (Chalip, 2006; O'Brien & Chalip, 2008).

During the EWC, an emerging action, intended to generate economic and environmental benefits (i.e., sale of a reusable bottle), revealed discrepancy in the event organizers' practice of distributing thousands of disposable plastic water bottles. As a result, the action generated a perception of inconsistency among stakeholders, and the message of the action suffered from a lack of transparency. This finding corroborates the argument advanced by Mascarenhas et al. (2021), highlighting that the incongruity between the action and environmental message of organizations results in consumer inaction.

To encourage sport participation and increase the local community's interest in one of the portfolio events (i.e., EWC), since 2017 (i.e., the first edition of this event in Nazaré), it has been mandatory to offer a wild-card for the main competition to a local team. This action has substantially reinforced the interest of local participants in beach soccer. Also to boost sport participation, an interaction session with elite athletes and the school community was held, operationalizing the recommendations of Chalip et al. (2017) to alleviate the perception of a gap in sport performance that exists between potential practitioners and elite athletes. The provision of small beach soccer fields to encourage the informal practice of beach soccer among younger people, with the aim of increasing the number of people practicing this sport, was another action implemented to strengthen sport participation. However, the inability to establish the necessary links with local clubs that develop beach soccer at youth levels was clear, since the implementation of the action did not involve any of these clubs. The strategic leverage process requires a wide range of resources, which are often centralized in the event itself and/or organizing entities. Consequently, as highlighted by Chalip et al. (2017), the centralization of the entire strategic leverage process in a single entity does not favour the optimization of the intended benefits for the local community. Therefore, a combined approach between local beach soccer clubs, in order to develop the action themselves, additionally creating sporting offers for their participants, could have provided even more effective results for increasing sport participation.

As for the political dimension of strategic leverage, in all events, throughout the process of implementing actions, it was clear that there are underlying decisions that have to be taken and controlled by the people who are responsible for the strategic apex of a given organization (Mintzberg, 2010). This dynamic reveals the exercise of power by decision-makers, whose influence has also been acknowledged in sport events (Chen et al., 2022). Furthermore, the connection of events to local government also appeared to be inevitable, implying a political component that determines a set of actions that aim to reinforce the political advantage of decision-makers. These actions ensured the visibility of local government actors, thus reinforcing the political advantage arising from the media exposure provided by the beach soccer events hosted in Nazaré (Pereira et al., 2020).

The analysis of the strategic leverage process of the portfolio of beach soccer events hosted in 2021 allowed the consultation of different stakeholders involved in defining the strategic objectives and corresponding actions, and even more, it encouraged their joint reflection. In this light, focus groups proved to be fundamental tools for activating this reflective, participatory and democratic method (Chevalier & Buckles, 2019; Hennink, 2014), which is essential given the purpose of contemplating a set of heterogeneous interests of different stakeholders. The planning of strategic leverage objectives and respective actions proposed by the focus groups generated a very positive effect, for the following reasons: (a) the number of leverage actions implemented in 2021 increased considerably (an additional 11 actions - seven planned and four emerging), compared to actions already developed in previous editions of the same events; and (b) the spectrum of dimensions of strategic leverage was also amplified (five dimensions of strategic leverage, implying the addition of strategic objectives in the social and environmental areas, to the three areas already covered in previous editions, namely, economic, sport participation and political). This positive result was achieved due to the innovation provided by the action research developed in the process of strategically leveraging the portfolio of beach soccer events hosted in 2021 in Nazaré.

The selection of different action proposals to strategically leverage a portfolio of sport events is more effective if its fundamental premise is cross-leveraging, both through actions capable of achieving different strategic objectives (or dimensions of strategic leverage) within the same sport event. (i.e., intra-events), as well as actions whose implementation crosses more than one of the events in the portfolio (i.e., inter-events). In this way, the strategic leverage process can achieve multiple strategic objectives in the different dimensions of leverage (i.e., economic, social, environmental, sport participation and political) with the lowest possible expenditure of efforts / resources (Antchak et al., 2019; Ziakas & Getz, 2020).

In this investigation, although the strategic leverage actions that were implemented in sport events revealed a clearly economic bias, corroborating the results of Schulenkorf et al. (2022), the implementation of actions aimed at strategic multi-leveraging was experienced, through intra-event and inter-event cross-leverage. This result supports other recent studies that have highlighted the primacy of heterogeneous approaches, which enable the production of multiple benefits in the short, medium and long term in the communities that host sport events (Mascarenhas et al., 2023; Pereira et al., 2015, 2020; Sant et al., 2019; Ziakas, 2019).

Due to the limited availability of some key actors, the individualized interviews were not applied to all stakeholders involved in the focus groups, so the application criterion was limited to those present at the HEZ. As such, this is a limitation of the present investigation. Additionally, the essence of action-research is the sequence of the process, which, after planned and implemented, integrates the evaluation of the result with a view to future application in the new planning and implementation cycle (Chevalier & Buckles, 2019). Despite the longitudinal panorama provided by the analysis of past editions of events, the action-research carried out in this study involved only a single cycle of planning, implementation and analysis, which must be repeated in future research.

## 6. CONCLUSION

This research allowed immersion in a specific context using a holistic approach to stimulate and understand the process of strategic leverage of a portfolio of sport events composed of three beach soccer events, two international competitions and one national competition. In the process of planning the strategic leverage of this portfolio of events, strategic objectives and corresponding actions were defined to pursue them in five different dimensions, namely economic, social, environmental, sport participation and political. Strategic economic leverage pointing to the long-term benefits arising from the intensification of the destination's image was the most developed dimension. This study also verified the increased concern of organizers in inducing visitor spending, both in terms of publicizing the destination's attractions, and at the same time, through the increase in entertainment activities and the offering of souvenirs, both imbued with local culture, to provide greater retention of tourist revenue associated with beach soccer events in the destination.

Although fewer in number, the implementation of actions aimed at cross-leveraging was also verified, both between the different leverage dimensions and between the sporting events that made up the portfolio. For greater effectiveness in the process of strategically leveraging sport events in Nazaré (and other similar destinations), it is urgent to stimulate strategic collaboration between the various types of stakeholders. In this way, different interests can be combined from the planning phase of the destination's strategic objectives and design of the corresponding leverage actions, and subsequently, efficiently operationalized in the implementation phase of these actions.

In conclusion, this investigation contributed to the sport event tourism literature by validating the application of the conceptualization of strategic leverage, in its various dimensions. The contributions of this study also extend to a set of practical implications for the management of sporting events, assessing the importance of the multidisciplinary process inherent to the strategic leverage of sport events, as well as the unavoidable need to involve all individuals / organizations with specific interests in the events and/or the destination, so that the process promotes maximization of their concertation. As leverageable resources, these beach soccer events must continue to be the target of strategic leverage processes to also promote the resolution of the most pressing weaknesses of the local community, and bring to light issues on the global agenda, such as the immensity of plastic in the oceans, which entails increased responsibilities for coastal destinations whose tourism is developed through surfing, beach soccer or other sport events.

## ACKNOWLEDGMENT

This paper is financed by National Funds provided by FCT- Foundation for Science and Technology through project UIDB/04020/2020 and with DOI 10.54499/UIDB/04020/2020 (https://doi.org/10.54499/UIDB/04020/2020).

## REFERENCES

E24. (2021). *Walter Chicharro sublinha importância dos grandes eventos para a Nazaré*. https://www.youtube.com/watch?v=-7TP7FAjsco

Aicher, T. J., Buning, R. J., & Newland, B. L. (2020). Running Through Travel Career Progression: Social Worlds and Active Sport Tourism. *Journal of Sport Management*, 34(6), 542–553. 10.1123/jsm.2019-0256

Antchak, V. (2017). Portfolio of major events in Auckland: Characteristics, perspectives and issues. *Journal of Policy Research in Tourism, Leisure & Events*, 9(3), 280–297. 10.1080/19407963.2017.1312421

Antchak, V., Ziakas, V., & Getz, D. (2019). *Event Portfolio Management: Theory and methods for event management and tourism* (1st ed.). Goodfellow Publishers., 10.23912/978-1-911396-91-8-4260

Bardin, L. (2018). *Análise de Conteúdo* (A. Rego & A. Pinheiro, Trans.). Edições 70.

Chalip, L. (2004). Beyond impact: A general model for sport event leverage. In B. W. Ritchie & D. Adair (Eds.), *Sport tourism: Interrelationships, impacts and issues* (pp. 226–252). Blue Ridge Summit: Channel View Publications. 10.21832/9781873150672-014

Chalip, L. (2006). Towards social leverage of sport events. *Journal of Sport & Tourism*, 11(2), 109–127. 10.1080/14775080601155126

Chalip, L. (2018). Trading legacy for leverage. In I. Brittain (Ed.), *Legacies and mega events: Fact or fairy tales?* (pp. 25–41). Routledge, Taylor & Francis Group.

Chalip, L., Green, B. C., Taks, M., & Misener, L. (2017). Creating sport participation from sport events: Making it happen. *International Journal of Sport Policy*, 9(2), 257–276. 10.1080/19406940.2016.1257496

Chen, S., Xing, X., & Chalip, L. (2022). Planning and implementation of event leveraging strategy: China's legacy pledge to motivate 300 million people to be involved in winter sport. *Sport Management Review*, 25(5), 771–790. 10.1080/14413523.2021.1987737

Chersulich Tomino, A., Perić, M., & Wise, N. (2020). Assessing and considering the wider impacts of sport-tourism events: A research agenda review of sustainability and strategic planning elements. *Sustainability (Basel)*, 12(11), 4473. 10.3390/su12114473

Chevalier, J. M., & Buckles, D. J. (2019). *Participatory Action Research: Theory and Methods for Engaged Inquiry*. Routledge. 10.4324/9781351033268

Creswell, J. W., & Creswell, J. D. (2018). *Research design: Qualitative, quantitative, and mixed methods approaches*. SAGE Publications.

Duglio, S., & Beltramo, R. (2017). Estimating the economic impacts of a small-scale sport tourism event: The case of the Italo-Swiss mountain trail Collon trek. *Sustainability (Basel)*, 9(3), 343. 10.3390/su9030343

Duignan, M. B., Parent, M. M., & McGillivray, D. (2023). Accommodating (Global–Glocal) Paradoxes Across Event Planning. *Event Management*, 27(1), 149–154. 10.3727/152599522X16419948694991

Getz, D., & Page, S. (2016). Progress and prospects for event tourism research. *Tourism Management*, 52, 593–631. 10.1016/j.tourman.2015.03.007

Gibson, H. J., Kaplanidou, K., & Kang, S. J. (2012). Small-scale event sport tourism: A case study in sustainable tourism. *Sport Management Review*, 15(2), 160–170. 10.1016/j.smr.2011.08.013

Hennink, M. M. (2014). *Focus Group Discussions: Understanding qualitative research*. Oxford University Press. 10.1093/acprof:osobl/9780199856169.001.0001

Lu, H.-F., & Lin, H.-W. (2020). Does the legacy governance of major sporting events affect urban development? The stakeholder's perspectives on the Taipei 2017 Universiade. *International Journal of Sports Marketing & Sponsorship*, 22(1), 87–106. 10.1108/IJSMS-02-2020-0022

Mainolfi, G., & Marino, V. (2020). Destination beliefs, event satisfaction and post-visit product receptivity in event marketing. Results from a tourism experience. *Journal of Business Research*, 116, 699–710. 10.1016/j.jbusres.2018.03.001

Malchrowicz-Mośko, E., Botiková, Z., & Poczta, J. (2019). "Because We Don't Want to Run in Smog": Problems with the Sustainable Management of Sport Event Tourism in Protected Areas (A Case Study of National Parks in Poland and Slovakia). *Sustainability (Basel)*, 11(2), 325. 10.3390/su11020325

Malchrowicz-Mośko, E., & Poczta, J. (2018). A small-scale event and a big impact—Is this relationship possible in the world of sport? The meaning of heritage sporting events for sustainable development of tourism—Experiences from Poland. *Sustainability (Basel)*, 10(11), 4289. Advance online publication. 10.3390/su10114289

Martins, R., Pereira, E., Rosado, A., Marôco, J., McCullough, B., & Mascarenhas, M. (2022). Understanding spectator sustainable transportation intentions in international sport tourism events. *Journal of Sustainable Tourism*, 30(8), 1972–1991. 10.1080/09669582.2021.1991936

Mascarenhas, M., Pereira, E., Rosado, A., & Martins, R. (2021). How has science highlighted sports tourism in recent investigation on sports' environmental sustainability? A systematic review. *Journal of Sport & Tourism*, 25(1), 42–65. 10.1080/14775085.2021.1883461

Mascarenhas, M., Vieira, H., & Martins, R. (2023). *Factors shaping spectators' consumption patterns: Controversial arisings from an international sport tourism event*. Consumer Behavior in Tourism and Hospitality., 10.1108/CBTH-08-2023-0108

McCullough, B. P., Pfahl, M. E., & Nguyen, S. N. (2016). The green waves of environmental sustainability in sport. *Sport in Society*, 19(7), 1040–1065. 10.1080/17430437.2015.1096251

Melovic, B., Rogic, S., Cerovic Smolovic, J., Dudic, B., & Gregus, M. (2019). The Impact of Sport Sponsorship Perceptions and Attitudes on Purchasing Decision of Fans as Consumers—Relevance for Promotion of Corporate Social Responsibility and Sustainable Practices. *Sustainability (Basel)*, 11(22), 6389. 10.3390/su11226389

Mhanna, R., Blake, A., & Jones, I. (2017). Challenges facing immediate tourism leveraging: Evidence from the London 2012 Olympic Games. *Managing Sport and Leisure*, 22(2), 147–165. 10.1080/23750472.2017.1412269

Mintzberg, H. (2010). *Estrutura e Dinâmica das Organizações* (4th ed.). Publicações Dom Quixote.

Misener, L., Lu, L. D., & Carlisi, R. (2020). Leveraging events to develop collaborative partnerships: Examining the formation and collaborative dynamics of the Ontario Parasport Legacy Group. *Journal of Sport Management*, 34(5), 447–461. 10.1123/jsm.2019-0283

Musikavanhu, R. R., Ladkin, A., & Sadd, D. (2021). The lasting social value of mega events: Experiences from green point community in Cape Town, South Africa. *Journal of Sustainable Tourism*, 29(11–12), 1832–1849. 10.1080/09669582.2021.1874395

Newland, B. L., & Yoo, J. J.-E. (2021). Active sport event participants' behavioural intentions: Leveraging outcomes for future attendance and visitation. *Journal of Vacation Marketing*, 27(1), 32–44. 10.1177/1356766720948249

O'Brien, D., & Chalip, L. (2008). Sport events and strategic leveraging: Pushing towards the triple bottom line. In Woodside, A. G., & Martin, D. (Eds.), *Tourism management: Analysis, behaviour and strategy* (1st ed., pp. 318–338). CAB International. 10.1079/9781845933234.0318

Pereira, E., Mascarenhas, M., Flores, A., Chalip, L., & Pires, G. (2020). Strategic leveraging: Evidences of small-scale sport events. *International Journal of Event and Festival Management*, 11(1), 69–88. 10.1108/IJEFM-07-2018-0046

Pereira, E., Mascarenhas, M., Flores, A., & Pires, G. (2015). Nautical small-scale sports events portfolio: A strategic leveraging approach. *European Sport Management Quarterly*, 15(1), 27–47. 10.1080/16184742.2015.1007883

Potwarka, L. R., Snelgrove, R., Drewery, D., Bakhsh, J., & Wood, L. (2020). From intention to participation: Exploring the moderating role of a voucher-based event leveraging initiative. *Sport Management Review*, 23(2), 302–314. 10.1016/j.smr.2019.03.002

Richardson, K., & Fletcher, T. (2020). Community sport development events, social capital and social mobility: A case study of Premier League Kicks and young black and minoritized ethnic males in England. *Soccer and Society*, 21(1), 79–95. 10.1080/14660970.2018.1506334

Richelieu, A. (2018). A sport-oriented place branding strategy for cities, regions and countries. *Sport, Business and Management*, 8(4), 354–374. 10.1108/SBM-02-2018-0010

Rocha, C. (2017). Rio 2016 Olympic Games and diplomatic legacies. *International Journal of Sport Policy*, 9(2), 277–294. 10.1080/19406940.2017.1287762

Salgado-Barandela, J., Barajas, A., & Sanchez-Fernandez, P. (2022). Sports event portfolios: An innovative tool and a new management paradigm. *International Journal of Sports Marketing & Sponsorship*, 23(5), 920–933. 10.1108/IJSMS-02-2021-0024

Sant, S.-L., Misener, L., & Mason, D. S. (2019). Leveraging sport events for tourism gain in host cities: A regime perspective. *Journal of Sport & Tourism*, 23(4), 203–223. 10.1080/14775085.2019.1711444

Schulenkorf, N., Giannoulakis, C., & Blom, L. (2019). Sustaining commercial viability and community benefits: Management and leverage of a sport-for-development event. *European Sport Management Quarterly*, 19(4), 502–519. 10.1080/16184742.2018.1546755

Schulenkorf, N., Peachey, J., Chen, G., & Hergesell, A. (2022). Event leverage: A systematic literature review and new research agenda. *European Sport Management Quarterly*, 1–25. 10.1080/16184742.2022.2160477

Soares, J. P., Antunes, H. L., Bárbara, A., Escórcio, C., & Saldanha, P. (2016). O interesse público do desporto das organizações desportivas sem fins lucrativos que são apoiadas pela administração pública. *Revista Brasileira de Educação Física e Esporte*, 30(3), 675–688. 10.1590/1807-55092016000300675

Sroka, R. (2021). Mega-projects and mega-events: Evaluating Vancouver 2010 stadium and convention infrastructure. *Journal of Sport & Tourism*, 25(3), 183–200. 10.1080/14775085.2021.1881590

Taks, M., Green, B. C., Misener, L., & Chalip, L. (2014). Evaluating sport development outcomes: The case of a medium-sized international sport event. *European Sport Management Quarterly*, 14(3), 213–237. 10.1080/16184742.2014.882370

Tavakkoli, M. H. (2016). Impact of City Image, Event Fit, and Word-of-Mouth Attempt Through the Host City of an International Sporting Event. *American Journal of Sports Science*, 4(6), 112–119. 10.11648/j.ajss.20160406.13

Weed, M. (2017). Sport participation legacies of mega sporting events. In Brittain, I., Bocarro, J., Byers, T., & Swart, K. (Eds.), *Legacies and Mega Events: Fact or Fairy Tales?* (1st ed., pp. 79–97). Routledge. 10.4324/9781315558981-6

Xing, X., & Chalip, L. (2006). Effects of Hosting a Sport Event on Destination Brand: A Test of Co-branding and Match-up Models. *Sport Management Review*, 9(1), 49–78. 10.1016/S1441-3523(06)70019-5

Yin, R. K. (2018). *Case study research and applications: Design and methods* (6th ed.). SAGE Publications, Inc.

Ziakas, V. (2016). Fostering the social utility of events: An integrative framework for the strategic use of events in community development. *Current Issues in Tourism*, 19(11), 1136–1157. 10.1080/13683500.2013.849664

Ziakas, V. (2019). Issues, patterns and strategies in the development of event portfolios: Configuring models, design and policy. *Journal of Policy Research in Tourism, Leisure & Events*, 11(1), 121–158. 10.1080/19407963.2018.1471481

Ziakas, V. (2023). Leveraging sport events for tourism development: The event portfolio perspective. *Journal of Global Sport Management*, 8(1), 43–72. 10.1080/24704067.2020.1731700

Ziakas, V., & Getz, D. (2020). Shaping the event portfolio management field: Premises and integration. *International Journal of Contemporary Hospitality Management*, 32(11), 3523–3544. 10.1108/IJCHM-05-2020-0486

# Conclusion

In the selection of works presented in this book, we were able to see that scientific innovations – alongside the high importance given to sea/land interface zones – are contributing to the reshaping of the panorama of coastal resource management, promoting an integrated approach in the sense of balancing ecological integrity, economic viability and social equity. This somewhat comprehensive conclusion highlights, through a broad lens, the multifaceted impact of these innovations on the management and conservation of marine resources, fishing, tourism and various coastal human activities. By exploring key areas such as artificial reefs, ecohydrology and science communication, we highlight the essential but critical role of scientific advances in promoting sustainable coastal management. Even though these advances are sometimes small on a global scale, they still mark their importance, and make a difference on a more local scale.

## Management and Conservation of Marine Resources

The management and conservation of marine resources, although not always carried out completely in an effective manner, are fundamental to ensuring the vigor and resilience of coastal ecosystems. Scientific innovations have gradually introduced advanced tools and methodologies that not only enable, but also improve our ability to monitor and mitigate environmental impacts. The concept of an Ecological Constitution, for example, integrates recognized ecological principles that are reflected in legal frameworks, providing a solid basis for marine conservation efforts. This legal innovation is largely based on ensuring that conservation measures are applicable and sustainable, protecting marine biodiversity for future generations.

Marine debris represents a relevant threat that needs to be resolved, with a visible impact on coral reefs, especially in regions of high pressure from this threat, such as Indonesia. In this sense, important decisions incorporating innovative mitigation strategies, raising awareness among populations in community cleanups and, in parallel, technological interventions using specific sensors to locate debris, have proven to be effective in reducing the impact of marine litter. These joint efforts, which involve raising awareness among affected populations, are important not only because they safeguard coral ecosystems, but also because they support the livelihoods of communities that depend on these vital resources.

## Fisheries Management

Artificial reefs in essence emerged as a fundamental innovation mainly in supporting local fisheries. By providing habitats for marine life, these structures generally improve fish stocks and, if properly monitored and managed, promote sustainable fishing practices. In São Tomé and Príncipe – benefiting from positive results in other parts of the world – the relatively recent implementation of artificial reefs will allow for an increase in fish populations, benefiting local communities economically and nutritionally. In places where protein shortages are still an issue, this type of innovation with local implementation, although seemingly a minuscule contribution at a global level, is expected to have a substantial impact on local communities.

Another aspect that allows us to embark on the avenues of scientific innovation is the understanding and integration of socioeconomic variables used in fisheries management, as this point is essential for the development of appropriate and effective policies. There are several scientific studies that examine the market dynamics of fisheries (and related) products, community dependence and regulatory frameworks, many of them of fundamental importance as they provide critical information on the complexities of adequate fisheries management. The appropriate integration of these variables into management practices provides greater guarantees that fisheries are ecologically sustainable and socially equitable.

## Tourism and Recreational Activities

The role of artificial reefs goes beyond fishing and also extends into the realm of tourism. The artificial reef of Vilamoura, in the south of Portugal, for example, implemented more than two decades ago, although not per se a significant attraction for tourists, allows for the enhancement of recreational activities carried out by tour operators who are able to expand their portfolio of points of interest. The mere presence of these structures will contribute as further support to local economies. Blue tourism, which emphasizes sustainable marine and coastal tourism, benefits enormously from these scientific innovations, promoting the conservation of species and sites, while boosting economic growth.

Sporting events linked to coastal environments, such as beach soccer and big wave surfing, benefiting from places of great demand and importance for relaxation and vacations, also play a strategic role in stimulating coastal tourism. These events take advantage of natural resources, promoting recreation to attract diverse audiences, increasing the visibility and economic benefits of coastal destinations. The synergy between recreational use and conservation efforts is fundamental, not always evident, but can be leveraged by promoting the potential of innovative approaches in tourism development.

## Multiple Use of Coastal Resources

The multifaceted use of coastal resources requires a balanced scope to ensure sustainability. Innovations with an ecohydrology approach exploring perspectives to offer nature-based solutions to improve ecosystem health and quality of life, appear highly rewarding and sustainable if well pursued. In Luanda Bay, Angola, ecohydrological interventions have led to significant improvements in water quality and habitat restoration, demonstrating the effectiveness of integrating ecological and hydrological knowledge into coastal management.

Technological advances in recent decades, such as the Automatic Identification System (AIS), have contributed to the improvement of maritime management, and have also proven to be a very important contribution to the safety of navigation and monitoring of activities at sea. These technologies support the sustainable use of resources, ensuring that maritime activities are conducted in a responsible and transparent manner.

## Coastal Science Communication

Effective communication of scientific knowledge, taking into account the use of various channels and targeting different audiences (e.g., of varying ages and levels of education) is crucial to addressing coastal and marine issues that would otherwise not succeed. Science communication strategies that engage stakeholders and the public through accessible narratives, visual tools, and community engagement are vital to not only enlighten and empower, but also to help promote informed decision-making. Several case studies demonstrate the impact of well-designed communication efforts in raising awareness and promoting action on different aspects related to coastal conservation.

## Final Reflections

The integration of scientific innovations into coastal local management represents a transformative shift towards sustainable practices that harmonize environmental conservation with economic development. By leveraging technology, enhancing legal frameworks, and promoting community involvement, it can be ensured the resilience and health of coastal and marine ecosystems.

The comprehensive approach outlined in this conclusion underscores the importance of interdisciplinary collaboration and the continuous advancement of scientific knowledge to address the complex challenges facing coastal regions. Through collective efforts and innovative strategies, we can achieve a sustainable balance that supports both human well-being and ecological integrity, ensuring that coastal resources are preserved and enhanced for future generations.

By fostering a holistic understanding of coastal systems and implementing adaptive management practices, scientific innovations offer a promising pathway towards a sustainable and thriving coastal future.

# Compilation of References


Abdillah. (2023, August 15). *Reducing Marine Debris, Deputy Jodi Expects Innovation to Fight ALDFG*. Kemenko Bidang Kemaritiman dan Investasi. https://maritim.go.id:443/

Abecasis, M. F. (1957). Les flèches de sable de la côte d'Angola [Sand spit formation on the coast of Angola]. *Journées de l'hydraulique, 4*(1), 100-108.

Agardy, T. S. (2010). *Ocean zoning: making marine management more effective*. Routledge. 10.4324/9781849776462

Agmour, Bentounsi, Achtaich, & Fountayeni. (2018). Carrying capacity influence on the incomes of seiners exploiting marine species in the Atlantic coast of Morocco. *Mathematical Biosciences*, 305, 10-17. 10.1016/j.mbs.2018.08.012

Aguila, Y., & Viñuales, J. E. (2019). A global pact for the environment: Conceptual foundations. *Review of European, Comparative & International Environmental Law*, 28(1), 3–12. 10.1111/reel.12277

Aicher, T. J., Buning, R. J., & Newland, B. L. (2020). Running Through Travel Career Progression: Social Worlds and Active Sport Tourism. *Journal of Sport Management*, 34(6), 542–553. 10.1123/jsm.2019-0256

Alder, J., Campbell, B., Karpouzi, V., Kashner, K., & Pauly, D. (2008). Forage Fish: From Ecosystems to Markets. *Annual Review of Environment and Resources*, 33(1), 33, 153–166. 10.1146/annurev.environ.33.020807.143204

Alder, J., & Pauly, D. (2008). A comparative assessment of biodiversity, fisheries and aquaculture in 53 countries' exclusive economic zones. *Fisheries Centre Research Reports*, 16, 90.

Alves, F., Costa, P. M., Novelli, L., & Vidal, D. G. (2023). The rights of nature and the human right to nature: An overview of the European legal system and challenges for the ecological transition. *Frontiers in Environmental Science*, 11, 1175143. 10.3389/fenvs.2023.1175143

Amri, R., Kholifiyanti, C., Wijayanti, E. S., Bayan, S., Hidayat, R. R., & Hidayati, N. V. (2023). Composition and Distribution of Marine Debris at Pasir Putih Losari Beach, Brebes, Central Java. *Jurnal Kelautan Tropis, 26*(1), 135–147. 10.14710/jkt.v26i1.15770

Anderson, L. G., & Seijo, J. C. (2010). *Bioeconomics of Fisheries Management*. Wiley-Blackwell.

Andrade, E. (2022). *Zooplâncton como indicador de qualidade ambiental na Baía de Luanda* [Zooplankton as an indicator of environmental quality in Luanda Bay] [Unpublished BSc degree dissertation]. Universidade Agostinho Neto, Luanda, LD, Angola.

Antchak, V. (2017). Portfolio of major events in Auckland: Characteristics, perspectives and issues. *Journal of Policy Research in Tourism, Leisure & Events*, 9(3), 280–297. 10.1080/19407963.2017.1312421

Antchak, V., Ziakas, V., & Getz, D. (2019). *Event Portfolio Management: Theory and methods for event management and tourism* (1st ed.). Goodfellow Publishers., 10.23912/978-1-911396-91-8-4260

Antsulevich, A. E. (1994). Artificial reefs project for improvement of water quality and environmental enhancement of Neva Bay (St.-Petersburg County Region). *Bulletin of Marine Science*, 55(2-3), 1189–1192.

Aoki. (2018). *Fish population dynamics, monitoring, and management—sustainable fisheries in the eternal ocean*. Springer Nature.

ARE. (n.d.). Algarve Real Estate. Retrieved from https://algarverealestate.com/

Arias-González, J. E., Baums, I. B., Banaszak, A. T., Prada, C., Rossi, S., Hernández-Delgado, E. A., & Rinkevich, B. (2022). Editorial: Coral Reef Restoration in a Changing World: Science-Based Solutions. *Frontiers in Marine Science*, 9, 919603. 10.3389/fmars.2022.919603

Arlinghaus, R., & Cooke, S. J. (2009). Recreational fisheries: socioeconomic importance, conservation issues and management challenges. *Recreational hunting, conservation and rural livelihoods: science and practice*, 39-58

Arteaga, M., Ernst, B., Vásquez, S., & Gatica, C. (2014). Bases conceptuales para la aplicación de una evaluación de estrategias de manejo (EEM) en sardina común (Strangomera bentincki) y anchoveta (Engraulis ringens) en la zona centro-sur de Chile. *Latin American Journal of Aquatic Research*, 42(3), 3, 445–467. 10.3856/vol42-issue3-fulltext-6

Asim, T., Islam, S. Z., Hemmati, A., & Khalid, M. S. U. (2022). A Review of Recent Advancements in Offshore Wind Turbine Technology. *Energies*, 15(2), 579. 10.3390/en15020579

Ba, Chaboud, Brehmer, & Schmidt. (2022). Are subsidies still relevant in West African artisanal small pelagic fishery? Insights from long run bioeconomic scenarios. *Marine Policy*, 146, 105294. 10.1016/j.marpol.2022.105294

Ba, A., Chaboud, C., Schmidt, J., Diouf, M., Fall, M., Dème, M., & Brehmer, P. (2019). The potential impact of marine protected areas on the Senegalese sardinella fishery. *Ocean and Coastal Management*, 169, 239–246. 10.1016/j.ocecoaman.2018.12.020

Baine, M. (2001). Artificial reefs: A review of their design, application, management and performance. *Ocean and Coastal Management*, 44(3-4), 241–259. 10.1016/S0964-5691(01)00048-5

Ballesteros & Dickey-Collas. (2023). Managing participation across boundaries: A typology for stakeholder engagement in the International Council for the Exploration of the Sea. *Marine Policy*, 147, 105389. 10.1016/j.marpol.2022.105389

Baptista, V., Encarnação, J., Serrão, E. A., Wirtz, P., Pestana, L. B., Faria, S., & Teodósio, M. A. (2021). New Records of Fish Species from the Coast of Luanda, Angola. *Thalassas*, 37(2), 803–811. 10.1007/s41208-021-00297-1

Baptista, V., Leitão, F., Morais, P., Teodósio, M. A., & Wolanski, E. (2020). Modelling the ingress of a temperate fish larva into a nursery coastal lagoon. *Estuarine, Coastal and Shelf Science*, 235, 106601. 10.1016/j.ecss.2020.106601

Barange, M. A., Bernal, M., Cergole, M. C., Cubillos, L. A., Cunningham, C. L., Daskalov, M., & De Oliveira, G. (2009). Current trend in the assessment and management of stocks. In *Climate Change and Small Pelagic Fish*. Cambirdge University Press.

Barbier, E. B. (2017). Marine ecosystem services. *Current Biology*, 27(11), R507–R510. 10.1016/j.cub.2017.03.02028586688

Bardin, L. (2018). *Análise de Contéudo* (A. Rego & A. Pinheiro, Trans.). Edições 70.

Bauer, M. W., Allum, N., & Miller, S. (2007). What can we learn from 25 years of PUS survey research? Liberating and expanding the agenda. *Public Understanding of Science (Bristol, England)*, 16(1), 79–95. 10.1177/0963662506071287

Baumann, J. H., Ries, J. B., Rippe, J. P., Courtney, T. A., Aichelman, H. E., Westfield, I., & Castillo, K. D. (2019). Nearshore coral growth declining on the Mesoamerican Barrier Reef System. *Global Change Biology*, 25(11), 3932–3945. 10.1111/gcb.1478431456305

Bax, N., Novaglio, C., Maxwell, K. H., Meyers, K., McCann, J., Jennings, S., Frusher, S., Fulton, E. A., Nursey-Bray, M., Fischer, M., Anderson, K., Layton, C., Emad, G. R., Alexander, K. A., Rousseau, Y., Lunn, Z., & Carter, C. G. (2022). Ocean resource use: Building the coastal blue economy. *Reviews in Fish Biology and Fisheries*, 32(1), 189–207. 10.1007/s11160-021-09636-033679009

Beddington, Agnew, & Clark. (2007). Current problems in the management of marine fisheries. *Science*, 316(5832), 1713–1716. 10.1126/science.1137362

Beeden, R. J., Turner, M. A., Dryden, J., Merida, F., Goudkamp, K., Malone, C., Marshall, P. A., Birtles, A., & Maynard, J. A. (2014). Rapid survey protocol that provides dynamic information on reef condition to managers of the Great Barrier Reef. *Environmental Monitoring and Assessment*, 186(12), 8527–8540. 10.1007/s10661-014-4022-025179944

Belekubun, R. A. (2022, April 10). *Traditional Markets in Jakarta Cannot Get Rid of Plastic Bags*. kompas.id. https://www.kompas.id/baca/humaniora/2022/10/04/pasar-tradisional-di-jakarta-belum-bisa-lepas-dari-kantong-plastik

Belhassen, Y., Rousseau, M., Tynyakov, J., & Shashar, N. (2017). Evaluating the attractiveness and effectiveness of artificial coral reefs as a recreational ecosystem service. *Journal of Environmental Management*, 203, 448–456. 10.1016/j.jenvman.2017.08.02028837911

Bellou, N., Gambardella, C., Karantzalos, K., Monteiro, J. G., Canning-Clode, J., Kemna, S., Arrieta-Giron, C. A., & Lemmen, C. (2021). Global assessment of innovative solutions to tackle marine litter. *Nature Sustainability*, 4(6), 516–524. 10.1038/s41893-021-00726-2

Bennett, D. J., & Jennings, R. C. (2011). *Successful science communication: Telling it like it is*. Cambridge University Press.

Benson & Stephenson. (2018). Options for integrating ecological, economic, and social objectives in evaluation and management of fisheries. *Fish and Fisheries*, 19(1), 40–56. 10.1111/faf.12235

Bettim, M., Krelling, A. P., Di Domenico, M., Cornwell, T. O., & Turra, A. (2021). Daily environmental variation influences temporal patterns of marine debris deposition along an estuarine outlet in southern Brazil. *Marine Pollution Bulletin*, 172, 112859. 10.1016/j.marpolbul.2021.11285934418710

Bideci, C., & Cater, C. (2019). In search of underwater atmosphere: A new diving world on artificial reefs. In *Atmospheric turn in culture and tourism: Place, design and process impacts on customer behaviour, marketing and branding* (pp. 245–257). Emerald Publishing Limited. 10.1108/S1871-317320190000016023

Blackmore, A. C. (2018). Rediscovering the origins and inclusion of the public trust doctrine in South African environmental law: A speculative analysis. *Review of European, Comparative & International Environmental Law*, 27(2), 187–198. 10.1111/reel.12239

Blanco, J., Livramento, F., & Rangel, I. M. (2010). Amnesic shellfish poisoning (ASP) toxins in plankton and molluscs from Luanda Bay. Angola. *Toxicon*, 55(2–3), 541–546. 10.1016/j.toxicon.2009.10.00819825382

Blount, C., Komyakova, V., Barnes, L., Smith, M. L., Zhang, D., Reeds, K., McPhee, D., Taylor, M. D., Macbeth, W., & Needham, E. (2021). Using ecological evidence to refine approaches to deploying offshore artificial reefs for recreational fisheries. *Bulletin of Marine Science*, 97(4), 665–698. 10.5343/bms.2020.0059

Bonnevie, I. M., Hansen, H. S., & Schrøder, L. (2019). Assessing use-use interactions at sea: A theoretical framework for spatial decision support tools facilitating co-location in maritime spatial planning. *Marine Policy*, 106, 103533. 10.1016/j.marpol.2019.103533

Boström-Einarsson, L., Babcock, R. C., Bayraktarov, E., Ceccarelli, D., Cook, N., Ferse, S. C. A., Hancock, B., Harrison, P., Hein, M., Shaver, E., Smith, A., Suggett, D., Stewart-Sinclair, P. J., Vardi, T., & McLeod, I. M. (2020). Coral restoration – A systematic review of current methods, successes, failures and future directions. *PLoS One*, 15(1), e0226631. 10.1371/journal.pone.022663131999709

Boswell, R. (2021). Art and the senses for ocean conservation. *Journal of Marine and Island Cultures*, 10(1), 39–56. 10.21463/jmic.2021.10.1.03

Bowater, L., & Yeoman, K. (2013). *Science communication: a practical guide for scientists*. Wiley-Blackwell.

Boxman, R., & Boxman, E. (2017). *Communicating science: a practical guide for engineers and physical scientists*. World Scientific Publishing.

Bradley, D., Merrifield, M., Miller, K. M., Lomonico, S., Wilson, J. R., & Gleason, M. G. (2019). Opportunities to improve fisheries management through innovative technology and advanced data systems. *Fish and Fisheries*, 20(3), 564–583. 10.1111/faf.12361

Brown, K. (2016). Oceanic geographies. *Performance Research*, 21(2), 118–124. 10.1080/13528165.2016.1162506

Brundtland, G. H. (1987). Brundtland report. Our common future. *Comissão Mundial*, 4(1), 17–25.

Bullard, R. (2005). Environmental justice in the 21st century. *Debating the earth*, 3222-3356.

Buonocore, E., Grande, U., Franzese, P. P., & Russo, G. F. (2021). Trends and evolution in the concept of marine ecosystem services: An overview. *Water (Basel)*, 13(15), 2060. 10.3390/w13152060

Burdon, P. D. (2013). The earth community and ecological jurisprudence. *Oñati Socio-Legal Series*, 3(5).

Butler, C. M., Rudershausen, P. J., & Buckel, J. A. (2010). Atlantic Bluefin Tuna (Thunnus thynnus) feeding ecology and potential ecosystem effects during winter in North Carolina. *U.S. National Marine Fisheries Service. Fish Bulletin*, 108, 56–69.

Butterworth, D. S., Johnston, S. J., & Brandão, A. (2010). Pretesting the likely efficacy of suggested management approaches to data-poor fisheries. *Marine and Coastal Fisheries*, 2(1), 131–145. 10.1577/C08-038.1

Caddy, J. F., & Mahon, R. (1995). *Reference points for fisheries management. FAO Fisheries Technical Paper. 347*. FAO.

Canda, R. (2022). *Contribuição para o conhecimento da ictiofauna da Baía de Luanda* [A contribution to the knowledge of the ichthyofauna of the Luanda Bay] [Unpublished BSc degree dissertation]. Universidade Agostinho Neto, Luanda, LD, Angola.

Carapuço, M. M., Taborda, R., Silveira, T. M., & Andrade, C. (2017). Upstream public engagement on coastal issues: Audience response to a science-based exhibition. *Ocean and Coastal Management*, 144, 83–89. 10.1016/j.ocecoaman.2017.04.008

Carey, J. (2021, June). The complex case of Chesapeake Bay restoration, PNAS. https://www.pnas.org/doi/10.1073/pnas.2108734118

Carr, M. H., & Hixon, M. A. (1997). Artificial reefs: The importance of comparisons with natural reefs. *Fisheries (Bethesda, Md.)*, 22(4), 28–33. 10.1577/1548-8446(1997)022<0028:ARTIOC>2.0.CO;2

Chalip, L. (2004). Beyond impact: A general model for sport event leverage. In B. W. Ritchie & D. Adair (Eds.), *Sport tourism: Interrelationships, impacts and issues* (pp. 226–252). Blue Ridge Summit: Channel View Publications. 10.21832/9781873150672-014

Chalip, L. (2018). Trading legacy for leverage. In I. Brittain (Ed.), *Legacies and mega events: Fact or fairy tales?* (pp. 25–41). Routledge, Taylor & Francis Group.

Chalip, L. (2006). Towards social leverage of sport events. *Journal of Sport & Tourism*, 11(2), 109–127. 10.1080/14775080601155126

Chalip, L., Green, B. C., Taks, M., & Misener, L. (2017). Creating sport participation from sport events: Making it happen. *International Journal of Sport Policy*, 9(2), 257–276. 10.1080/19406940.2016.1257496

Chen, Liu, & Yin. (2023). Evaluation of Yellow Sea anchovy (Engraulis japonicus) fishery for management strategies using a data-limited management toolkit. *Heliyon*, 9, e18058. 10.1016/j.heliyon.2023.e18058

Chen, Zhang, Sun, Xu, Xue, Ren, & Chen. (2018). The impact of natural mortality variations on the performance of management procedures for Spanish mackerel (Scomberomorus niphonius) in the Yellow Sea, China. *Acta Oceanologica Sinica*, 37(8), 21–30. 10.1007/s13131-018-1234-0

Chen, J. L., Chuang, C. T., Jan, R. Q., Liu, L. C., & Jan, M. S. (2013). Recreational benefits of ecosystem services on and around artificial reefs: A case study in Penghu, Taiwan. *Ocean and Coastal Management*, 85, 58–64. 10.1016/j.ocecoaman.2013.09.005

Chen, J. L., Hsu, K., & Chuang, C. T. (2020). How do fishery resources enhance the development of coastal fishing communities: Lessons learned from a community-based sea farming project in Taiwan. *Ocean and Coastal Management*, 184, 105015. 10.1016/j.ocecoaman.2019.105015

Chen, L., Hu, Y., Wang, R., Li, X., Chen, Z., Hua, J., Osman, A. I., Farghali, M., Huang, L., Li, J., Dong, L., Rooney, D. W., & Yap, P.-S. (2024). Green building practices to integrate renewable energy in the construction sector: A review. *Environmental Chemistry Letters*, 22(2), 751–784. 10.1007/s10311-023-01675-2

Chen, S., Xing, X., & Chalip, L. (2022). Planning and implementation of event leveraging strategy: China's legacy pledge to motivate 300 million people to be involved in winter sport. *Sport Management Review*, 25(5), 771–790. 10.1080/14413523.2021.1987737

Chersulich Tomino, A., Perić, M., & Wise, N. (2020). Assessing and considering the wider impacts of sport-tourism events: A research agenda review of sustainability and strategic planning elements. *Sustainability (Basel)*, 12(11), 4473. 10.3390/su12114473

Chevalier, J. M., & Buckles, D. J. (2019). *Participatory Action Research: Theory and Methods for Engaged Inquiry*. Routledge. 10.4324/9781351033268

Chicharo, L., Ben Hamadou, R., Amaral, A., Range, P., Mateus, C., Piló, D., Marques, R., Morais, P., & Alexandra Chícharo, M. (2009). Application and demonstration of the Ecohydrology approach for the sustainable functioning of the Guadiana estuary (South Portugal) UNESCO-IHP Demonstration Projects in Ecohydrology. *Ecohydrology & Hydrobiology*, 9(1), 55–71. 10.2478/v10104-009-0039-3

Cho, J. H., Gates, J. M., Logan, P., Kitts, A., & Soboil, M. (2005). The Economic Values of Atlantic Herring in the Northeast Shelf Large Marine Ecosystem. *Large Marine Ecosystems*, 13, 215–228. 10.1016/S1570-0461(05)80034-5

Clarke, M., & Egan, A. (2017). Good luck or good governance? The recovery of Celtic Sea herring. *Marine Policy*, 78, 78, 163–170. 10.1016/j.marpol.2016.10.025

Claudet, J., & Pelletier, D. (2004). Marine protected areas and artificial reefs: A review of the interactions between management and scientific studies. *Aquatic Living Resources*, 17(2), 129–138. 10.1051/alr:2004017

Cochet, Y., & Gilly, B. (1990). *Computer programmes for fisheries bioeconomic analysis. BEAM 3: Bioeconomic analytical simulation model, for tropical shrimp fisheries with fixed or random recruitment. Fisheries Technical Papers, no: 310.2.* FAO.

Collins, L. (2021). *The ecological constitution: reframing environmental law.* Routledge. 10.4324/9780429277320

Cordova, M. R., Bernier, N., Yogaswara, D., Subandi, R., Wibowo, S. P. A., Kaisupy, M. T., & Haulussy, J. (2023). Land-derived litter load to the Indian Ocean: A case study in the Cimandiri River, southern West Java, Indonesia. *Environmental Monitoring and Assessment*, 195(10), 1251. 10.1007/s10661-023-11831-437768383

Costa, S. (2022). *Incidência do fitoplâncton tóxico ou nocivo na Baía de Luanda com impacto na saúde pública e na qualidade do ambiente* [Incidence of toxic or harmful phytoplankton in Luanda Bay with an impact on public health and environmental quality] [Unpublished Master's dissertation]. Universidade Agostinho Neto, Luanda, LD, Angola.

Costanza, R., d'Arge, R., De Groot, R., Farber, S., Grasso, M., Hannon, B., Limburg, K., Naeem, S., O'Neill, R. V., Paruelo, J., Raskin, R. G., Sutton, P., & Van Den Belt, M. (1997). The value of the world's ecosystem services and natural capital. *Nature*, 387(6630), 253–260. 10.1038/387253a0

Craig, R. K., & Ruhl, J. B. (2010). Governing for Sustainable Coasts: Complexity, Climate Change, and Coastal Ecosystem Protection. *Sustainability (Basel)*, 2(5), 1361–1388. 10.3390/su2051361

Creswell, J. W., & Creswell, J. D. (2018). *Research design: Qualitative, quantitative, and mixed methods approaches*. SAGE Publications.

Croese, S. (2018). Global urban policymaking in Africa: A view from Angola through the redevelopment of the Bay of Luanda. *International Journal of Urban and Regional Research*, 42(2), 198–209. 10.1111/1468-2427.12591

Cullinan, C. (2011). *Wild law: A manifesto for earth justice*. Bloomsbury Publishing.

Cury, P., Bakun, A., Crawford, R.J.M., Jarre, A., Quinones, R.A., Shannon, L.J., & Verheye, H.M. (2000). Small pelagics in upwelling systems: Patterns of interaction and structural changes in "wasp-waist" ecosystems. *Ices Journal of Marine Science*, 57(3), 603–618. 10.1006/jmsc.2000.0712

d'Eon-Eggertson, F., Dulvy, N. K., & Peterman, R. (2015). Reliable Identification of Declining Populations in an Uncertain World. *Conservation Letters*, 8(2), 86–96. 10.1111/conl.12123

Da Costa, J. P., Mouneyrac, C., Costa, M., Duarte, A. C., & Rocha-Santos, T. (2020). The role of legislation, regulatory initiatives and guidelines on the control of plastic pollution. *Frontiers in Environmental Science*, 8, 104. 10.3389/fenvs.2020.00104

da Fonte, M. M. A. (2007). *Urbanismo e aquitectura em Angola - de Norton de Matos à Revolução* [Urbanism and architecture in Angola - from Norton de Matos to the Revolution]. [Doctoral thesis, Universidade de Lisboa]. ULisboa Repository.

Da Rocha, J.M., & Gutiérrez, M.J. (2011). Lessons from the long-term management plan for northern hake: could the economic assessment have accepted it? *ICES Journal of Marine Science*, 68(9), 1937–1941. 10.1093/icesjms/fsr105

Da Rocha, J.M., Cerviño, S., & Gutiérrez, M.J. (2010). An endogenous bioeconomic optimization algorithm to evaluate recovery plans: an application to southern hake. *ICES Journal of Marine Science*, 67(9), 1957–1962. 10.1093/icesjms/fsq116

Daily, G. C. (Ed.). (1997). *Nature's Services: Societal Dependence on Natural Ecosystems*. Island Press.

Danovaro, R., Aronson, J., Cimino, R., Gambi, C., Snelgrove, P. V., & Van Dover, C. (2021). Marine ecosystem restoration in a changing ocean. *Restoration Ecology*, 29(S2), e13432. 10.1111/rec.13432

Day, J. C., & Dobbs, K. (2013). Effective governance of a large and complex cross-jurisdictional marine protected area: Australia's Great Barrier Reef. *Marine Policy*, 41, 14–24. 10.1016/j.marpol.2012.12.020

De Anda-Montañez, J. A., Salas, S., & Galindo-Cortes, G. (2017). Dealing with dynamics and uncertainty of small pelagic fisheries: bioeconomic analysis of manager's responses to alternative management strategies. *Revista de Biología Marina y Oceanografía*, 52(1), 51-65. https://www.redalyc.org/articulo.oa?id=47950143004

De Labra, G., Vilela López, B., Prieto Porriños, G., Blanc, P. P., Vasta, A., & Anibal, O. (2023). *A cadeia de valor dos pelágicos costeiros em São Tomé e Príncipe: Relatório de síntese*. FAO.

de Lima, R. L. P., Boogaard, F. C., & de Graaf-van Dinther, R. E. (2020). Innovative Water Quality and Ecology Monitoring Using Underwater Unmanned Vehicles: Field Applications, Challenges and Feedback from Water Managers. *Water (Basel)*, 12(4), 1196. 10.3390/w12041196

De Lucia, V. (2013). Towards an ecological philosophy of law: A comparative discussion. *Journal of Human Rights and the Environment*, 4(2), 167–190. 10.4337/jhre.2013.02.03

De Moor, C. L., Butterworth, D. S., & De Oliveira, J. A. A. (2011). Is the management procedure approach equipped to handle short-lived pelagic species with their boom and bust dynamics? The case of the South African fishery for sardine and anchovy. *ICES Journal of Marine Science*, 68(10), 2075–2085. 10.1093/icesjms/fsr165

de Vargas, C., Le Bescot, N., Pollina, T., Henry, N., Romac, S., Colin, S., Haëntjens, N., Carmichael, M., Berger, C., Le Guen, D., Decelle, J., Mahé, F., Poulain, J., Malpot, E., Beaumont, C., Hardy, M., Guiffant, D., Probert, I., Gruber, D. F., & Prakash, M. (2022). Plankton Planet: A frugal, cooperative measure of aquatic life at the planetary scale. *Frontiers in Marine Science*, 9(August), 1–17. 10.3389/fmars.2022.936972

Dean, L. (1983, May). Undersea oases made by man: Artificial reefs create new fishing grounds. *Oceans*, 26(May–June), 27–29.

Delacámara, G., O'Higgins, T. G., Lago, M., & Langhans, S. (2020). Ecosystem-Based Management: Moving from Concept to Practice. In O'Higgins, T., Lago, M., & DeWitt, T. (Eds.), *Ecosystem-Based Management, Ecosystem Services and Aquatic Biodiversity*. Springer., 10.1007/978-3-030-45843-0_3

Delgado Fernandez, I. (2022). Coast for Kids (C4K): A transdisciplinary science communication effort. *EGU General Assembly*, 2022, EGU22–EGU13427. 10.5194/egusphere-egu22-13427

Department of Housing. Settlement Areas and Land. (2020, August 1). *Bali Becomes the First Province in Indonesia to Implement Plastic Waste Free Rules*. Dinas Perumahan, Kawasan Permukiman dan Pertanahan | Pemerintah Kabupaten Buleleng. https://disperkimta.bulelengkab.go.id/informasi/detail/artikel/bali-menjadi-provinsi-pertama-di-indonesia-yang-menerapkan-aturan-bebas-sampah-plastik-13

Department of the Environment. (2022, March 21). *Plastic Bag Free Jakarta Policy*. Portal Resmi Provinsi DKI Jakarta. https://jakarta.go.id/kebijakan-jakarta-bebas-kantong-plastik

Díaz-Vega, J.P., & Flores-Arévalo, Y. (2021). Evaluating marine reserves as a management policy in the central-southern anchovy (Engraulis ringens) fishery of Chile. *Latin American Journal of Aquatic Research,* 49(2), 212-226. 10.3856/vol49-issue2-fulltext-2614

Dichmont, C.M., & Brown, I.W. (2010). A case study in successful management of a data-poor fishery using simple decision rules: the Queensland spanner crab fishery. *Marine and Coastal Fisheries: Dynamics, Management, and Ecosystem Science*, 2, 1–13. 10.1577/C08-034.1

Dichmont, C. M., Ellis, N., Bustamante, R. H., Deng, R., Tickell, S., Pascual, R., Lozano-Montes, H., & Griffiths, S. (2013). Evaluating marine spatial closures with conflicting fisheries and conservation objectives. *Journal of Applied Ecology*, 50(4), 1060–1070. 10.1111/1365-2664.12110

Dijkstra, A. M., de Bakker, L., van Dam, F., & Jensen, E. A. (2020). Setting the Scene. In van Dam, F., de Bakker, L., Dijkstra, A. M., & Jensen, E. A. (Eds.), *Science Communication. An Introduction* (pp. 1–16). World Scientific Publishing. 10.1142/9789811209888_0001

Dimmock, K. (2007). SCUBA diving, snorkeling, and free diving. In *Water-based tourism, sport, leisure, and recreation experiences* (pp. 128–147). Routledge. 10.1016/B978-0-7506-6181-2.50013-0

Directorate General of Waste. Waste and B3 Management. (2022, May 24). *National Dialogue: Waste Reduction by Producers, Opportunities and Challenges towards Sustainable Business*. Direktorat Jenderal Pengelolaan Sampah, Limbah Dan B3 Kementrian Lingkungan Hidup Dan Kehutanan. https://pslb3.menlhk.go.id/portal/read/dialog-nasional-pengurangan-sampah-oleh-produsen-peluang-dan-tantangan-menuju-bisnis-yang-berkelanjutan

Divan, S., & Rosencranz, A. (2001). *Environmental law and policy in India: cases. Materials and Statistics* (2nd ed.). OUP.

Domínguez-Torreiro, M., & Surís-Regueiro, J.C. (2007). Cooperation and non-cooperation in the Ibero-atlantic sardine shared stock fishery. *Fisheries Research*, 83, 1-10. 10.1016/j.fishres.2006.07.012

Duarte, C. M. (2000). Marine biodiversity and ecosystem services: An elusive link. *Journal of Experimental Marine Biology and Ecology*, 250(1-2), 117–131. 10.1016/S0022-0981(00)00194-510969166

Dudley, R.G. (2008). A basis for understanding fishery management dynamics. *System Dynamics Review*, 24(1), 1-29. 10.1002/sdr.392

Duglio, S., & Beltramo, R. (2017). Estimating the economic impacts of a small-scale sport tourism event: The case of the Italo-Swiss mountain trail Collon trek. *Sustainability (Basel)*, 9(3), 343. 10.3390/su9030343

Duhec, A. V., Jeanne, R. F., Maximenko, N., & Hafner, J. (2015). Composition and potential origin of marine debris stranded in the Western Indian Ocean on remote Alphonse Island, Seychelles. *Marine Pollution Bulletin*, 96(1), 76–86. 10.1016/j.marpolbul.2015.05.04226024564

Duignan, M. B., Parent, M. M., & McGillivray, D. (2023). Accommodating (Global–Glocal) Paradoxes Across Event Planning. *Event Management*, 27(1), 149–154. 10.3727/152599522X16419948694991

Dupont, J. M. (2008). Artificial Reefs as restoration tools: A case study on West Florida Shelf. *Coastal Management*, 36(5), 495–507. 10.1080/08920750802395558

E24. (2021). *Walter Chicharro sublinha importância dos grandes eventos para a Nazaré*. https://www.youtube.com/watch?v=-7TP7FAjsco

EC. (2023). *European Commission. The EU Blue Economy Report. 2023.* Publications Office of the European Union.

Edmunds, P. J., Tsounis, G., Boulon, R., & Bramanti, L. (2018). Long-term variation in light intensity on a coral reef. *Coral Reefs*, 37(3), 955–965. 10.1007/s00338-018-1721-y

Edward, J. K. P., Mathews, G., Raj, K. D., Laju, R. L., Bharath, M. S., Kumar, P. D., Arasamuthu, A., & Grimsditch, G. (2020). Marine debris—An emerging threat to the reef areas of Gulf of Mannar, India. *Marine Pollution Bulletin*, 151, 110793. 10.1016/j.marpolbul.2019.11079332056588

EEA. (2009). European Environment Agency. Common international classification of ecosystem services (CICES). European Environment Agency. https://cices.eu/

Ehrlich, P. R., & Ehrlich, A. H. (1981). *Extinction: the causes and consequences of the disappearance of species/Paul and Anne Ehrlich* (1st ed.). Random House.

El Gohary, G. (2013). Wet landscape: Using sculptures to form underwater landscape. *Democratic Transition and Sustainable Communities*, 6, 351.

Elliott, M., Mander, L., Mazik, K., Simenstad, C., Valesini, F., Whitfield, A., & Wolanski, E. (2016). Ecoengineering with Ecohydrology: successes and failures in estuarine restoration. *Estuarine, Coastal and Shelf Science 176*, 12-35.

El-Mohtady Bellah, A., & Chiha, I. (2017). Sustainable Development in International and Comparative Constitutional Law.

Emery, T. J., Gardner, C., Hartmann, K., & Cartwright, I. (2017). Incorporating economics into fisheries management frameworks in Australia. *Marine Policy*, 77, 77, 136–143. 10.1016/j.marpol.2016.12.018

Entradas, M., Junqueira, L., & Pinto, B. (2020). Portugal: The late bloom of (modern) science communication. In T. Gascoigne, B. Schiele, J. Leach, M. Riedlinger, B. V Lewenstein, L. Massarani, & P. Broks (Eds.), *Communicating science - a global perspective* (pp. 693–741). Australian National University Press. 10.22459/CS.2020

Eriksen, T., Høye, G., Narheim, B., & Meland, B. J. (2006). Maritime traffic monitoring using a space-based AIS receiver. *Acta Astronautica*, 58(10), 537–549. 10.1016/j.actaastro.2005.12.016

Essington, T. E., Sanchirico, J. N., & Baskett, M. L. (2018). Economic value of ecological information in ecosystem-based natural resource management depends on exploitation history. *Proceedings of the National Academy of Sciences of the United States of America*, 115(7), 1658–1663. 10.1073/pnas.171685811529378966

European Commission. (2017). Attitudes of European citizens towards the environment. In *Eurobarometer*. European Commission.

European Commission. (2021). *European citizens' knowledge and attitudes towards science and technology*. 10.2775/071577

Eurostat. (n.d.). GISCO - Geographic Information System for Cooperation. Retrieved from https://ec.europa.eu/eurostat/web/gisco/geodata/reference-data

Fabi, G., Spagnolo, A., Bellan-Santini, D., Charbonnel, E., Çiçek, B. A., García, J. J. G., Jensen, A. C., Kallianiotis, A., & Santos, M. N. D. (2011). Overview on artificial reefs in Europe. *Brazilian Journal of Oceanography*, 59(spe1), 155–166. 10.1590/S1679-87592011000500017

Faculty of Fisheries and Marine Sciences. (2018). Faculty of Fisheries and Marine Sciences. https://fpik.ub.ac.id/?page_id=3452

Falcão, M., Santos, M. N., Vicente, M., & Monteiro, C. C. (2007). Biogeochemical processes and nutrient cycling within an artificial reef off Southern Portugal. *Marine Environmental Research*, 63(5), 429–444. 10.1016/j.marenvres.2006.12.00117239434

Fan, J., Kuang, C., Liu, H., Wang, D., Liu, J., Wang, G., & Zou, Q. (2023). Impact of Artificial Islands and Reefs on Water Quality in Jinmeng Bay, China. *Water (Basel)*, 15(5), 959. 10.3390/w15050959

FAO (Food and Agriculture Organization of the United Nations). (2016). *The State of World Fisheries and Aquaculture 2016. Contributing to Food Security and Nutrition for All*. Food and Agriculture Organization of the United Nations.

FAO. (2012). *Food and Agriculture Organization. Technical Guidelines for Responsible Fisheries, No. 13*. FAO.

Faria, S., Macuéria, M., Mosley, B. A., Teodósio, M. A., & Baptista, V. (2021). Characterization of small-scale fishing activity in Luanda Bay (Angola). *Journal of Spatial and Organizational Dynamics*, 9(3), 225–238.

Fecher, B., Kuper, F., Sokolovska, N., Fenton, A., Hornbostel, S., & Wagner, G. G. (2021). Understanding the Societal Impact of the Social Sciences and Humanities: Remarks on Roles, Challenges, and Expectations. *Frontiers in Research Metrics and Analytics*, 6(July), 1–14. 10.3389/frma.2021.69680434278206

Feigenbaum, D., Bushing, M., Woodward, J., & Friedlander, A. (1989). Artificial reefs in Chesapeake Bay and nearby coastal waters. *Bulletin of Marine Science*, 44(2), 734–742.

Fernandes, M., Fernandes, C., Barroqueiro, T., Agostinho, P., Martins, N., & Alonso-Martirena, A. (2018). Extreme wave height events in Algarve (Portugal): Comparison between HF radar systems and wave buoys. *Proceedings of the 5th Jornadas Engenharia Hidrográfica*, 19-21.

Fernando, R., & Rathnayake, S. (2020). Potentials of Underwater Cultural Heritage Towards Tourism in Galle, Sri Lanka: Perspectives of Tourism Service Providers and Maritime Archaeologists.

Firth, L. B., Farnworth, M., Fraser, K. P., & McQuatters-Gollop, A. (2023). Make a difference: Choose artificial reefs over natural reefs to compensate for the environmental impacts of dive tourism. *The Science of the Total Environment*, 901, 165488. 10.1016/j.scitotenv.2023.16548837524181

Fischer, L., Barata, G., Scheu, A. M., & Ziegler, R. (2024). Connecting science communication research and practice: Challenges and ways forward. *Journal of Science Communication, 23*(02). *JCOM, Journal of Science Communication*, 23(2). Advance online publication. 10.22323/2.23020501

Ford, H. V., Jones, N. H., Davies, A. J., Godley, B. J., Jambeck, J. R., Napper, I. E., Suckling, C. C., Williams, G. J., Woodall, L. C., & Koldewey, H. J. (2022). The fundamental links between climate change and marine plastic pollution. *The Science of the Total Environment*, 806(Pt 1), 150392. 10.1016/j.scitotenv.2021.15039234583073

Fosse, J., Tonazzini, D., Morales, E., Gonzalez, A., Klarwein, S., Moukaddem, K., & Louveau, O. (2019). Blue Tourism: The Transition Towards Sustainable Coastal and Maritime Tourism in World Marine Regions.

Francis, R. I. C. C., & Shotton, R. (1997). Risk in fisheries management: A review. *Canadian Journal of Fisheries and Aquatic Sciences*, 54(8), 1699–1715. 10.1139/f97-100

Frangovska, A. (2020). Inter/Trans-Disciplinarity as a Platform for Seeking Universal Common Ground: From Speculation to Pragmatism Through Kristina Pulejkova's Wedding Route Art Project. *AM Journal of Art and Media Studies*, 21(21), 147–158. 10.25038/am.v0i21.366

Fréon, P., Bouchon, M., Mullon, C., García, C., & Ñiquen, M. (2008). Interdecadal variability of anchoveta abundance and overcapacity of the fishery in Peru. *Progress in Oceanography*, 79(2), 401–412. 10.1016/j.pocean.2008.10.011

Freon, P., Cury, P., Shannon, L., & Roy, C. (2005). Sustainable exploitation of small pelagic fish stocks challenged by environmental and ecosystem changes: A review. *Bulletin of Marine Science*, 76(2), 385–462.

Fujioka, S. N. (2015). *The Effect of Inorganic Waste on the Condition of Hard Corals*. Universitas Hasanuddin

Fulton, E.A., Smith, A.D.M., Smith, D.C., & Johnson, P. (2014). An Integrated Approach Is Needed for Ecosystem Based Fisheries Management: Insights from Ecosystem-Level Management Strategy Evaluation. *PLoS One*, 9(1), e84242. 10.1371/journal.pone.0084242

Fulton, E. A., Link, J. S., Kaplan, I. C., Savina-Rolland, M., Johnson, P., Ainsworth, C., Horne, P., Gorton, R., Gamble, R. J., Smith, A. D. M., & Smith, D. C. (2011). Lessons in modelling and management of marine ecosystems: The Atlantis experience. *Fish and Fisheries*, 12(2), 12, 171–188. 10.1111/j.1467-2979.2011.00412.x

Furness, R. W. (2003). Impacts of fisheries on seabird communities. *Scientia Marina*, 67(S2), 33–45. 10.3989/scimar.2003.67s233

Gelcich, S., Buckley, P., Pinnegar, J. K., Chilvers, J., Lorenzoni, I., Terry, G., Guerrero, M., Castilla, J. C., Valdebenito, A., & Duarte, C. M. (2014). Public awareness, concerns, and priorities about anthropogenic impacts on marine environments. *Proceedings of the National Academy of Sciences of the United States of America*, 111(42), 15042–15047. 10.1073/pnas.141734411125288740

Getz, D., & Page, S. (2016). Progress and prospects for event tourism research. *Tourism Management*, 52, 593–631. 10.1016/j.tourman.2015.03.007

Ghosh, T. (2012). Sustainable coastal tourism: Problems and management options. *Journal of Geography and Geology*, 4(1), 163–169. 10.5539/jgg.v4n1p163

Gibson, H. J., Kaplanidou, K., & Kang, S. J. (2012). Small-scale event sport tourism: A case study in sustainable tourism. *Sport Management Review*, 15(2), 160–170. 10.1016/j.smr.2011.08.013

Giyanto. (2014). *Guide to Monitoring Coral Reef Health: Coral Reefs, Reef Fish, Megabenthos, and Report Writing*. Pusat Penelitian Oseanografi, Lembaga Ilmu Pengetahuan Indonesia.

Goethel, D. R., Omori, K. L., Punt, A. E., Lynch, P. D., Berger, A. M., de Moor, C. L., Plagányi, É. E., Cope, J. M., Dowling, N. A., McGarvey, R., Preece, A. L., Thorson, J. T., Chaloupka, M., Gaichas, S., Gilman, E., Hesp, S. A., Longo, C., Yao, N., & Methot, R. D. (2023). Oceans of plenty? Challenges, advancements, and future directions for the provision of evidence-based fisheries management advice. *Reviews in Fish Biology and Fisheries*, 33(2), 375–410. 10.1007/s11160-022-09726-736124316

Gormsen, E. (1997). The impact of tourism on coastal areas. *GeoJournal*, 42(1), 39–54. 10.1023/A:1006840622450

Gouveia, L. (1970). Luanda. Estudo de Geografia Urbana: Ílidio do Amaral [Luanda. Study of urban geography: Ílidio do Amaral]. *Finisterra*, 5(10). Advance online publication. 10.18055/Finis2446

Gray, M. (2012). Valuing geodiversity in an 'ecosystem services' context. *Scottish Geographical Journal*, 128(3-4), 177–194. 10.1080/14702541.2012.725858

Griggs, G., & Reguero, B. G. (2021). Coastal Adaptation to Climate Change and Sea-Level Rise. *Water (Basel)*, 13(16), 2151. 10.3390/w13162151

Grorud-Colvert, K., Lester, S. E., Airamé, S., Neeley, E., & Gaines, S. D. (2010). Communicating marine reserve science to diverse audiences. *Proceedings of the National Academy of Sciences of the United States of America*, 107(43), 18306–18311. 10.1073/pnas.0914292107 20427745

Guan, S., Qu, F., & Qiao, F. (2023). United Nations Decade of Ocean Science for Sustainable Development (2021-2030): From innovation of ocean science to science-based ocean governance. *Frontiers in Marine Science*, 9(January), 1–8. 10.3389/fmars.2022.1091598

Guerra, F., Grilo, C., Pedroso, N. M., & Cabral, H. (2015). Environmental Impact Assessment in the marine environment: A comparison of legal frameworks. *Environmental Impact Assessment Review*, 55, 182–194. 10.1016/j.eiar.2015.08.003

Gunn, C. M., Amerson, A. M., Adkisson, K. L., & Haxel, J. H. (2022). A Framework for Effective Science Communication and Outreach Strategies and Dissemination of Research Findings for Marine Energy Projects. *Journal of Marine Science and Engineering*, 10(2), 130. Advance online publication. 10.3390/jmse10020130

Gunningham, N. (2009). Environment law, regulation and governance: Shifting architectures. *Journal of Environmental Law*, 21(2), 179–212. 10.1093/jel/eqp011

Guzman, R. (2022, March 7). *Marine Plastic Pollution and Climate Change: Regional and National Action in Southeast Asia*. Heinrich Böll Foundation | Southeast Asia Regional Office. https://th.boell.org/en/2022/03/07/marine-plastic-pollution-seasia

Haines-Young, R., & Potschin, M. (2012). *Common International classification of ecosystem services*. Centre for Environmental Management, University of Nottingham.

Halpern, B. S., Frazier, M., Afflerbach, J., Lowndes, J. S., Micheli, F., O'Hara, C., Scarborough, C., & Selkoe, K. A. (2019). Recent pace of change in human impact on the world's ocean. *Scientific Reports*, 9(1), 11609. 10.1038/s41598-019-47201-931406130

Hamilton, L., Otterstad, O., & Ogmundardottir, H. (2006). *Rise and fall of the herring towns: impacts of climate and human teleconnections. Climate Change and the Economics of the World's Fisheries*. Edward Elgar Publishing.

Hamka, A. (2020). *Impact of Plastic Waste on Bleaching Intensity and Recovery Efforts on Acropora sp coral*. Universitas Hasanuddin.

Hanafiah, M. (2022). Framing The Future Agenda Of Blue Tourism In Sustainable Coastal Tourism Destinations. *Tourism and Hospitality Management*, 28(2), 465–470. 10.20867/thm.28.2.6

Handjar, H., Guendouzi, Y., Niar, A., Kara, H., & Benhalima, M. (2022). Stock status of the bogue Boops boops (Walbaum, 1792) in Algerian West Coast (Southwestern Mediterranean Sea). *Regional Studies in Marine Science*, 56, 102727. 10.1016/j.rsma.2022.102727

Harati-Mokhtari, A., Wall, A., Brooks, P., & Wang, J. (2007). Automatic Identification System (AIS): data reliability and human error implications. The Journal of Navigation, 60(3), 373-389.

Hariram, N. P., Mekha, K. B., Suganthan, V., & Sudhakar, K. (2023). Sustainalism: An Integrated Socio-Economic-Environmental Model to Address Sustainable Development and Sustainability. *Sustainability (Basel)*, 15(13), 10682. 10.3390/su151310682

Harley, M. D., & Kinsela, M. A. (2022). CoastSnap: A global citizen science program to monitor changing coastlines. *Continental Shelf Research*, 245(June), 104796. 10.1016/j.csr.2022.104796

Harris, L. E. (2009). Artificial reefs for ecosystem restoration and coastal erosion protection with aquaculture and recreational amenities. *The R Journal*, 1(1), 235–246.

Harrison, J. (2017). *Saving the oceans through law: the international legal framework for the protection of the marine environment*. Oxford University Press.

Harrison, S., & Rousseau, M. (2020). Comparison of artificial and natural reef productivity in Nantucket Sound, MA, USA. *Estuaries and Coasts*, 43(8), 2092–2105. 10.1007/s12237-020-00749-6

Harvey, N. (1999). Australian Integrated Coastal Management: A Case Study of the Great Barrier Reef. In Salomons, W., Turner, R. K., de Lacerda, L. D., & Ramachandran, S. (Eds.), *Perspectives on Integrated Coastal Zone Management. Environmental Science*. Springer. 10.1007/978-3-642-60103-3_16

Hassan, D. (2017). *Protecting the marine environment from land-based sources of pollution: towards effective international cooperation*. Routledge. 10.4324/9781315245829

Heffernan, J. P. (2014). Dealing with Mediterranean bluefin tuna: A study in international environmental management. *Marine Policy*, 50, 81–88. 10.1016/j.marpol.2014.05.014

Helbert, M. (2024). The Space Between Motherhood and Mother Earth: An Ecofeminist Analysis of the Post-Development Model in Bolivia. In *Ecofeminism on the Edge: Theory and Practice* (pp. 201-214). Emerald Publishing Limited.

Hennink, M. M. (2014). *Focus Group Discussions: Understanding qualitative research*. Oxford University Press. 10.1093/acprof:osobl/9780199856169.001.0001

Hermawan, S., Abiyudo, B., & Isnawanjati, G. (2022). Potential Use of Social Media Platforms to Reduce Plastic Waste in Indonesian Oceans. *Kosmik Hukum*, 22(2), 132. 10.30595/kosmikhukum.v22i2.14140

Hermawan, S., & Astuti, W. (2021). Analysing several ASEAN countries' policy for combating marine plastic litter. *Environmental Law Review*, 23(1), 9–22. 10.1177/1461452921991731

Holmlund, C. M., & Hammer, M. (1999). Ecosystem services generated by fish populations. *Ecological Economics*, 29(2), 253–268. 10.1016/S0921-8009(99)00015-4

Hoshino, E., Milner-Gulland, E. J., & Hillary, R. M. (2012). Bioeconomic adaptive management procedures for short-lived species: A case study of Pacific saury (Cololabis saira) and Japanese common squid (Todarodes pacificus). *Fisheries Research*, 121, 17–30. 10.1016/j.fishres.2012.01.007

Hsui, C. Y., & Wang, C. C. (2013). Synergy between fractal dimension and lacunarity index in design of artificial habitat for alternative SCUBA diving site. *Ecological Engineering*, 53, 6–14. 10.1016/j.ecoleng.2013.01.014

humasft. (2023, March 27). *Caring for the Environment, BEM FT Beach Cleanup*. Prasetya UB. https://prasetya.ub.ac.id/peduli-lingkungan-bem-ft-bersih-pantai/

Hung, L.-Y., Wang, S.-M., & Yeh, T.-K. (2022). Collaboration between the government and environmental non-governmental organisations for marine debris policy development: The Taiwan experience. *Marine Policy*, 135, 104849. 10.1016/j.marpol.2021.104849

Hurtado-Ferro, F., Hiramatsu, K., & Shirakihara, K. (2010). Allowing for environmental effects in a management strategy evaluation for Japanese sardine. *ICES Journal of Marine Science*, 67(9), 2012–2017. 10.1093/icesjms/fsq126

Ibrahim, L. A., Shaghaleh, H., El-Kassar, G. M., Abu-Hashim, M., Elsadek, E. A., & Alhaj Hamoud, Y. (2023). Aquaponics: A Sustainable Path to Food Sovereignty and Enhanced Water Use Efficiency. *Water (Basel)*, 15(24), 4310. 10.3390/w15244310

ICES 2015. *Report of the Working Group on Southern Horse Mackerel, Anchovy and Sardine* (WGHANSA) ICES CM 2015/ACOM:16

ICES 2021. Working group on economics (WGECON; outputs from 2020 meeting. *ICES Scientific Report* (3(4)). 10.17895/ices.pub.7652

ICES. (2023). Working Group on Southern Horse Mackerel, Anchovy and Sardine (WGHANSA). ICES Scientific Reports. Report. 10.17895/ices.pub.23507922.v1

Idris, I., Fakhrurrozi, F., Suhendar, D. B., Rochyadi, A., Wahyudi, R., & Larasati, C. E. (2023). Distribution of Marine debris in Coral Reef Ecosystems on Mules Island, Sawu Marine National Park. *Jurnal Biologi Tropis*, 23(2), 2. Advance online publication. 10.29303/jbt.v23i2.4618

Ikegwu, A. C., Nweke, H. F., Mkpojiogu, E., Anikwe, C. V., Igwe, S. A., & Alo, U. R. (2024). Recently emerging trends in big data analytic methods for modeling and combating climate change effects. *Energy Informatics*, 7(1), 6. 10.1186/s42162-024-00307-5

Ilieva, I., Jouvet, L., Seidelin, L., Best, B. D., Aldabet, S., da Silva, R., & Conde, D. A. (2019). A global database of intentionally deployed wrecks to serve as artificial reefs. *Data in Brief*, 23, 103584. 10.1016/j.dib.2018.12.02330809558

Illingworth, S., & Allen, G. (2020). *Effective science communication: a practical guide to surviving as a scientist* (2nd ed.). IOP Publishing. 10.1088/978-0-7503-2520-2

INE. (2015). Anuário Estatístico Regional do Algarve -INE. Available at: https://www.cm-albufeira.pt/sites/default/files/public/RepositorioDocumentos/plano_alinhamento_intermunicipal_algarve2020_0.pdf

INE. (2020). *Perfil da Pobreza com base no Inquérito de Orçamento Familiar de 2017*. Cidade de São Tomé, República Democrática de São Tomé e Príncipe.

INE. (2023). Instituto Nacional de Estatística. Estatísticas do Turismo: 2022. Lisboa.

Irawati, D. (2020, September 8). *Nestle Immediately Replace Plastic Straws with Paper*. kompas.id. https://www.kompas.id/baca/nusantara/2020/09/08/nestle-segera-ganti-sedotan-plastik-dengan-kertas

Irwin, A. (2008). Risk, science and public communication: Third-order thinking about scientific culture. In Bucchi, M., & Trench, B. (Eds.), *Handbook of Public Communication of Science and Technology*. Routledge. https://doi.org/https://doi.org/10.4324/9780203928240

Issifu, I., Dahmouni, I., García-Lorenzo, I., & Sumaila, U. R. (2024). Economics in Marine Spatial Planning: A Review of Issues in British Columbia and Similar Jurisdictions. *Sustainability (Basel)*, 16(3), 1210. 10.3390/su16031210

Ives, M.C., & Scandol, J.P. (2013). BIOMAS A bioeconomic modelling and assessment system for fisheries management strategy evaluation. *Ecological Modelling*, 249, 42-49. 10.1016/j.ecolmodel.2012.07.006

Jakšić, S., Stamenković, I., & Đorđević, J. (2013). Impacts of artificial reefs and diving tourism. *Turizam*, 17(4), 155–165. 10.5937/Turizam1304155J

Jang, H. G., Yamazaki, S., & Hoshino, E. (2019). Profit and equity trade-offs in the management of small pelagic fisheries: The case of the Japanese sardine fishery. *The Australian Journal of Agricultural and Resource Economics*, 63(3), 549–574. 10.1111/1467-8489.12317

Johansen, K. (2013). SCUBA diving education and training. In *SCUBA diving tourism* (pp. 81–96). Routledge.

Jones, P. J., Qiu, W., & De Santo, E. M. (2013). Governing marine protected areas: Social–ecological resilience through institutional diversity. *Marine Policy*, 41, 5–13. 10.1016/j.marpol.2012.12.026

Kaharto, S. W., & Hermansyah, Y. (2023). Analysis of Public Awareness of the Impact of Pollution on Losari Beach, Makassar City. *INNOVATIVE: Journal Of Social Science Research, 3*(4), 3510–3517. https://j-innovative.org/index.php/Innovative

Kanojia, S. (2024). Guardians of Atolls: Examining the Paris Agreement's Role in Climate Change Mitigation and Resilience in the Maldives. In Bridging Health, Environment, and Legalities: A Holistic Approach (pp. 271-291). IGI Global.

Keller, K., Steffe, A. S., Lowry, M., Murphy, J. J., & Suthers, I. M. (2016). Monitoring boat-based recreational fishing effort at a nearshore artificial reef with a shore-based camera. *Fisheries Research*, 181, 84–92. 10.1016/j.fishres.2016.03.025

Khandu, F., Forootan, E., Schumacher, M., Awange, J. L., & Müller Schmied, H. (2016). Exploring the influence of precipitation extremes and human water use on total water storage (TWS). *Water Resources Research*, 52(3), 1541–2375. 10.1002/2015WR018113

Kibel, P. S. (2007). Grasp on water: A natural resource that eludes NAFTA's notion of investment. *Ecology Law Quarterly*, 34, 655.

Kim, H., Kang, H., & Zhang, C.I. (2022). Ecosystem-based fisheries risk assessment and forecasting considering a spatio-temporal component in Korean waters. *Ocean and Coastal Management*, 230, 106356. 10.1016/j.ocecoaman.2022.106356

Kim, S. (2010). Fisheries development in northeastern Asia in conjunction with changes in climate and social systems. *Marine Policy*, 34(4), 803–809. 10.1016/j.marpol.2010.01.028

Kjærsgaard, J. (2007). Multiple criteria decision making in fisheries management: experiences and implications. *Acta Agric. Scand. Sect. C - Food Econ.*, 4, 49–64. 10.1080/16507540701362831

Knowlton, N., Brainard, R. E., Fisher, R., Moews, M., Plaisance, L., & Caley, M. J. (2010). Coral reef biodiversity. *Life in the world's oceans: diversity distribution and abundance*, 65-74.

Kompas Cyber Media. (2020, March 9). *Study of Huntete Wakatobi Coastal Waste, Much of it from Neighboring Countries*. KOMPAS.com. https://www.kompas.com/sains/read/2020/03/09/181000123/kajian-sampah-pesisir-huntete-wakatobi-banyak-dari-negara-tetangga

Kompas Cyber Media. (2023, March 5). *Kresek Project, Handicraft Business from Recycling Kresek Waste*. KOMPAS.com. https://umkm.kompas.com/read/2023/03/05/192824183/kresek-project-bisnis-kerajinan-tangan-dari-daur-ulang-sampah-kresek

Kompas, T. V. Bengkulu. (2023, March 17). *Utilizing Waste into Valuable Items*. KOMPAS.tv. https://www.kompas.tv/regional/388999/pemanfaatan-sampah-menjadi-barang-bernilai

Kompasiana.com. (2023, January 10). *SeaBin Ocean Cleaning Technology, Available in Indonesia?* KOMPASIANA. https://www.kompasiana.com/yudiramid0862/63bb9e8d4addee77b65bfe12/seabin-teknologi-permbersih-laut-adakah-di-indonesia

Kos, S., Vojković, L., & Brčić, D. (2014, May). Development of AIS and its influence on marine traffic control. In *8 Proceedings of the 8th Annual Baška GNSS Conference* (pp. 7-9).

Kotzé, L. J. (2012). Arguing global environmental constitutionalism. *Transnational Environmental Law*, 1(1), 199–233. 10.1017/S2047102511000094

Kusumawardhani, A. (2016, November 25). *Tangerang City Government-style Garbage Pickup Application*. Bisnis.com. https://jakarta.bisnis.com/read/20161125/383/606245/aplikasi-jemput-sampah-ala-pemkot-tangerang

Kusumawati, I., Nasution, M. A., & Alamsyah, A. (2019). Distribution and Composition Coastal Marine Garbage in The Sub-District Nagan Raya District. *Jurnal Laot Ilmu Kelautan*, 1(1). Advance online publication. 10.35308/jlaot.v1i1.1073

Kvamsdal, S.F., Eide, A., Ekerhovd, N.A., Enberg, K., Gudmundsdottir, A., Hoel, A.H., Mills, K.E., Mueter, F.J., Ravn-Jonsen, L., Sandal, L.K., Stiansen, J.E., & Vestergaard, N. (2016). Harvest control rules in modern fisheries management. *Elementa: Science of the Anthropocene*, 4, 000114. 10.12952/journal.elementa.000114

Laine, C., Ollikainen, M., Kankainen, M., Setälä, J., & Vielma, J. (2023). Social net benefits from aquaculture production: A comparison of net cage cultivation and recirculating aquaculture systems. *Aquaculture Economics & Management*, 28(1), 1–31. 10.1080/13657305.2023.2222681

LaJeunesse, T. C. (2020). Zooxanthellae. *Current Biology*, 30(19), R1110–R1113. 10.1016/j.cub.2020.03.05833022248

Lam, M.E., Pitcher, T.J., Surma, S., Scott, J., Kaiser, M., White, A.S.J., Pakhomov, E.A., & Ward, L.M. (2019). Value- and ecosystem-based management approach: the Pacific herring fishery conflict. *Marine Ecology Progress Series*, 617-618, 341-364. 10.3354/meps12972

Lamb, J. B., Willis, B. L., Fiorenza, E. A., Couch, C. S., Howard, R., Rader, D. N., True, J. D., Kelly, L. A., Ahmad, A., Jompa, J., & Harvell, C. D. (2018). Plastic waste associated with disease on coral reefs. *Science*, 359(6374), 460–462. 10.1126/science.aar332029371469

Laszlo, P. (2006). *Communicating science: a practical guide*. Springer Berlin Heidelberg New York., 10.1007/3-540-31920-4

Leach, A., Levontin, P., Holt, J., Kell, L., & Mumford, J. (2014). Identification and prioritization of uncertainties for management of eastern Atlantic bluefin tuna (Thunnus thynnus). *Marine Policy*, 48, 48, 84–92. 10.1016/j.marpol.2014.03.010

Lebreton, L. C. M., Van Der Zwet, J., Damsteeg, J.-W., Slat, B., Andrady, A., & Reisser, J. (2017). River plastic emissions to the world's oceans. *Nature Communications*, 8(1), 15611. 10.1038/ncomms1561128589961

Lee, E., Mokashi, A. J., Moon, S. Y., & Kim, G. (2019). The maturity of automatic identification systems (AIS) and its implications for innovation. *Journal of Marine Science and Engineering*, 7(9), 287. 10.3390/jmse7090287

Lee, M. O., Otake, S., & Kim, J. K. (2018). Transition of artificial reefs (ARs) research and its prospects. *Ocean and Coastal Management*, 154, 55–65. 10.1016/j.ocecoaman.2018.01.010

Leitão, A., Santos, A. M., & Boaventura, R. A. R. (2016). Complexation of lead by organic matter in Luanda Bay, Angola. *Environmental Monitoring and Assessment*, 188(10), 563. 10.1007/s10661-016-5557-z27624745

Leitão, F. (2013). Artificial reefs: From ecological processes to fishing enhancement tools. *Brazilian Journal of Oceanography*, 61(1), 77–81. 10.1590/S1679-87592013000100009

Leitão, F., Santos, M. N., Erzini, K., & Monteiro, C. C. (2008). Fish assemblages and rapid colonisation after enlargement of an artificial reef off the Algarve coast (Southern Portugal). *Marine Ecology (Berlin)*, 29(4), 435–448. 10.1111/j.1439-0485.2008.00253.x

Leitão, F., Santos, M. N., & Monteiro, C. C. (2007). Contribution of artificial reefs to the diet of the white sea bream (Diplodus sargus). *ICES Journal of Marine Science*, 64(3), 473–478. 10.1093/icesjms/fsm027

Leitão, F., Santos, M. N., & Monteiro, C. C. (2009). Diplodus spp. assemblages on artificial reefs: Importance for near shore fisheries. *Fisheries Management and Ecology*, 16(2), 88–99. 10.1111/j.1365-2400.2008.00646.x

León, I. (2012). The tourist project of the SETAP Group for the International Bid of Ideas for Maspalomas Costa Canaria. In *6th Conference of the International Forum on Urbanism (IFoU): TOURBANISM, Barcelona, 25-27 gener* (pp. 1-10). International Forum on Urbanism.

Leposa, N. (2020). Problematic blue growth: A thematic synthesis of social sustainability problems related to growth in the marine and coastal tourism. *Sustainability Science*, 15(4), 1233–1244. 10.1007/s11625-020-00796-9

Lewenstein, B. V. (2011). Changing our ideas. *International Journal of Science Education. Part B*, 1(1), 17–21.

Liberati, A., Altman, D.G., Tetzlaff, J., Mulrow, C., Gøtzsche, P.C., Ioannidis, J.P.A., Clarke, M., Devereaux, P.J., Kleijnen, J., & Moher, D. (2010). The PRISMA statement for reporting systematic reviews and meta-analyses of studies that evaluate health care interventions: explanation and elaboration, *Journal of Clinical Epidemiology*, 62, 1-34. 10.1016/j.jclinepi.2009.06.006

Licandeo, R., de la Puente, S., Christensen, V., Hilborn, R., & Walters, C. (2023). A delay-differential model for representing small pelagic fish stock dynamics and its application for assessing alternative management strategies under environmental uncertainty. *Fish and Fisheries*, 24(4), 24, 544–566. 10.1111/faf.12743

Lima, J. S., & Zalmon, I. R. (2022) Impacts of Artisanal Reefs on Artisanal Fishers: Southeastern Brazil. 10.4018/978-1-6684-2344-8.ch005

Lima, J. S., Zalmon, I. R., & Love, M. (2019). Overview and trends of ecological and socioeconomic research on artificial reefs. *Marine Environmental Research*, 145, 81–96. 10.1016/j.marenvres.2019.01.01030837123

Lincoln, S., Andrews, B., Birchenough, S. N. R., Chowdhury, P., Engelhard, G. H., Harrod, O., Pinnegar, J. K., & Townhill, B. L. (2022). Marine litter and climate change: Inextricably connected threats to the world's oceans. *The Science of the Total Environment*, 837, 155709. 10.1016/j.scitotenv.2022.15570935525371

Liputan6.com. (2018, November 21). *3 Cases of Throwing Garbage into the Sea that Go Viral on Social Media*. liputan6.com. https://www.liputan6.com/lifestyle/read/3697231/3-kasus-buang-sampah-ke-laut-yang-viral-di-media-sosial

Little, D. C., Newton, R. W., & Beveridge, M. C. M. (2016). Aquaculture: A rapidly growing and significant source of sustainable food? Status, transitions and potential. *The Proceedings of the Nutrition Society*, 75(3), 274–286. 10.1017/S002966511600066527476856

Little, P. (2013). Making Law Matter: Environmental Protection & Legal Institutions in Brazil by Lesley K. McAllister. *Luso-Brazilian Review*, 50(2), 145–146. 10.1353/lbr.2013.0042

Liverman, D., & Jaramillo, M. (2011). Communicating environmental geoscience - An international survey. *Episodes*, 34(1), 25–31. 10.18814/epiiugs/2011/v34i1/004

Lobo, S. (2010). Sun, Sand, Sea and Bikini. Arquitectura e turismo: Portugal anos 60. [in Portuguese]. *Revista Critica de Ciencias Sociais*, (91), 91–106. 10.4000/rccs.4170

Loo, M. J., Demirel, M. C., Garg, A., & Jayasinghe, R. P. (2009). The Impact of Marinas on the Coastal Zone. INUAF STUDIA-Instituto Universitário Dom Afonso III. Suplemento, 16, 25-27.

Lopes, L. F., & Eugénio, K. (2012, Aug 29). Inaugurada nova marginal da baía de Luanda [New waterfront inaugurated in Luanda Bay]. *Expresso*. https://expresso.pt/economia/inaugurada-nova-marginal-da-baia-de-luanda=f749660

Lu, H.-F., & Lin, H.-W. (2020). Does the legacy governance of major sporting events affect urban development? The stakeholder's perspectives on the Taipei 2017 Universiade. *International Journal of Sports Marketing & Sponsorship*, 22(1), 87–106. 10.1108/IJSMS-02-2020-0022

Mackinson, S., Sumaila, U. R., & Pitcher, T. J. (1997). Bioeconomics and catchability: Fish and fishers behaviour during stock collapse. *Fisheries Research*, 31(1-2), 11–17. 10.1016/S0165-7836(97)00020-9

Madricardo, F., Ghezzo, M., Nesto, N., Mc Kiver, W. J., Faussone, G. C., Fiorin, R., Riccato, F., Mackelworth, P. C., Basta, J., De Pascalis, F., Kruss, A., Petrizzo, A., & Moschino, V. (2020). How to Deal With Seafloor Marine Litter: An Overview of the State-of-the-Art and Future Perspectives. *Frontiers in Marine Science*, 7, 505134. 10.3389/fmars.2020.505134

Maia, H. A., Morais, R. A., Siqueira, A. C., Hanazaki, N., Floeter, S. R., & Bender, M. G. (2018). Shifting baselines among traditional fishers in São Tomé and Príncipe islands, Gulf of Guinea. *Ocean and Coastal Management*, 154, 133–142. 10.1016/j.ocecoaman.2018.01.006

Mainolfi, G., & Marino, V. (2020). Destination beliefs, event satisfaction and post-visit product receptivity in event marketing. Results from a tourism experience. *Journal of Business Research*, 116, 699–710. 10.1016/j.jbusres.2018.03.001

Ma, J., Ma, R., Pan, Q., Liang, X., Wang, J., & Ni, X. (2023). A Global Review of Progress in Remote Sensing and Monitoring of Marine Pollution. *Water (Basel)*, 15(19), 3491. 10.3390/w15193491

Makino, M. (2018). Rebuilding and full utilization of alternating pelagic species around Japan: A social-ecological approach. *FAO Fisheries and Aquaculture Technical Paper*, 630(2), 61–73.

Malchrowicz-Mośko, E., Botiková, Z., & Poczta, J. (2019). "Because We Don't Want to Run in Smog": Problems with the Sustainable Management of Sport Event Tourism in Protected Areas (A Case Study of National Parks in Poland and Slovakia). *Sustainability (Basel)*, 11(2), 325. 10.3390/su11020325

Malchrowicz-Mośko, E., & Poczta, J. (2018). A small-scale event and a big impact—Is this relationship possible in the world of sport? The meaning of heritage sporting events for sustainable development of tourism—Experiences from Poland. *Sustainability (Basel)*, 10(11), 4289. Advance online publication. 10.3390/su10114289

Malve, H. (2016). Exploring the ocean for new drug developments: Marine pharmacology. *Journal of Pharmacy & Bioallied Sciences*, 8(2), 83. 10.4103/0975-7406.17170027134458

Mamun, A. A., & Natcher, D. C. (2023). The promise and pitfalls of community-based monitoring with a focus on Canadian examples. *Environmental Monitoring and Assessment*, 195(4), 445. 10.1007/s10661-022-10841-y36877375

Manfreda, S., McCabe, M. F., Miller, P. E., Lucas, R., Pajuelo Madrigal, V., Mallinis, G., Ben Dor, E., Helman, D., Estes, L., Ciraolo, G., Müllerová, J., Tauro, F., De Lima, M. I., De Lima, J. L. M. P., Maltese, A., Frances, F., Caylor, K., Kohv, M., Perks, M., & Toth, B. (2018). On the Use of Unmanned Aerial Systems for Environmental Monitoring. *Remote Sensing (Basel)*, 10(4), 641. 10.3390/rs10040641

Maouel, D., Maynou, F., & Bedrani, S. (2014). Bioeconomic Analysis of Small Pelagic Fishery in Central Algeria. *Turkish Journal of Fisheries and Aquatic Sciences*, 14(4), 14, 897–904. 10.4194/1303-2712-v14_4_08

Ma, Q., Ding, J., Xi, Y., Song, J., Liang, S., & Zhang, R. (2022). An evaluation method for determining the optimal structure of artificial reefs based on their flow field effects. *Frontiers in Marine Science*, 9, 962821. Advance online publication. 10.3389/fmars.2022.962821

Marin-Diaz, B., Fivash, G. S., Nauta, J., Temmink, R. J., Hijner, N., Reijers, V. C., Cruijsen, P. P. M. J. M., Didderen, K., Heusinkveld, J. H. T., Penning, E., Maldonado-Garcia, G., van Belzen, J., de Smit, J. C., Christianen, M. J. A., van der Heide, T., van der Wal, D., Olff, H., Bouma, T. J., & Govers, L. L. (2021). On the use of large-scale biodegradable artificial reefs for intertidal foreshore stabilization. *Ecological Engineering*, 170, 106354. 10.1016/j.ecoleng.2021.106354

Marois, D. E., & Mitsch, W. J. (2015). Coastal protection from tsunamis and cyclones provided by mangrove wetlands – a review. *The International Journal of Biodiversity Science, Ecosystem Services & Management*, 11(1), 71–83. 10.1080/21513732.2014.997292

Marshall, C. T. (2009). Implementing Information on Stock Reproductive Potential in Fisheries Management: The Motivation, Challenges and Opportunities. In Jakobsen, T., Fogarty, M. J., Megrey, B. A., & Moksness, E. (Eds.), *Fish Reproductive Biology*., 10.1002/9781444312133.ch11

Martins, R., Pereira, E., Rosado, A., Marôco, J., McCullough, B., & Mascarenhas, M. (2022). Understanding spectator sustainable transportation intentions in international sport tourism events. *Journal of Sustainable Tourism*, 30(8), 1972–1991. 10.1080/09669582.2021.1991936

Mascarenhas, M., Pereira, E., Rosado, A., & Martins, R. (2021). How has science highlighted sports tourism in recent investigation on sports' environmental sustainability? A systematic review. *Journal of Sport & Tourism*, 25(1), 42–65. 10.1080/14775085.2021.1883461

Mascarenhas, M., Vieira, H., & Martins, R. (2023). *Factors shaping spectators' consumption patterns: Controversial arisings from an international sport tourism event.* Consumer Behavior in Tourism and Hospitality., 10.1108/CBTH-08-2023-0108

Masifundise Development Trust. (2013). *Assessment report on small-scale fisheries in Africa*. J. Cox.

Matias, A., Carrasco, A. R., Pinto, B., & Reis, J. (2023). The role of art in coastal and marine sustainability. *Cambridge Prisms: Coastal Futures, 1*(e25), 1–10. https://doi.org/https://doi.org/10.1017/cft.2023.13

Matias, A., Carrasco, A. R., Ramos, A. A., & Borges, R. (2020). Engaging children in geosciences through storytelling and creative dance. *Geoscience Communication*, 3(2), 167–177. 10.5194/gc-3-167-2020

Matias, A., Delgado-Fernandez, I., Costas, S., & Loureiro, C. (2022). Adaptation of 'Coasts for Kids' animations: an exercise of cultural diversity in coastal communication. *5th International Marine Science Communication Conference - CommOcean22*, 6–7.

Matias, A., Pinto, B., Areia, N. P., & Carrasco, A. R. (2024). How do coastal experts communicate science? Insights from an international survey. *Egusphere, EGU24-1703*. https://doi.org/https://doi.org/10.5194/egusphere-egu24-17037

McCullough, B. P., Pfahl, M. E., & Nguyen, S. N. (2016). The green waves of environmental sustainability in sport. *Sport in Society*, 19(7), 1040–1065. 10.1080/17430437.2015.1096251

McGlennon, D., & Branden, K. L. (1994). Comparison of catch and recreational anglers fishing on artificial reefs and natural seabed in Gulf St. Vincent, South Australia. *Bulletin of Marine Science*, 55(2-3), 510–523.

Mega, V. P. (2018). *Eco-responsible cities and the global ocean: geostrategic shifts and the sustainability trilemma*. Springer.

Megdal, S. B., Gerlak, A. K., Varady, R. G., & Huang, L. Y. (2015). Groundwater governance in the United States: Common priorities and challenges. *Groundwater*, 53(5), 677–684. 10.1111/gwat.1229425406503

Meijer, L. J. J., Van Emmerik, T., Van Der Ent, R., Schmidt, C., & Lebreton, L. (2021). More than 1000 rivers account for 80% of global riverine plastic emissions into the ocean. *Science Advances*, 7(18), eaaz5803. Advance online publication. 10.1126/sciadv.aaz580333931460

Meisel-Lusby, C., & Cottrell, S. (2008). Understanding motivations and expectations of SCUBA divers. *Tourism in Marine Environments*, 5(1), 1–14. 10.3727/154427308785855279

Mejjad, N., Rossi, A., & Pavel, A. B. (2022). The coastal tourism industry in the Mediterranean: A critical review of the socio-economic and environmental pressures and impacts. *Tourism Management Perspectives*, 44, 101007. 10.1016/j.tmp.2022.101007

Melanidis, M. S., & Hagerman, S. (2022). Competing narratives of nature-based solutions: Leveraging the power of nature or dangerous distraction? *Environmental Science & Policy*, 132, 273–281. 10.1016/j.envsci.2022.02.028

Melovic, B., Rogic, S., Cerovic Smolovic, J., Dudic, B., & Gregus, M. (2019). The Impact of Sport Sponsorship Perceptions and Attitudes on Purchasing Decision of Fans as Consumers—Relevance for Promotion of Corporate Social Responsibility and Sustainable Practices. *Sustainability (Basel)*, 11(22), 6389. 10.3390/su11226389

Metcalfe, J. (2022). Science communication: A messy conundrum of practice, research and theory. *JCOM, Journal of Science Communication*, 21(7), 1–8. 10.22323/2.21070307

Mhanna, R., Blake, A., & Jones, I. (2017). Challenges facing immediate tourism leveraging: Evidence from the London 2012 Olympic Games. *Managing Sport and Leisure*, 22(2), 147–165. 10.1080/23750472.2017.1412269

Millennium Ecosystem Assessment, M. E. A. (2005). *Ecosystems and Human Wellbeing: Biodiversity Synthesis*. Island Press.

Mill, R. C. (2008). *Resorts: management and operation*. John Wiley and Sons.

Milon, J. W., Holland, S. M., & Whitmarsh, D. J. (2000). *Social and economic evaluation methods*. CRC Press LLC.

Ministry of Fisheries. (2004). *Decreto-Lei n.º 6-A/04, de 8 de outubro* [Decree-Law No 6-A/04, of 8 October] (I series, Diário da República No. 81). Angolan National Government Press.

Ministry of Fisheries. (2005). *Decreto-Lei n.º 41/05, de 13 de Junho* [Decree-Law No 41/05, of 13 June] (I series Diário da República No. 70). Angolan National Government Press.

Ministry of Fisheries. (2015). *Decreto Presidencial n.º 28/15, de 13 de Janeiro* [Presidential Decree no. 28/15 of 13 January] (I series Diário da República No. 8). Angolan National Government Press.

Mintzberg, H. (2010). *Estrutura e Dinâmica das Organizações* (4th ed.). Publicações Dom Quixote.

Misener, L., Lu, L. D., & Carlisi, R. (2020). Leveraging events to develop collaborative partnerships: Examining the formation and collaborative dynamics of the Ontario Parasport Legacy Group. *Journal of Sport Management*, 34(5), 447–461. 10.1123/jsm.2019-0283

Monteiro, C. C., & Santos, M. N. (2000). Portuguese artificial reefs. In *Artificial Reefs in European Seas* (pp. 249–261). Springer Netherlands. 10.1007/978-94-011-4215-1_15

Mooney, H., Larigauderie, A., Cesario, M., Elmquist, T., Hoegh-Guldberg, O., Lavorel, S., Mace, G. M., Palmer, M., Scholes, R., & Yahara, T. (2009). Biodiversity, climate change, and ecosystem services. *Current Opinion in Environmental Sustainability*, 1(1), 46–54. 10.1016/j.cosust.2009.07.006

Moreno, A., & Amelung, B. (2009). Climate change and coastal and marine tourism: Review and analysis. *Journal of Coastal Research*, 1140–1144.

Moura, A., Fonseca, L. C. D., Boaventura, D., Santos, M. N., & Monteiro, C. C. (2011). Estimation of secondary production of the Faro/Ancão artificial reefs. *Brazilian Journal of Oceanography*, 59(spe1), 91–94. 10.1590/S1679-87592011000500011

MT. (n.d.). MarineTraffic - Global Ship Tracking Intelligence. https://www.marinetraffic.com/

Mueller, J. S., Bill, N., Reinach, M. S., Lasut, M. T., Freund, H., & Schupp, P. J. (2022). A comprehensive approach to assess marine macro litter pollution and its impacts on corals in the Bangka Strait, North Sulawesi, Indonesia. *Marine Pollution Bulletin*, 175, 113369. 10.1016/j.marpolbul.2022.11336935144214

Mulazzani, L., Curtin, R., Andrés, M., & Malorgio, G. (2013). Multilevel governance and management of shared stocks with integrated markets: The European anchovy case. *Marine Policy*, 38, 407–416. 10.1016/j.marpol.2012.06.020

Mulazzani, L., & Malorgio, G. (2013). Bioeconomic equilibrium and supply regime for a multi-species multi-fleet fishery: An analysis in the Mediterranean Sea. *New Medit*, 3.

Mullon, C., Mittaine, J. F., Thébaud, O., Péron, G., Merino, G., & Barenge, M. (2009). Modeling the global fishmeal and fish oil markets. *Natural Resource Modeling*, 22(4), 564–609. 10.1111/j.1939-7445.2009.00053.x

Musikavanhu, R. R., Ladkin, A., & Sadd, D. (2021). The lasting social value of mega events: Experiences from green point community in Cape Town, South Africa. *Journal of Sustainable Tourism*, 29(11–12), 1832–1849. 10.1080/09669582.2021.1874395

Musolino, E. (2018). A. Vercelli, Crisis and Sustainability. The Delusion of Free Markets, 2017. *Rassegna Italiana di Sociologia*, 59(2), 408–409.

Nakamura, M. (1985). Evolution of artificial fishing reef concepts in Japan. *Bulletin of Marine Science*, 37(1), 271–278.

Nama, S., Shanmughan, A., Nayak, B. B., Bhushan, S., & Ramteke, K. (2023). Impacts of marine debris on coral reef ecosystem: A review for conservation and ecological monitoring of the coral reef ecosystem. *Marine Pollution Bulletin*, 189, 114755. 10.1016/j.marpolbul.2023.11475536905864

National Coordination Team for Handling Marine Debris. (2024, February 2). *Tim Koordinasi Nasional Penanganan Sampah Laut | TKN PSL*. Tim Koordinasi Nasional Penanganan Sampah Laut. https://sampahlaut.id/

Nauta, J., Christianen, M. J., Temmink, R. J., Fivash, G. S., Marin-Diaz, B., Reijers, V. C., Didderen, K., Penning, E., Borst, A. C., Heusinkveld, J. H., Zwarts, M., Cruijsen, P. M., Hijner, N., Lengkeek, W., Lamers, L. P., van der Heide, T., Bouma, T. J., van der Wal, D., Olff, H., & Govers, L. L. (2023). Biodegradable artificial reefs enhance food web complexity and biodiversity in an intertidal soft-sediment ecosystem. *Journal of Applied Ecology*, 60(3), 541–552. 10.1111/1365-2664.14348

Nerghes, A., Mulder, B., & Lee, J. S. (2022). Dissemination or participation? Exploring scientists' definitions and science communication goals in the Netherlands. *PLoS ONE, 17*(12), 1–29. 10.1371/journal.pone.0277677

Nesci, O., & Valentini, L. (2020). Science, poetry, and music for landscapes of the Marche region, Italy: Communicating the conservation of natural heritage. *Geoscience Communication*, 3(2), 393–406. 10.5194/gc-3-393-2020

Newland, B. L., & Yoo, J. J.-E. (2021). Active sport event participants' behavioural intentions: Leveraging outcomes for future attendance and visitation. *Journal of Vacation Marketing*, 27(1), 32–44. 10.1177/1356766720948249

Nguyen, Q. H. (2021). Impact of Investment in Tourism Infrastructure Development on Attracting International Visitors: A Nonlinear Panel ARDL Approach Using Vietnam's Data. *Economies*, 9(3), 131. 10.3390/economies9030131

Ni, Y., Steinshamn, S.I., & Kvamsdal, C.F. (2022). Negative shocks in an age-structured bioeconomic model and how to deal with them. *Economic Analysis and Policy*, 76, 15-30. 10.1016/j.eap.2022.07.009

Nicolau, S. (2016). *Caracterização da macrofauna bentónica da Baía de Luanda – Angola* [Characterisation of the benthic macrofauna of the Luanda Bay – Angola] [Master's dissertation, Universidade do Algarve]. Sapientia

Nielsen, J.R., Thunberg, E., Holland, D.S., Schmidt, J.O., Fulton, E.A., Bastardie, F., & Bethke, E. (2018). Integrated ecological–economic fisheries models—evaluation, review and challenges for implementation. *Fish and Fisheries,* 19(1), 1–29. 10.1111/faf.12232

Nisbet, M. C., & Scheufele, D. A. (2009). What's next for science communication? Promising directions and lingering distractions. *American Journal of Botany*, 96(10), 1767–1778. 10.3732/ajb.090004121622297

Nogueira, P., Faria, S., Mosley, B. A., & Domingues, R. B. (2021). Environmental risks and quality of life of a coastal community (Luanda, Angola). *Journal of Spatial and Organizational Dynamics*, 9(3), 205–224.

Nonet, P., Selznick, P., & Kagan, R. A. (2017). *Law and society in transition: Toward responsive law*. Routledge. 10.4324/9780203787540

Nunes, L. J., Raposo, M. A., & Gomes, C. J. P. (2020). The impact of tourism activity on coastal biodiversity: A case study at praia da cova redonda (Algarve—Portugal). *Environments (Basel, Switzerland)*, 7(10), 88. 10.3390/environments7100088

Nuno, A., Matos, L., Metcalfe, K., Godley, B. J., & Broderick, A. C. (2021). Perceived influence over marine conservation: Determinants and implications of empowerment. *Conservation Letters*, 14(3), e12790. 10.1111/conl.12790

Nurdin, M., Litaay, M., Priosambodo, D., & Moka, W. (2019). Coral Conditions on Baranglompo and Bone Batang Islands Based on Coral Health Tables. *Jurnal Ilmu Alam Dan Lingkungan, 10*(1), 15–28. https://journal.unhas.ac.id/index.php/jai2/article/view/6384

O'Brien, D., & Chalip, L. (2008). Sport events and strategic leveraging: Pushing towards the triple bottom line. In Woodside, A. G., & Martin, D. (Eds.), *Tourism management: Analysis, behaviour and strategy* (1st ed., pp. 318–338). CAB International. 10.1079/9781845933234.0318

O'Donnell, E., Poelina, A., Pelizzon, A., & Clark, C. (2020). Stop burying the Lede: The essential role of indigenous law (s) in creating rights of nature. *Transnational Environmental Law*, 9(3), 403–427. 10.1017/S2047102520000242

OECD. (n.d.). Country note on national fisheries management systems – Portugal. Available at: https://www.oecd.org/portugal/34431028.pdf

Okubo, H., & Kakimoto, H. (1991). Changes in the community composition around artificial reefs. in Recent advances in aquatic habitat technology. Proc. Japan-U.S. Symposium on artificial habitat for fisheries, 161-168.

Okubo, A. (1971). Oceanic diffusion diagrams. *Deep-Sea Research and Oceanographic Abstracts*, 18(8), 789–802. 10.1016/0011-7471(71)90046-5

Oliveira, M. J., Farinha, F., da Silva, E. M., Lança, R., Pinheiro, M. D., & Miguel, C. (2019). Observatory of sustainability of the Algarve Region for tourism: Proposal for environmental and sociocultural indicators. *International Journal of Humanities and Social Science*, 13, 1237–1244.

Olsen, A. (2018). Dancing in wild places: Seaweed and ocean health. *Coreographic Practices*, 9(1), 97–118. 10.1386/chor.9.1.97_1

OR. (n.d.) Ocean Revival. Retrieved from http://www. oceanrevival .pt/pt/

Ortega, C., Nogueira, C., & Pinto, H. (2013). Sea and littoral localities' economy: Exploring potentialities for a maritime cluster-An integrated analysis of Huelva, Spain and Algarve, Portugal. *Journal for Maritime Research*, 10(2), 35–42.

Ortiz, J.-C., Wolff, N. H., Anthony, K. R. N., Devlin, M., Lewis, S., & Mumby, P. J. (2018). Impaired recovery of the Great Barrier Reef under cumulative stress. *Science Advances*, 4(7), eaar6127. Advance online publication. 10.1126/sciadv.aar612730035217

OSPAR Commission. (2013). OSPAR Guidelines on artificial reefs in relation to living marine resources. London: OSPAR 99/15/1-E, Annex 6.

Overholtz, W. J., Link, J. S., & Suslowicz, L. E. (2000). Consumption of important pelagic fish and squid by predatory fish in the northeastern USA shelf ecosystem with some fishery comparisons. *ICES Journal of Marine Science*, 57(4), 1147–1159. 10.1006/jmsc.2000.0802

Özgül, A., & Lök, A. (2022). Artificial Reef Applications for Diving Tourism: Artificial Wreck Reefs in Turkey. In *Impact of Artificial Reefs on the Environment and Communities* (pp. 151-168). IGI Global.

Palme, A. C., Nguyen, V. M., Gutowsky, L. F., & Cooke, S. J. (2016). Do fishing education programs effectively transfer 'catch-and-release'best practices to youth anglers yielding measurable improvements in fish condition and survival? *Knowledge and Management of Aquatic Ecosystems*, (417), 42. 10.1051/kmae/2016029

Pascoe, S., Paredes, S., & Coglan, L. (2023). The Indirect Economic Contribution of Fisheries to Coastal Communities through Tourism. *Fishes*, 8(3), 138. 10.3390/fishes8030138

Pauly, D., & Zeller, D. (2017). Comments on FAOs state of world fisheries and aquaculture (SOFIA 2016). *Marine Policy*, 77, 176–181. 10.1016/j.marpol.2017.01.006

Pawar, P. R., Shirgaonkar, S. S., & Patil, R. B. (2016). Plastic marine debris: Sources, distribution and impacts on coastal and ocean biodiversity. *PENCIL Publication of Biological Sciences, 3*(1), 40–54. www.pencilacademicpress.org/ppbs

Pêgo, A. (n.d.-a). *Plasticus maritimus*. Facebook Page @plasticusmaritimus. Retrieved 10 May 2024, from https://www.facebook.com/plasticusmaritimus/

Pêgo, A. (n.d.-b). *Plasticus maritimus*. Instagram Page @plasticus_maritimus. Retrieved 10 May 2024, from https://www.instagram.com/plasticus_maritimus/

Pelich, R., Chini, M., Hostache, R., Matgen, P., Lopez-Martinez, C., Nuevo, M., Ries, P., & Eiden, G. (2019). Large-scale automatic vessel monitoring based on dual-polarization sentinel-1 and AIS data. *Remote Sensing (Basel)*, 11(9), 1078. 10.3390/rs11091078

Pereira, E., Mascarenhas, M., Flores, A., Chalip, L., & Pires, G. (2020). Strategic leveraging: Evidences of small-scale sport events. *International Journal of Event and Festival Management*, 11(1), 69–88. 10.1108/IJEFM-07-2018-0046

Pereira, E., Mascarenhas, M., Flores, A., & Pires, G. (2015). Nautical small-scale sports events portfolio: A strategic leveraging approach. *European Sport Management Quarterly*, 15(1), 27–47. 10.1080/16184742.2015.1007883

Pestana, L. B., Dias, G. M., & Marques, A. C. (2020). Spatial and temporal diversity of non-native biofouling species associated with marinas in two Angolan bays. *African Journal of Marine Science*, 42(4), 413–422. 10.2989/1814232X.2020.1831954

Picken, F. (2023). Tourism and the blue economy. *Tourism Geographies*, 1–9. 10.1080/14616688.2023.2291821

Pickering, H. (1996). Artificial reefs of bulk waste materials: A scientific and legal review of the suitability of using the cement stabilised by-products of coal-fired power stations. *Marine Policy*, 20(6), 483–497. 10.1016/S0308-597X(96)00036-X

Pickering, H., Whitmarsh, D., & Jensen, A. (1999). Artificial reefs as a tool to aid rehabilitation of coastal ecosystems: Investigating the potential. *Marine Pollution Bulletin*, 37(8-12), 505–514. 10.1016/S0025-326X(98)00121-0

Picou, J. S., & Martin, C. G. (2007). *Long-term community impacts of the Exxon Valdez oil spill: patterns of social disruption and psychological stress seventeen years after the disaster*. Department of Sociology, Anthropology and Social Work, University of South Alabama.

Pikitch, E. K., Rountos, K. J., Essington, T. E., Santora, C., Pauly, D., Watson, R., Sumaila, U. R., Boersma, P. D., Boyd, I. L., Conover, D. O., Cury, P., Heppell, S. S., Houde, E. D., Mangel, M., Plagányi, É., Sainsbury, K., Steneck, R. S., Geers, T. M., Gownaris, N., & Munch, S. B. (2014). The global contribution of forage fish to marine fisheries and ecosystems. *Fish and Fisheries*, 15(1), 43–64. 10.1111/faf.12004

Pinto, B., & Costa, J. L. (2018). What are the practitioners' views about past marine science communication to the public? An example from Portugal. *Regional Studies in Marine Science*, 18, 27–34. 10.1016/j.rsma.2017.12.013

Pinto, B., & Matias, A. (2023a). European journalists and the sea: Contexts, motivations, and difficulties. *Public Understanding of Science (Bristol, England)*, 32(4), 459–469. 10.1177/0963662522113703636448498

Pinto, B., & Matias, A. (2023b). How European journalists cover marine issues. *JCOM, Journal of Science Communication*, 22(5), 1–11. 10.22323/2.22050802

Polovina, J. J., & Sakai, I. (1989). Impacts of artificial reefs on fishery production in Shimamaki, Japan. *Bulletin of Marine Science*, 44(2), 997–1003.

Popescu, S. M., Mansoor, S., Wani, O. A., Kumar, S. S., Sharma, V., Sharma, A., Arya, V. M., Kirkham, M. B., Hou, D., Bolan, N., & Chung, Y. S. (2024). Artificial intelligence and IoT driven technologies for environmental pollution monitoring and management. *Frontiers in Environmental Science*, 12, 1336088. 10.3389/fenvs.2024.1336088

Potts, T., Pita, C., O'Higgins, T., & Mee, L. (2016). Who cares? European attitudes towards marine and coastal environments. *Marine Policy*, 72, 59–66. 10.1016/j.marpol.2016.06.012

Potwarka, L. R., Snelgrove, R., Drewery, D., Bakhsh, J., & Wood, L. (2020). From intention to participation: Exploring the moderating role of a voucher-based event leveraging initiative. *Sport Management Review*, 23(2), 302–314. 10.1016/j.smr.2019.03.002

Prajanti, A., Berlianto, M., Simamora, R. L., Imansari, M. B., & Sari, N. (2020). *Guidelines for Monitoring Marine Debris: Beach Debris, Floating Trash, and Seabed Debris*. Direktorat Pengendalian Pencemaran dan Kerusakan Pesisir dan Laut. https://www.ppkl.menlhk.go.id

Prakash, S., & Verma, A. K. (2022). Anthropogenic activities and Biodiversity threats. *International Journal of Biological Innovations, IJBI*, 4(1), 94–103. 10.46505/IJBI.2022.4110

Prasetiawan, T. (2018). Efforts to Overcome Plastic Waste in the Ocean. *PUSLIT BKD, 10*(10).

Prellezo, R., Accadia, P., Andersen, J.L., Andersen, B.S., Buisman, E., Little, A., & Röckmann, C. (2012). A review of EU bioeconomic models for fisheries: the value of a diversity of models. *Marine Policy,* 36(2), 423–431. 10.1016/j.marpol.2011.08.003

Punt, A. E., Butterworth, D. S., de Moor, C. L., De Oliveira, J. A. A., & Haddon, M. (2016). Management strategy evaluation: Best practices. *Fish and Fisheries*, 17(2), 303–334. 10.1111/faf.12104

Punt, A. E., Trinnie, F., Walker, T. I., McGarvey, R., Feenstra, J., Linnane, A., & Hartmann, K. (2013). The performance of a management procedure for rock lobsters,Jasus edwardsii, off western Victoria, Australia inthe face on non-stationary dynamics. *Fisheries Research*, 137, 116–128. 10.1016/j.fishres.2012.09.017

Purba, N. P., Handyman, D. I. W., Pribadi, T. D., Syakti, A. D., Pranowo, W. S., Harvey, A., & Ihsan, Y. N. (2019). Marine debris in Indonesia: A review of research and status. *Marine Pollution Bulletin*, 146, 134–144. 10.1016/j.marpolbul.2019.05.05731426141

Purwanto, F.E.C., White, A., & Mardiani, A.R. (2023). Multiple-goal bioeconomic programming to address conflicting management objectives in Indonesian small pelagic fisheries. *Marine Policy*, 150, 105519. 10.1016/j.marpol.2023.105519

Putra, M. G. A., Zamani, N. P., Natih, N. M., & Harahap, S. A. (2021). Relationship Between Characteristics of Marine Debris and Impact to Coral Reef. *Jurnal Ilmiah Perikanan Dan Kelautan*, 13(1), 11–19. 10.20473/jipk.v13i1.18896

Putri, R. A., & Hudaya, M. (2022). The Establishment of ASEAN Framework of Action on Marine Debris: The Role of Shared Knowledge. *Global: Jurnal Politik Internasional*, 24(1), 63–84. 10.7454/global.v24i1.668

Rahman, A. N. P., Fitria, D., & Sentika, H. T. (2022). *Effectiveness of Plastic Waste Management Policies in Indonesian Seas. Presidensi G-20 Indonesia Tahun 2022 'Recover Together, Recover Stronger"*. Transformasi Administrasi Dan Kebijakan Public Di Era Post Truth Dalam Mewujudkan Indonesia Tumbuh, Indonesia Tangguh.

Rahman, A., Farrok, O., & Haque, M. M. (2022). Environmental impact of renewable energy source based electrical power plants: Solar, wind, hydroelectric, biomass, geothermal, tidal, ocean, and osmotic. *Renewable & Sustainable Energy Reviews*, 161, 112279. Advance online publication. 10.1016/j.rser.2022.112279

Rallis, I., Chatzigeorgiou, G., Florido, M., Sedano, F., Procopiou, A., Chertz-Bynichaki, M., Vernadou, E., Plaiti, W., Koulouri, P., Dounas, C., Gerovasileiou, V., & Dailianis, T. (2022). Early succession patterns of benthic assemblages on artificial reefs in the oligotrophic eastern mediterranean basin. *Journal of Marine Science and Engineering*, 10(5), 620. 10.3390/jmse10050620

Ramadhani, M., & Maharani, E. (2018, November 22). *Plastic Waste Damages Bali's Coral Reef Ecosystem*. Republika Online. https://republika.co.id/share/pildnu335

Ramm, L. A., Florisson, J. H., Watts, S. L., Becker, A., & Tweedley, J. R. (2021). Artificial reefs in the Anthropocene: A review of geographical and historical trends in their design, purpose, and monitoring. *Bulletin of Marine Science*, 97(4), 699–728. 10.5343/bms.2020.0046

Ramos, J. (2022). Portuguese Artificial Reef Modules: Are They Suitable for Dive Tourism? In *Impact of Artificial Reefs on the Environment and Communities* (pp. 127-150). IGI Global.

Ramos, J., Santos, M. N., Whitmarsh, D., & Monteiro, C. C. (2007). Stakeholder perceptions regarding the environmental and socio-economic impacts of the Algarve artificial reefs. In *Biodiversity in Enclosed Seas and Artificial Marine Habitats: Proceedings of the 39th European Marine Biology Symposium, held in Genoa, Italy, 21–24 July 2004* (pp. 181-191). Springer Netherlands. 10.1007/978-1-4020-6156-1_16

Ramos, J., & dos Santos, M. (2022). Efficiency of Artificial Reefs for Local Communities Over Time. In *Impact of Artificial Reefs on the Environment and Communities* (pp. 1–20). IGI Global. 10.4018/978-1-6684-2344-8.ch001

Ramos, J., & Drakeford, B. (2021). Blue economy and tourism: is there a potential to create sustainable jobs? In *Handbook of research on human capital and people Management in the Tourism Industry* (pp. 243–264). IGI Global. 10.4018/978-1-7998-4318-4.ch012

Ramos, J., & Santos, M. N. (2015). Facilitating fishing decisions in an artificial reef area off southern Portugal: A case study using generalized additive models. *Journal of Applied Ichthyology*, 31, 24–34. 10.1111/jai.12948

Ramos, J., Santos, M. N., Whitmarsh, D., & Monteiro, C. C. (2006a). Patterns of use in an artificial reef system: A case study in Portugal. *Bulletin of Marine Science*, 78(1), 203–211.

Ramos, J., Santos, M. N., Whitmarsh, D., & Monteiro, C. C. (2006b). The usefulness of the analytic hierarchy process for understanding reef diving choices: A case study. *Bulletin of Marine Science*, 78(1), 213–219.

Ramos, J., Tuaty-Guerra, M., Almeida, M., Raposo, A. C., Gaudêncio, M. J., Silva, A. D., Rodrigues, N., Leandro, S. M., & Caetano, M. (2021). An artificial reef at the edge of the deep: An interdisciplinary case study. *Ocean and Coastal Management*, 210, 105729. 10.1016/j.ocecoaman.2021.105729

Rangel, I., & Silva, S. (2006). First records of *Gymnodinium catenatum, Gambierdiscus toxicus* and *Pyrodinium bahamense* on northern Luanda coast, Angola. *Harmful Algae News*, 32, 10–11.

Rangel, I., & Silva, S. (2007). *Pseudo-nitzschia* spp and *Prorocentrum micans* blooms in Luanda Bay, Angola. *Harmful Algae News*, 33, 8–9.

Rangel, I., Silva, S., & Muai, J. (2004). Mortality caused by dinoflagellates bloom in Luanda Bay. *Harmful Algae News*, 26, 8–9.

Rapada, M. Z., Yu, D. E., & Yu, K. D. (2021). Do Social Media Posts Influence Consumption Behavior towards Plastic Pollution? *Sustainability (Basel)*, 13(22), 12334. 10.3390/su132212334

Raven, J., Caldeira, K., Elderfield, H., Hoegh-Guldberg, O., Liss, P., Riebesell, U., & Watson, A. (2005). *Ocean acidification due to increasing atmospheric carbon dioxide*. The Royal Society.

Relini, G., Fabi, G., dos Santos, M. N., Moreno, I., & Charbonnel, E. (2008). Fisheries and their management using artificial reefs in the northwestern Mediterranean Sea and southern Portugal. *American Fisheries Society Symposium*, 49(1), 891.

Rendle, E. J., & Rodwell, L. D. (2014). Artificial surf reefs: A preliminary assessment of the potential to enhance a coastal economy. *Marine Policy*, 45, 349–358. 10.1016/j.marpol.2013.09.004

Richardson, K., & Fletcher, T. (2020). Community sport development events, social capital and social mobility: A case study of Premier League Kicks and young black and minoritized ethnic males in England. *Soccer and Society*, 21(1), 79–95. 10.1080/14660970.2018.1506334

Richelieu, A. (2018). A sport-oriented place branding strategy for cities, regions and countries. *Sport, Business and Management*, 8(4), 354–374. 10.1108/SBM-02-2018-0010

Ridha, D. M., Hendriarti, N., Koropitan, A. F., & Lestari, S. P. (2019, March 30). *Marine Debris and Its Impact on Climate Change*. Pojok Iklim. http://pojokiklim.menlhk.go.id/read/marine-debris-and-its-impact-on-climate-change

Riesch, H., Potter, C., & Davies, L. (2016). What Is Public Engagement, and What Is It for? A Study of Scientists' and Science Communicators' Views. *Bulletin of Science, Technology & Society*, 36(3), 179–189. 10.1177/0270467617690057

Rifai, H., Quevedo, J. M. D., Lukman, K. M., Sondak, C. F. A., Risandi, J., Hernawan, U. E., Uchiyama, Y., Ambo-Rappe, R., & Kohsaka, R. (2023). Potential of seagrass habitat restorations as nature-based solutions: Practical and scientific implications in Indonesia. *Ambio*, 52(3), 546–555. 10.1007/s13280-022-01811-236484926

Riksfardini, M. (2023). Analysis of the Implementation of Marine Waste Management Policies in the Muara Angke Coastal Area, North Jakarta. *PENTAHELIX*, 1(2), 217. 10.24853/penta.1.2.217-236

Rocha, C. (2017). Rio 2016 Olympic Games and diplomatic legacies. *International Journal of Sport Policy*, 9(2), 277–294. 10.1080/19406940.2017.1287762

Rodrigues, J., Castro, C., Costa, E., Silva, E., & Pereira, D. I. (2023). Geoscientists' views about science communication: Predicting willingness to communicate geoscience. *Geoscience Communication*, 6(1), 15–25. 10.5194/gc-6-15-2023

Rosa, A. S. M. (2017). *Modelação Hidrodinâmica da Baía de Luanda*. [Hydrodynamic modelling of the Luanda Bay] [Master's thesis, Instituto Superior Técnico-Universidade de Lisboa]. Scholar.

Rosa, R., Costa, T., & Mota, P.R. (2022). Incorporating economics into fishery policies: Developing integrated ecological-economics harvest control rules. *Ecological Economics*, 169, 107418. 10.1016/j.ecolecon.2022.107418

Rosenberg, A. A., & Restrepo, V. R. (1994). Uncertainty and risk evaluation in stock assessment advice for U.S. marine fisheries. *Canadian Journal of Fisheries and Aquatic Sciences*, 51(12), 2715–2720. 10.1139/f94-271

Rotter, A., Barbier, M., Bertoni, F., Bones, A. M., Cancela, M. L., Carlsson, J., Carvalho, M. F., Cegłowska, M., Chirivella-Martorell, J., Conk Dalay, M., Cueto, M., Dailianis, T., Deniz, I., Díaz-Marrero, A. R., Drakulovic, D., Dubnika, A., Edwards, C., Einarsson, H., Erdo an, A., & Vasquez, M. I. (2021). The Essentials of Marine Biotechnology. *Frontiers in Marine Science*, 8, 629629. 10.3389/fmars.2021.629629

RTA. (2021). Região de Turismo do Algarve. Plano de Marketing Estratégico do Turismo do Algarve: 2020-2023.

Rühs, N., & Jones, A. (2016). The implementation of earth jurisprudence through substantive constitutional rights of nature. *Sustainability (Basel)*, 8(2), 174. 10.3390/su8020174

Ruiz, J., Garcia-Isarch, E., Emma Huertas, I., Prieto, L., Juárez, A., Muñoz, J. L., Sánchez-Lamadrid, A., Rodríguez-Gálvez, S., Naranjo, J. M., & Baldó, F. (2006). Meteorological and oceanographic factors influencing Engraulis encrasicolus early life stages and catches in the Gulf of Cádiz. *Deep-sea Research. Part II, Topical Studies in Oceanography*, 53(11–13), 1363–1376. 10.1016/j.dsr2.2006.04.007

Ruiz, J., Rincon, M. M., Castilla, D., Ramos, F., & del Hoyo, J. J. G. (2017). Biological and economic vulnerabilities of fixed TACs in small pelagics: An analysis of the European anchovy (Engraulis encrasicolus) in the Gulf of Cádiz. *Marine Policy*, 78, 171–180. 10.1016/j.marpol.2017.01.022

Sabine, C. L., Heimann, M., Artaxo, P., Bakker, D. C., Chen, C. T. A., Field, C. B., ... Valentini, R. (2004). Current status and past trends of the global carbon cycle. *Scope-scientific committee on problems of the environment international council of scientific unions, 62*, 17-44.

Salgado-Barandela, J., Barajas, A., & Sanchez-Fernandez, P. (2022). Sports event portfolios: An innovative tool and a new management paradigm. *International Journal of Sports Marketing & Sponsorship*, 23(5), 920–933. 10.1108/IJSMS-02-2021-0024

Sánchez, S., Ibaibarriaga, L., Uriarte, A., Prellezo, R., Andrés, M., Abaunza, P., Jardim, E., Lehuta, S., Pawlowski, L., & Roel, B. (2019). Challenges of management strategy evaluation for small pelagic fish: the Bay of Biscay anchovy case study. *Marine Ecology Progress Series*, 617-618, 245-263. 10.3354/meps12602

Sánchez, S., Uriarte, A., Andrés, M., & Ibaibarriaga, L. (2014). Developments of the MSE methodology for the Bay of Biscay anchovy. Final Report for the European Commission, Directorate General for Maritime Affairs and Fisheries. Presented to STECF EWG 14-03.

Sanchez, E., Vinueza, R., Izurieta, X., & Rey, N. (2020). Use of muralism to promote awareness about aquatic ecosystems and wise water consumption in northwestern Ecuador. *Ocean and Coastal Management*, 190, 105165. 10.1016/j.ocecoaman.2020.105165

Sánchez-Jerez, P., & Ramos-Esplá, A. (2000). Changes in fish assemblages associated with the deployment of an antitrawling reef in seagrass meadows. *Transactions of the American Fisheries Society*, 129(5), 1150–1159. 10.1577/1548-8659(2000)129<1150:CIFAAW>2.0.CO;2

Sandberg, P., Bogstad, B., & Røttingen, I. (1998). Bioeconomic advice on TAC–the state of the art in the Norwegian fishery management. *Fisheries Research*, 37(1-3), 37, 259–274. 10.1016/S0165-7836(98)00142-8

Santos, A. M. P. A. (2012). *Estudo da qualidade da água da Baía de Luanda, distribuição de metais pesados na água, sólidos suspensos e sedimentos* [Study of the water quality of Luanda Bay, distribution of heavy metals in water, suspended solids and sediments] [Unpublished doctoral thesis]. Universidade Agostinho Neto, Luanda, LD, Angola.

Santos, M. N., & Monteiro, C. C. (2007). A fourteen-year overview of the fish assemblages and yield of the two oldest Algarve artificial reefs (southern Portugal). In *Biodiversity in Enclosed Seas and Artificial Marine Habitats:Proceedings of the 39th European Marine Biology Symposium, held in Genoa,Italy,21–24 July 2004* (pp. 225-231). Springer Netherlands.

Santos, M. N., Monteiro, C. C., & Leitão, F. (2011). The role of artificial reefs in the sustainability of artisanal fisheries. *Artificial reefs in Fisheries Management, S. Bortone, FP Brandini, G. Fabi and S. Otake, CRC Press, Taylor & Francis Group*, 221-237.

Santos, M. N., Monteiro, C. C., & Leitão, F. (2011). The role of artificial reefs in the sustainability of artisanal fisheries. Artificial reefs in Fisheries Management. CRC Press, Taylor and Francis Group.

Santos, M. N., Lino, P. G., Pousão-Ferreira, P., & Monteiro, C. C. (2006). Preliminary results of hatchery-reared seabreams released at artificial reefs off the Algarve coast (southern Portugal): A pilot study. *Bulletin of Marine Science*, 78(1), 177–184.

Santos, M. N., & Monteiro, C. C. (1997). The Olhao artificial reef system (south Portugal): Fish assemblages and fishing yield. *Fisheries Research*, 30(1-2), 33–41. 10.1016/S0165-7836(96)00567-X

Santos, M. N., & Monteiro, C. C. (1998). Comparison of the catch and fishing yield from an artificial reef system and neighbouring areas off Faro (Algarve, south Portugal). *Fisheries Research*, 39(1), 55–65. 10.1016/S0165-7836(98)00169-6

Santos, M. N., Monteiro, C. C., & Lasserre, G. (2005). Observations and trends on the intra-annual variation of the fish assemblages on two artificial reefs in Algarve coastal waters (southern Portugal). *Scientia Marina*, 69(3), 415–426. 10.3989/scimar.2005.69n3415

Sant, S.-L., Misener, L., & Mason, D. S. (2019). Leveraging sport events for tourism gain in host cities: A regime perspective. *Journal of Sport & Tourism*, 23(4), 203–223. 10.1080/14775085.2019.1711444

Sari, F. O., Bulut, C., & Pirnar, I. (2016). Adaptation of hospitality service quality scales for marina services. *International Journal of Hospitality Management*, 54, 95–103. 10.1016/j.ijhm.2016.02.004

Sarker, P. K. (2023). Microorganisms in Fish Feeds, Technological Innovations, and Key Strategies for Sustainable Aquaculture. *Microorganisms*, 11(2), 439. 10.3390/microorganisms1102043936838404

Sbragaglia, V., Morroni, L., Bramanti, L., Weitzmann, B., Arlinghaus, R., & Azzurro, E. (2018). Spearfishing modulates flight initiation distance of fishes: The effects of protection, individual size, and bearing a speargun. *ICES Journal of Marine Science*, 75(5), 1779–1789. 10.1093/icesjms/fsy059

Schruben, L., & Singham, D. (2010). *Simulating multivariate time series using flocking. Proceedings of the 2010 Winter Simulation Conference*10.1109/WSC.2010.5679085

Schulenkorf, N., Giannoulakis, C., & Blom, L. (2019). Sustaining commercial viability and community benefits: Management and leverage of a sport-for-development event. *European Sport Management Quarterly*, 19(4), 502–519. 10.1080/16184742.2018.1546755

Schulenkorf, N., Peachey, J., Chen, G., & Hergesell, A. (2022). Event leverage: A systematic literature review and new research agenda. *European Sport Management Quarterly*, 1–25. 10.1080/16184742.2022.2160477

Scotti, M., Opitz, S., MacNeil, L., Kreutle, A., Pusch, C., & Froese, R. (2022). Ecosystem-based fisheries management increases catch and carbon sequestration through recovery of exploited stocks: The western Baltic Sea case study. *Frontiers in Marine Science*, 9, 879998. 10.3389/fmars.2022.879998

Seys, J., Cox, L., Yücel, E. Ş., Ezgeta-Balić, D., Faimali, M., Garaventa, F., García-Martínez, M. C., Gili, C., Kopke, K., Petrera, F., Simon, D., Villwock, A., Zielinski, T., Moreau, K., & Piniella, A. M. (2022). *Marine Science Communication in Europe: A Way Forward.* Zenodo. 10.5281/zenodo.6444143

Shani, A., Polak, O., & Shashar, N. (2012). Artificial reefs and mass marine ecotourism. *Tourism Geographies*, 14(3), 361–382. 10.1080/14616688.2011.610350

Shrader-Frechette, K. (2002). *Environmental justice: Creating equality, reclaiming democracy*. Oxford University Press. 10.1093/0195152034.001.0001

Sidik, H. (2016, March 5). *Bantul residents process marine waste into crafts*. Antara News Yogyakarta. https://jogja.antaranews.com/berita/338113/warga-bantul-mengolah-sampah-laut-jadi-kerajinan

Sigee, D. C. (2005). *Freshwater microbiology: biodiversity and dynamic interactions of microorganisms in the aquatic environment*. John Wiley & Sons.

Sigurdardottir, S., & Shruben, L.W. (2014). A new approach to simulating fisheries data for policy making. *Natural Resource Modeling,* 27(3), 411-428. 10.1111/nrm.12041

Silvestri, S., & Maynou, F. (2009). Application of a bioeconomic model for supporting the management process of the small pelagic fishery in the Veneto region, northern Adriatic Sea, Italy. *Scientia Marina*, 73(3), 563–572. 10.3989/scimar.2009.73n3563

Siple, M.C., Koehn, L.E., Johnson, K.F., Punt, A.E., Canales, T.M., Carpi, P., & de Moor, C.L. (2021). Considerations for management strategy evaluation for small pelagic fishes, *Fish and Fisheries*, 22, 1167–1186. 10.1111/faf.12579

Smith, A. (2022). *Managing Marine Plastic Waste in Asia and the Pacific*. Economic and Social Commission for Asia and the Pacific (ESCAP).

Smith, A. D., Smith, D. C., Haddon, M., Knuckey, I. A., Sainsbury, K. J., & Sloan, S. R. (2014). Implementing harveststrategies in Australia: 5 years on. *ICES Journal Marine Science Journal Conservation*, 71, 195–203.

Smith, A., Hurtado, F., Klaer, N., & Fulton, E. (2015). *Review and update of harvest strategy settings for the Commonwealth Small Pelagic Fishery Single species and ecosystem considerations*.

Smith, M. D., Zhang, J., & Coleman, F. C. (2008). Econometric modeling of fisheries with complex life histories: Avoiding biological management failures. *Journal of Environmental Economics and Management*, 55(3), 265–280. 10.1016/j.jeem.2007.11.003

Soares, J. P., Antunes, H. L., Bárbara, A., Escórcio, C., & Saldanha, P. (2016). O interesse público do desporto das organizações desportivas sem fins lucrativos que são apoiadas pela administração pública. *Revista Brasileira de Educação Física e Esporte*, 30(3), 675–688. 10.1590/1807-55092016000300675

Sofian, H. (2023, June 12). *Economic value, plastic waste is processed into weaving crafts*. Rri.Co.Id - Portal Berita Terpercaya. https://www.rri.go.id/mataram/umkm/259600/bernilai-ekonomi-sampah-plastik-diolah-jadi-kerajinan-tenun

Soukissian, T. H., Denaxa, D., Karathanasi, F., Prospathopoulos, A., Sarantakos, K., Iona, A., Georgantas, K., & Mavrakos, S. (2017). Marine Renewable Energy in the Mediterranean Sea: Status and Perspectives. *Energies*, 10(10), 1512. 10.3390/en10101512

Spagnol, S., Wolanski, E., Deleersnijder, E., Brinkman, R., McAllister, F., Cushman-Roisin, R., & Hanert, E. (2002). An error frequently made in the evaluation of advective transport in two-dimensional Lagrangian models of advection-diffusion in coral reef waters. *Marine Ecology Progress Series*, 235, 299–302. 10.3354/meps235299

Sroka, R. (2021). Mega-projects and mega-events: Evaluating Vancouver 2010 stadium and convention infrastructure. *Journal of Sport & Tourism*, 25(3), 183–200. 10.1080/14775085.2021.1881590

Steins, N. A., Baker, M. R., Brooks, K., Mackinson, S., & Stephenson, R. L. (2023). Editorial: Co-creating knowledge with fishers: challenges and lessons for integrating fishers' knowledge contributions into marine science in well-developed scientific advisory systems. *Frontiers in Marine Science*, 10(December), 1–5. 10.3389/fmars.2023.1338271

Sumaila, , Ebrahim, N., Schuhbauer, A., Skerritt, D., Li, Y., Kim, H. S., Mallory, T. G., Lam, V. W. L., & Pauly, D. (2019). Sumaila U.R., Ebrahim N., Schuhbauer A., Skerritt D., Li Y., Kim H.S., Mallory T.G., Lam V.W.L. and Pauly D. (2019). Updated estimates and analysis of global fisheries subsidies. *Marine Policy*, 109, 103695. 10.1016/j.marpol.2019.103695

Supriyanto, E. E. (2022). Blue tourism: Treating marine ecosystems and increasing the potential of maritime tourism in Indonesia. *Jurnal Kepariwisataan Indonesia: Jurnal Penelitian dan Pengembangan Kepariwisataan Indonesia, 16*(2), 138-148.

Susanto, A., Syafrie, H., Nurdin, H. S., Irnawati, R., Supadminingsih, F. N., Hamzah, A., & Kurniawati, V. (2022). Lost Gear on Blue Swimming Crab Fisheries in Banten Bay: Case of Abandoned, Lost or Discarded Fishing Gear. *Marine Fisheries*, 13(2), 233–241. 10.29244/jmf.v13i2.41961

Suteja, Y., Atmadipoera, A. S., Riani, E., Nurjaya, I. W., Nugroho, D., & Purwiyanto, A. I. S. (2021). Stranded marine debris on the touristic beaches in the south of Bali Island, Indonesia: The spatiotemporal abundance and characteristic. *Marine Pollution Bulletin*, 173, 113026. 10.1016/j.marpolbul.2021.11302634628344

Sutton, S. G., & Bushnell, S. L. (2007). Socio-economic aspects of artificial reefs: Considerations for the Great Barrier Reef Marine Park. *Ocean and Coastal Management*, 50(10), 829–846. 10.1016/j.ocecoaman.2007.01.003

Sweet, M., Stelfox, M., & Lamb, J. (2019). *Plastics and shallow water coral reefs. Synthesis of the science for policy-makers*. UN Environment.

Szoboszlai, A. I., Thayer, J. A., Wood, S. A., Sydeman, W. J., & Koehn, L. E. (2015). Forage species in predator diets: Synthesis of data from the California Current. *Ecological Informatics*, 29, 29, 45–56. 10.1016/j.ecoinf.2015.07.003

Tabuenca, B., Kalz, M., & Löhr, A. (2019). Massive Open Online Education for Environmental Activism: The Worldwide Problem of Marine Litter. *Sustainability (Basel)*, 11(10), 2860. 10.3390/su11102860

Tacon, A.G.J., & Metian, M. (2009). Global overview on the use of fish meal and fish oil in industrially compounded aquafeeds: Trends and future prospects. *Aquaculture*, 285, 146-158. 10.1016/j.aquaculture.2008.08.015

Taks, M., Green, B. C., Misener, L., & Chalip, L. (2014). Evaluating sport development outcomes: The case of a medium-sized international sport event. *European Sport Management Quarterly*, 14(3), 213–237. 10.1080/16184742.2014.882370

Tang, J., Ni, X., Zhou, Z., Wang, L., & Lin, S. (2018). Acute microplastic exposure raises stress response and suppresses detoxification and immune capacities in the scleractinian coral Pocillopora damicornis. *Environmental Pollution*, 243, 66–74. 10.1016/j.envpol.2018.08.04530172125

Tavakkoli, M. H. (2016). Impact of City Image, Event Fit, and Word-of-Mouth Attempt Through the Host City of an International Sporting Event. *American Journal of Sports Science*, 4(6), 112–119. 10.11648/j.ajss.20160406.13

Tavares, A. I., Assis, J., Patrício, A. R., Ferreira, R., Cheikh, M. A. S., Bandeira, S., Regalla, A., Santos, I., Potouroglou, M., Nicolau, S., Teodósio, M. A., Almada, C., Santos, R., Pearson, G. A., & Serrao, E. A. (2022). Seagrass Connectivity on the West Coast of Africa Supports the Hypothesis of Grazer-Mediated Seed Dispersal. *Frontiers in Marine Science*, 9, 809721. 10.3389/fmars.2022.809721

Teodósio, M. A., Paris, C. B., Wolanski, E., & Morais, P. (2016). Biophysical processes leading to the ingress of temperate fish larvae into estuarine nursery areas: A review. *Estuarine, Coastal and Shelf Science*, 183, 187–202. 10.1016/j.ecss.2016.10.022

Teodósio, M. A., Velho, F., & Santos, C. (Eds.). (2019). Colaboração luso angolana para formação de excelência em ciências do mar em África [Luso-Angolan collaboration for excellence in marine science training in Africa]. In *Proceedings of the XXIX Meeting of the Association of Portuguese Language Universities Conference on Art and Culture in the Identity of Peoples*. AULP.

Terry, A., Lewis, K., & Bullimore, B. (2019). The impact of the Marine and Coastal Access Act (2009) on Welsh inshore fisheries and marine management. *Marine Policy*, 99, 359–368. 10.1016/j.marpol.2018.11.003

The Royal Society. (2006). *Survey of factors affecting science communication by scientists and engineers*. Author.

Tonazzini, D., Fosse, J., Morales, E., González, A., Klarwein, S., Moukaddem, K., & Louveau, O. (2019). *Blue Tourism. Towards a sustainable coastal and maritime tourism in world marine regions. Edited by eco-union.*

Torralba, J., & Besada, M. (2015). A stochastic model for the Iberoatlantic sardine fishery. Global warming and economic effects. *Ocean and Coastal Management*, 114, 114, 175–184. 10.1016/j.ocecoaman.2015.06.023

Treeck, P., & Schuhmacher, H. (1999). Mass diving tourism–a new dimension calls for new management approaches. *Marine Pollution Bulletin*, 37(8-12), 499–504. 10.1016/S0025-326X(99)00077-6

TYHA. (n.d.). The Yacht Harbour Association. Retrieved from https://www.tyha.co.uk/

Tymoshenko, Y., Kyslenko, D., Kuzmichova-Kyslenko, E., Leonenko, I., & Servetsky, I. (2022). Features of the Pre-trial Investigation of Air Pollution. Academic Press.

Uriarte, A., Ibaibarriaga, L., Sánchez-maroño, S., Abaunza, P., Andrés, M., Duhamel, E., Jardim, E., Pawlowski, L., Prellezo, R., & Roel, B.A. (2023). Lessons learnt on the management of short-lived fish from the Bay of Biscay anchovy case study: Satisfying fishery needs and sustainability under recruitment uncertainty. *Marine Policy*, 150, 105512. 10.1016/j.marpol.2023.105512

Utami, D. A., Reuning, L., Konechnaya, O., & Schwarzbauer, J. (2021). Microplastics as a sedimentary component in reef systems: A case study from the Java Sea. *Sedimentology*, 68(6), 2270–2292. 10.1111/sed.12879

Uusitalo, L., Blenckner, T., Puntila-Dodd, R., Skyttä, A., Jernberg, S., Voss, R., Müller-Karullis, B., Tomczak, M.T., Möllmann, C., & Pertonen, H. (2022). Integrating diverse model results into decision support for good environmental status and blue growth. *Science of the Total Environment*, 806, 150450. 10.1016/j.scitotenv.2021.150450

VA. (2021). *VisitAlgarve*. Boletim Conjuntura Turística.

Valadas-Monteiro, P. (2014). General insights of the Portuguese maritime economy and particularly of the Algarve region: Contributing towards a strategic vision. *Journal for Maritime Research*, 11(1), 3–9.

Valcic, B. (2009). Spatial policy and the behavior of fishermen. *Marine Policy*, 33(2), 215-222. 10.1016/j.marpol.2008.06.001

Vale, P., Rangel, I., Silva, B., Coelho, P., & Vilar, A. (2009). Atypical profiles of paralytic shellfish poisoning toxins in shellfish from Luanda and Mussulo bays, Angola. *Toxicon*, 53(1), 176–183. 10.1016/j.toxicon.2008.10.02919041661

Valladares, C., & Boelens, R. (2017). Extractivism and the rights of nature: Governmentality, 'convenient communities' and epistemic pacts in Ecuador. *Environmental Politics*, 26(6), 1015–1034. 10.1080/09644016.2017.1338384

Valle, P. O. D., Guerreiro, M., Mendes, J., & Silva, J. A. (2011). The cultural offer as a tourist product in coastal destinations: The Case of Algarve, Portugal. *Tourism and Hospitality Research*, 11(4), 233–247. 10.1177/1467358411420623

van Helmond, A. T. M., Mortensen, L. O., Plet-Hansen, K. S., Ulrich, C., Needle, C. L., Oesterwind, D., Kindt-Larsen, L., Catchpole, T., Mangi, S., Zimmermann, C., Olesen, H. J., Bailey, N., Bergsson, H., Dalskov, J., Elson, J., Hosken, M., Peterson, L., McElderry, H., Ruiz, J., & Poos, J. J. (2020). Electronic monitoring in fisheries: Lessons from global experiences and future opportunities. *Fish and Fisheries*, 21(1), 162–189. 10.1111/faf.12425

Viegas, S. L. (2016). Urbanisation and Peri-Urbanisation in Luanda: A Geopolitical and Socio-Spatial Perspective from the Late Colonial Period to the Present. *Journal of Southern African Studies*, 42(4), 595–618. 10.1080/03057070.2016.1192893

Voss, R., Quaas, M.F., Stoeven, M., Schmidt, J.O., Tomczak, M., & Möllmann, C. (2017). Ecological-economic fisheries management advice – quantification of potential benefits for the case of the eastern Baltic cod fishery. *Frontiers in Marine Science*, 4, 209. 10.3389/fmars.2017.00209

VW. (n.d.). Vilamoura World. Retrieved from https://www.vilamouraworld.com/

Watanuki, N., & Gonzales, B. J. (2006). The potential of artificial reefs as fisheries management tools in developing countries. *Bulletin of Marine Science*, 78(1), 9–19.

Weed, M. (2017). Sport participation legacies of mega sporting events. In Brittain, I., Bocarro, J., Byers, T., & Swart, K. (Eds.), *Legacies and Mega Events: Fact or Fairy Tales?* (1st ed., pp. 79–97). Routledge. 10.4324/9781315558981-6

Wenhai, L., Cusack, C., Baker, M., Tao, W., Mingbao, C., Paige, K., Xiaofan, Z., Levin, L., Escobar, E., Amon, D., Yue, Y., Reitz, A., Neves, A. A. S., O'Rourke, E., Mannarini, G., Pearlman, J., Tinker, J., Horsburgh, K. J., Lehodey, P., & Yufeng, Y. (2019). Successful Blue Economy Examples with an Emphasis on International Perspectives. *Frontiers in Marine Science*, 6, 261. 10.3389/fmars.2019.00261

Westman, W. E. (1977). How Much Are Nature's Services Worth? Measuring the social benefits of ecosystem functioning is both controversial and illuminating. *Science*, 197(4307), 960–964. 10.1126/science.197.4307.96017784115

Wibowo, T. W. (2021). Cleaning "Ghost Nets" in Karimunjawa National Park Waters: Pilot Study. *JECE - Journal of Empowerment Community and Education, 1*(4), 345–350. https://jurnalpengabdian.com/index.php/jece/article/view/651

Wilke, M. (2023). Comparing Public Participation in Coastal and Marine Planning in the Arctic: Lessons from Iceland and Norway. *Coasts*, 3(4), 345–369. 10.3390/coasts3040021

Wilks, J. (2021). Safety in coastal and marine tourism. In *Tourist health, safety and wellbeing in the new normal* (pp. 411–442). Springer Singapore. 10.1007/978-981-16-5415-2_17

Wise, L., Murta, A.G., Carvalho, J.P., & Mesquita, M. (2012). Qualitative modelling of fishermen's behaviour in a pelagic fishery. *Ecological Modelling*, 228, 112-122. 10.1016/j.ecolmodel.2011.12.008

Wolanski, E. & Elliott, M. (2024). The Physics-Biology Links in Suspended and Settled Estuarine Mud Dynamics: A Review. *Sediment Transport Research – Further Recent Advances*: 10.5772/intechopen.1005423

Wolanski, E. (1993). Water circulation in the Gulf of Carpentaria. *Journal of Marine Systems*, 4(5), 401–420. 10.1016/0924-7963(93)90024-G

Wolanski, E. (2006). Is harbour development ecologically sustainable. In Wolanski, E. (Ed.), *The environment in Asia Pacific harbours* (pp. 489–493). Springer. 10.1007/1-4020-3655-8_28

Wolanski, E. (2014). *Estuaries of Australia in 2050 and Beyond* (Wolanski, E., Ed.). Springer. 10.1007/978-94-007-7019-5

Wolanski, E., Chicharo, L., & Chicharo, M. A. (2008). Estuarine ecohydrology. In Jorgensen, S. E., & Fath, B. D. (Eds.), *Encyclopedia of Ecology* (pp. 1413–1422). Elsevier. 10.1016/B978-008045405-4.00044-6

Wolanski, E., & Elliott, M. (2015). *Estuarine Ecohydrology - An Introduction* (Wolanski, E., & Elliott, M., Eds.; 2nd ed.). Elsevier.

World Bank. (2021). *Plastic Waste Discharges from Rivers and Coastlines in Indonesia*. World Bank. 10.1596/35607

Xing, X., & Chalip, L. (2006). Effects of Hosting a Sport Event on Destination Brand: A Test of Co-branding and Match-up Models. *Sport Management Review*, 9(1), 49–78. 10.1016/S1441-3523(06)70019-5

Xiuzhen, L., Richard, B., Christopher, C. E., & Widney, S. (2018). Coastal wetland loss, consequences, and challenges for restoration. *Anthropocene Coasts.*, 1(1), 1–15. 10.1139/anc-2017-0001

Yatsu, A. (2019). Review of population dynamics and management of small pelagic fishes around the Japanese Archipelago. *Fisheries Science*, 85(4), 611–639. 10.1007/s12562-019-01305-3

Yingst, A. L. (2016), The Effect of Mangrove Development on Coral Reef Fish in Bocas Del Toro; Panama And Global Mangrove Management Method [Bachelor's Dissertation, University of Pittsburgh]. http://d-scholarship.pitt.edu/27790/1/Yingst_Bachelor_of_Philosophy.pdf

Yin, R. K. (2018). *Case study research and applications: Design and methods* (6th ed.). SAGE Publications, Inc.

Zacarias, W. B. M., Dai, X., Kindong, R., Sarr, O., & Moussa, A. H. (2022). Analysis of Fishery Resource Management Practices in São Tomé and Príncipe: Perception of the Dynamics of Catches from 1950–2020. Recommendations and Strategies for Future Research. *Sustainability (Basel)*, 14(20), 13367. 10.3390/su142013367

Zalewski, M., Wagner-Lotkowska, I., & Robarts, R. D. (2004). *Integrated watershed management - ecohydrology and phytotechnology – manual* (1st ed.). UNESCO - IHP, UNESCO - ROSTE, UNEP IETC.

Zhu, H., Liu, X., Cheng, S., & Wang, J. (2021). Effects of Artificial Reefs on Phytoplankton Community Structure in Baiyangdian Lake, China. *Water (Basel)*, 13(13), 1802. 10.3390/w13131802

Ziakas, V. (2016). Fostering the social utility of events: An integrative framework for the strategic use of events in community development. *Current Issues in Tourism*, 19(11), 1136–1157. 10.1080/13683500.2013.849664

Ziakas, V. (2019). Issues, patterns and strategies in the development of event portfolios: Configuring models, design and policy. *Journal of Policy Research in Tourism, Leisure & Events*, 11(1), 121–158. 10.1080/19407963.2018.1471481

Ziakas, V. (2023). Leveraging sport events for tourism development: The event portfolio perspective. *Journal of Global Sport Management*, 8(1), 43–72. 10.1080/24704067.2020.1731700

Ziakas, V., & Getz, D. (2020). Shaping the event portfolio management field: Premises and integration. *International Journal of Contemporary Hospitality Management*, 32(11), 3523–3544. 10.1108/IJCHM-05-2020-0486

Zurba, N. (2019). *Introduction to Coral Reefs as the Main Foundation of Our Ocean*. UNIMAL PRESS.

# About the Contributors

**Jorge Ramos** is an Integrated Researcher of the Research Centre for Tourism, Sustainability and Well-being (CinTurs) at the University of Algarve, Portugal. He has worked on socioeconomics related to coastal issues, particularly ecotourism, small-scale fisheries and offshore aquaculture. His research interests include aspects related to ecotourism, stakeholder analysis, artificial reefs, among other areas. He holds a PhD in Fisheries Economics and Management, a PgC in Social Research Methods, a MSc in Fisheries Economics (University of Portsmouth, UK) and a graduate degree in Marine Biology and Fisheries (University of Algarve, Portugal).

**Benjamin Drakeford** is an Associate Professor in Marine Resource Economics at the Centre for Blue Governance at the University of Portsmouth. His research interests lie in the broad field of blue economy development and governance. He has worked on such matters in developing countries (with the FAO on fisheries and aquaculture development in Central Asia) and developed countries (with DG Mare on revitalising the blue economy in the Atlantic Ocean). He has led several EU projects on blue economy matters, and national projects on blue economy development. He has more than 50 contributions in journal articles, books and conference proceedings.

* * *

**Alexandra Teodósio Anica Teodosio** (formerly Alexandra Chicharo) is a Full Professor of Marine Sciences at the University of the Algarve (UAlg) - Portugal, with a Ph.D. in Biological Oceanography in 1996. Alexandra developed her research on coastal ecology at CCMAR, where she coordinates the Ecoreach group - Ecology and Restoration of Riverine, Estuarine, and Coastal Habitats. She is responsible for the planktonic time series in the South Iberian Lower Guadiana Station (https://www.st.nmfs.noaa.gov/copepod/time-series/pt-30201/). She has been teaching and researching at UAlg for the last 35 years, with sabbaticals at the Institut de Ciències del Mar (ICM), Barcelona, Spain, in 2005/06, and at the University of Miami, USA, in 2015/16. In UAlg she has coordinated several study programmes, including bachelor's, master's and Ph.D. programmes. She is currently the Vice-Rector for Internationalisation and Sustainable Development at UAlg and at SEA_EU (European University of The Seas). ~ https://www.scopus.com/authid/detail.uri?authorId=57214851696.

**Vânia Baptista** is a Junior Researcher and Invited Assistant Professor at the Centre of Marine Sciences and the University of Algarve. She completed her PhD in Marine Sciences at the University of Algarve in 2019. Her work has been focused on fish larvae behaviour and ecology, and the effects of environmental and climate variability on the recruitment of fish species. Vânia has been coordinating the LittleFish-STP project, whose main objective is to study an endemic and economically exploited fish species in São Tomé.

**Sumanta Bhattacharya** is a Research Scholar at MAKAUT, a Public Policy Analyst, Chartered economist, Chartered Engineer, Cyber Crime Intervention officer and Zonal advisor at Consumer Rights Organization .He holds the record for maximum number of degree and certificate courses in Asia. He completed his M.Tech along with that he has MA in Development studies, LLB, Masters in Security and Defence Law, MA in Economics, MBA . He has done Diploma in International affairs and Diplomacy, Diploma in Governance and Government studies, Post graduation certificate course in Public policy and Advocacy, MPI from Oxford University followed by that Dr Bhattacharya did many courses from different International and National organization which includes ISRO, ABDI, UN, IIGL, NCAS, NIRDPR .Currently pursuing his PH.D-Tech, he has a PH.D in Human Rights also completed 108 courses in a span for 5 months along with that he is having 216 research papers, his research is versatile specialised in various topics, He has 46 book chapters and 70 national and International Patents .

**A. Rita Carrasco** is an Assistant Researcher at CIMA/ARNET - University of Algarve. RCarrasco holds a degree in Environmental Engineering (2003, University of Algarve), an MSc degree in Water and Coastal Management (2006, University of Algarve and University of Plymouth), a PhD in Marine, Earth, and Environmental Sciences, Coastal Dynamics Branch, Specializing in Geosciences (2012, University of Algarve). Currently, she is on the Coordination Board of CIMA, responsible for the scientific coordination of the objective 'Assessing and Mitigating Global Risks'. She teaches at the Bachelor's and Master's levels at the University of Algarve, covering subjects related to climate change and coastal geochemistry. Her current research is dedicated to unraveling the biogeomorphic adaptations and resilience mechanisms of coastal wetlands in the face of climate change.

**Andik Isdianto** is a Lecturer at Universitas Brawijaya Indonesia's Department of Fisheries and Marine Resources Utilization. Graduated from Universitas Brawijaya Indonesia with a bachelor's degree in regional and city planning, Institut Teknologi Sepuluh Nopember Indonesia with a master's degree in coastal management engineering, and Universitas Brawijaya Indonesia with a doctorate in environmental science. Teaches Marine and Fisheries Laws and Regulations, Marine Protected Areas, Marine Remote Sensing, Mapping of Marine Biological Resources, Marine Geographic Information Systems, Rehabilitation of Coastal Ecosystems, Management of Coastal and Marine Areas, Environmental Impact Analysis, and Conservation of Marine and Fisheries Resources. Has written various scientific papers on the topics of coastal management, coastal resilience, marine conservation, coastal environment, marine ecology, and ecological risk assessment.

**Siddharth Kanojia** is an Asst. Professor at O.P Jindal Global University. He is also associated with Indian Institute of Management Amritsar and Sirmaur as visiting faculty. He holds a Ph.D in Corporate Law & Governance, LL.M in International Business and Commercial Law, LL.B and BBS in Finance. Prior to his academic stints, he has worked at UK based Intellectual Property Management Company as Legal Consultant and Manchester based Start-up Company as Legal Advisor. He is also an active member in Human Development and Capability Association, USA and Development Studies Association, U.K.

**Rute Martins**, PhD in Sport Management, is an integrated researcher at Research Centre for Tourism, Sustainability and Well-being (CinTurs), University of Algarve. She holds a master's degree in Sport Management (Faculty of Human Kinetics, University of Lisbon) and a master's degree in Environmental Citizenship and Participation (Universidade Aberta). Her research interests are situated in nature-based sport, sport events, environmental sustainability, sport consumer behaviour.

**Margarida Mascarenhas**, PhD in Sport Sciences, is a professor at Faculty oh Human Kinetics, University of Lisbon. She is an integrated researcher at Research Centre for Tourism, Sustainability and Well-being (CinTurs), University of Algarve. She holds a master's degree in Industrial Management and Strategy (Lisbon School of Economics and Management, University of Lisbon) and graduated in Management (Business and Economics School, Lisbon). Her research interests are situated in the field of Sport Management, namely sport finance and economics, sport strategy, sport sustainability, sport events, sport tourism and sport consumer behaviour.

**Ana Matias** is a Research Fellow at the Center for Marine and Environmental Research (CIMA), University of Algarve. She has a Geology Degree, an MSc in Coastal and Marine Studies, a PhD in Marine Geology, and an MSc in Science Communication. She devoted most of her academic career to researching coastal geological processes and participating in projects that covered dune vulnerability, overwash, and coastal resilience. Since 2017, A. Matias's primary research focus has been Transdisciplinary Research on Geoscience, Art, and Communication. She is currently carrying out research in transdisciplinary approaches to coastal and marine environments; coastal risk communication; social inclusion in science communication; and experimental approaches to science communication formats.

**Édi Milhazes**, Master in Sport Management (Faculty of Human Kinetics, University of Lisbon) and graduated in Sports Sciences (University of Coimbra). He is a Senior Sports Technician. His research interests are situated in sport events and community development, sport tourism and sport public policies.

**Cristina Pallero** is currently research technician at the Institute of Marine Sciences of Andalusia (ICMAN-CSIC), since December 2021. He collaborates in several of the active projects of the Ecology and Coastal Management Group, such as FISHCLIM, CREMAR, among others. Previously, he has worked in environmental consulting and at the University of Cadiz, in the area of Regional Geographic Analysis at the Institute for Social and Sustainable Development (INDESS). Doctorate from the University of Cádiz (UCA, Spain) and the Federal University of Santa Catarina (UFSC, Brazil) in international co-supervision, with the thesis: "Management of complex natural and social realities: estuaries". Degree in Marine and Environmental Sciences from the UCA, where she completed a postgraduate course in Integrated Management of Coastal Areas (2013). Member of the Research Group on Integrated Management of Coastal Areas (UCA) 2010-2020 and of the Laboratorio de Gestão Costeira Integrada (UFSC, Brazil) 2014-2018. He has participated in several research projects on integrated management of coastal and marine areas, including the European project Macaronesia Marine Spatial Planning (MarSP). He participated in the coordination of the Ibero-American Network of Integrated Coastal Management (IBERMAR). I have several publications developed in the conduct of research in different projects and investigations, in addition to technical reports. In addition, it adds experience as a technician in public institution (City Council).

**Bruno Pinto** is currently an assistant researcher and a science communicator at the Faculty of Sciences of the University of Lisbon, and a member of MARE- Center for Marine and Environmental Sciences (FCUL) and ARNET- Aquatic Research Network. He has a special interest in themes related to biodiversity, climate change and marine sciences, and the use of diverse media such as comics, documentary or exhibition to communicate them. Previously, he was a science communicator at the Natural History and Science Museum (University of Lisbon).

**Berlania Mahardika Putri** completed her bachelor degree in Marine Science at the Brawijaya University, Indonesia with a final project entitled Differences Growth of Crustose Coralline Algae on Collector Substrates in South Malang Waters in 2021. During college, the author was active in campus non-academic activities such as the Fisheries Diving Club, member and division head of the Marine Science Student Association Universitas Brawijaya (HIMALAYA) and active as a teaching assistant of Marine Zoology and Marine Coralogy. She focused on research and publications related to the environmental science, especially marine science. Currently, she is pursuing a Master Program of Environmental Science at Gadjah Mada University, and interested in environmental issues, conservation and ecology.

**Arief Setyanto**, S.Pi., M.App.Sc., was born in Rembang, Central Java in 1971 and completed elementary to high school education in Rembang (SD Kutoharjo 2, SMP 3 and SMA 2 Rembang). In 1991 he studied at the Fisheries Resource Utilization study program, Faculty of Fisheries and Marine Science, Brawijaya University. This is where the author began to study the field of ecological systems in the management of capture fisheries resources. During the 1998 reform era, he became a lecturer at his alma mater. The author began to study aspects of fisheries and marine resource management, especially those related to biological aspects, since studying at James Cook University Townsville, Queensland-Australia. The author has also studied education regarding the collection and analysis of fisheries data at Wageningan International, the Netherlands. The author also had time to research and study fisheries management at the Australian National Center for Ocean Resources and Security (ANCORS), University of Wollongong, New South Wales-Australia. Has produced more than 30 scientific articles which were disseminated at various scientific meetings, and published in seminar proceedings, national and international scientific journals. The book written: Marine Conservation in a Biological Approach, and Tools and Methods for Catching Marine Fish with Gill Nets in East Java was published by UB Press and MNC Media Nusa Creative.

# Index

## W



## Z

www.ingramcontent.com/pod-product-compliance
Lightning Source LLC
LaVergne TN
LVHW080600300926
849162LV00005B/1011